T0206160

Chapman & Hall/CRC
Computer Science and Data Analysis Series

BAYESIAN REGRESSION MODELING WITH INLA

XIAOFENG WANG
Cleveland Clinic, Cleveland, Ohio

YU RYAN YUE
Baruch College, The City University of New York

JULIAN J. FARAWAY
University of Bath, UK

CRC Press is an imprint of the
Taylor & Francis Group, an **informa** business
A CHAPMAN & HALL BOOK

Chapman & Hall/CRC
Computer Science and Data Analysis Series

The interface between the computer and statistical sciences is increasing, as each discipline seeks to harness the power and resources of the other. This series aims to foster the integration between the computer sciences and statistical, numerical, and probabilistic methods by publishing a broad range of reference works, textbooks, and handbooks.

Proposals for the series should be sent directly to one of the series editors above, or submitted to:

Chapman & Hall/CRC
Taylor and Francis Group
3 Park Square, Milton Park
Abingdon, OX14 4RN, UK

Published Titles

Semisupervised Learning for Computational Linguistics
Steven Abney

Visualization and Verbalization of Data
Jörg Blasius and Michael Greenacre

Chain Event Graphs
Rodrigo A. Collazo, Christiane Görgen, and Jim Q. Smith

Design and Modeling for Computer Experiments
Kai-Tai Fang, Runze Li, and Agus Sudjianto

Microarray Image Analysis: An Algorithmic Approach
Karl Fraser, Zidong Wang, and Xiaohui Liu

R Programming for Bioinformatics
Robert Gentleman

Exploratory Multivariate Analysis by Example Using R
François Husson, Sébastien Lê, and Jérôme Pagès

Published Titles cont.

Bayesian Artificial Intelligence, Second Edition
Kevin B. Korb and Ann E. Nicholson

Computational Statistics Handbook with MATLAB®, Third Edition
Wendy L. Martinez and Angel R. Martinez

Exploratory Data Analysis with MATLAB®, Third Edition
Wendy L. Martinez, Angel R. Martinez, and Jeffrey L. Solka

Statistics in MATLAB®: A Primer
Wendy L. Martinez and MoonJung Cho

Clustering for Data Mining: A Data Recovery Approach, Second Edition
Boris Mirkin

Introduction to Machine Learning and Bioinformatics
Sushmita Mitra, Sujay Datta, Theodore Perkins, and George Michailidis

Introduction to Data Technologies
Paul Murrell

R Graphics
Paul Murrell

Correspondence Analysis and Data Coding with Java and R
Fionn Murtagh

Data Science Foundations: Geometry and Topology of Complex Hierarchic Systems and Big Data Analytics
Fionn Murtagh

Pattern Recognition Algorithms for Data Mining
Sankar K. Pal and Pabitra Mitra

Statistical Computing with R
Maria L. Rizzo

Statistical Learning and Data Science
Mireille Gettler Summa, Léon Bottou, Bernard Goldfarb, Fionn Murtagh, Catherine Pardoux, and Myriam Touati

Bayesian Regression Modeling With INLA
Xiaofeng Wang, Yu Ryan Yue, and Julian J. Faraway

Music Data Analysis: Foundations and Applications
Claus Weihs, Dietmar Jannach, Igor Vatolkin, and Günter Rudolph

Foundations of Statistical Algorithms: With References to R Packages
Claus Weihs, Olaf Mersmann, and Uwe Ligges

CRC Press
Taylor & Francis Group
6000 Broken Sound Parkway NW, Suite 300
Boca Raton, FL 33487-2742

First issued in paperback 2020

Version Date: 20180111

ISBN-13: 978-0-367-57226-6 (pbk)
ISBN-13: 978-1-4987-2725-9 (hbk)

Visit the Taylor & Francis Web site at
http://www.taylorandfrancis.com

and the CRC Press Web site at
http://www.crcpress.com

Dedicated to my parents, Huahui Wang and Xianzhang Zeng, and my lovely family.
– X.W.

For Amy, Annie and Evan.
– Y.Y.

Contents

Preface

INLA stands for Integrated Nested Laplace Approximations. It is a method for fitting a broad class of Bayesian models. Historically, it was difficult to fit anything but the most simple Bayesian models. Over the last twenty years, a class of Bayesian computational methods based on a simulation method called Markov chain Monte Carlo (MCMC) has been developed and has seen wide acceptance in statistics. Popular packages using these methods include BUGS, JAGS and STAN. Despite impressive improvements, these packages suffer from two problems. First, they are slow. For some more complex and/or larger data problems, MCMC can be infeasibly slow, even if you are prepared to wait days. But even for more modest problems, the ability to fit models quickly is crucial to exploratory data analyses. INLA is an approximation method. Typically, the approximations are more than adequate and remember that simulation methods are inevitably approximations also if they are to finish in finite time. INLA takes no more than a few seconds to fit the models found in this book.

The other practical difficulty with MCMC methods is they take substantial expertise to use. You need to learn a specialized programming language to specify the models. You also need to understand the diagnostics that determine whether the model has been fit correctly or whether the simulation process has failed. Some skill is necessary to make a success of this. Despite ongoing improvement in this area, this has been an obstacle to wider adoption of these methods in the scientific community.

There are some drawbacks to INLA too. Although you do not need to learn a specific programming language as the models can be specified and analyzed using R, it is still not that straightforward. Indeed, that is why you should read this book. Furthermore, INLA only applies to a class called latent Gaussian models. If you browse through the table of contents for this book, you will see that this class is very broad. Nevertheless, there will be some models that cannot be fit with INLA but can be done with MCMC.

We make some assumptions about you, the reader. We expect that you have a basic knowledge of statistical theory and practice. We expect you already know something about Bayesian methodology. This is not a theoretical book as we focus our presentation around examples. We hope that scientists, who already use some statistics, will find this book accessible. We also expect you have some knowledge of R. We do provide fully working code for all the examples so you may be able to adapt this to your own needs without proficiency in R. Even so, you will need some experience with R to draw the full benefit. If you need more introductory material, we direct you to more comprehensive and accessible texts.

This book is about regression models and we have not presented the extensive spatial data analysis capabilities of INLA. For those specifically interested in spatial data, we direct you to Blangiardo and Cameletti (2015).

We have gathered the data and additional functions we use in this text as an R package which you may find currently at:

```
https://github.com/julianfaraway/brinla
```

Our first thanks go to Håvard Rue and his coworkers for developing the theory and producing the software for INLA. Thanks also to Finn Lindgren, Daniel Simpson and Egil Ferkingstad for helpful advice.

1

Introduction

INLA can be quite complex so we warm up with a simple regression example to illustrate the most basic features. We do not pretend this as a comprehensive introduction to Bayesian methods. Instead we review the Bayesian theory that will arise later in the book. We also discuss why you might want to use INLA in preference to the alternatives.

1.1 Quick Start

1.1.1 Hubble's Law

Let's get started with INLA using a simple example. Suppose we want to know the age of the universe. In the beginning, there was the big bang. Galaxies moved away from the center of the universe at a constant velocity. If we measure the velocity of other galaxies relative to us and their distance from us, we can estimate the age of the universe. We use Hubble's law which states that:

$$y = \beta x$$

where y is the relative velocity between us and another galaxy and x is the distance from us to the other galaxy. β is called Hubble's constant. We can estimate the age of the universe as β^{-1}.

The Hubble Space Telescope has allowed the collection of the data vital to answering this question. We use data on 24 galaxies reported in Freedman et al. (2001). The same data is also analyzed in Wood (2006). Now refer to Appendix A for how to install INLA and the R package for this book. The data can be found in our R package, which is called `brinla`.

```
data(hubble, package = "brinla")
```

We highlight the R code throughout this book in the manner above. You should type this in at the R console or you may find it easier to copy and paste the code from R scripts for each chapter. See the appendix for how to obtain these scripts. Alternatively, you can gain access to the data and functions in our package with the command: `library(brinla)`. We have used the `data` command instead so that the source of the data is clear.

We plot the data as seen in Figure 1.1.

```
plot(y ~ x, xlab = "Distance(Mpc)", ylab = "Velocity(km/s)",
    data = hubble)
```

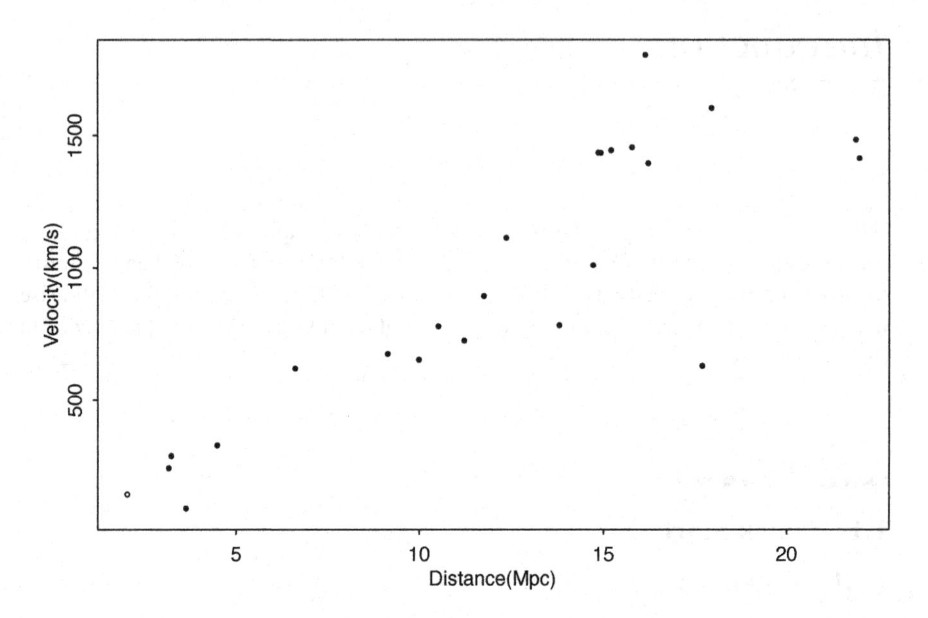

FIGURE 1.1
Distance and velocity of 24 galaxies relative to our location.

We see that the observations do not all lie on a line and so there must be some measurement error. This error is hardly surprising given the difficulty of observing such distant objects. Even so, we do see an approximately linear relationship between the variables, consistent with Hubble's law. We allow for the error by specifying a model:

$$y_i = \beta x_i + \varepsilon_i, \quad i = 1, \ldots, 24.$$

For now, we assume only that the errors have mean zero and variance σ^2.

1.1.2 Standard Analysis

The method of least squares is now almost 200 years old and can be applied to estimate β. We do not have an intercept term in this model so we put a `-1` in the model formula.

```
lmod <- lm(y ~ x - 1, data = hubble)
coef(lmod)
```

```
     x
76.581
```

This is our estimate of β, which is Hubble's constant. We need to convert the units of the data to something more convenient. We have 60 seconds, 60 minutes, 24 hours and 365.25 days in a year and one megaparsec (Mpc) is 3.09×10^{19} km. Here is a function to transform Hubble's constant into the age of the universe in billions of years. We apply it to our estimate:

```
hubtoage <- function(x) 3.09e+19/(x * 60^2 * 24 * 365.25 * 1e+09)
hubtoage(coef(lmod))
```

```
     x
12.786
```

Our point estimate of the age of the universe is 12.8 billion years old. We know there is some uncertainty in this estimate and so we may want a confidence interval to express this. Now we need to make a distributional assumption on the errors. We assume they follow a Gaussian distribution. There are some other assumptions. We assume the structural form of the model is correct. We take this from Hubble's law and the data appear to support this. We have also assumed that the errors have constant variance. This may not be quite true but we shall let this pass in the interest of simplicity.

We construct a 95% confidence interval for Hubble's constant:

```
(bci <- confint(lmod))
```

```
   2.5 % 97.5 %
x 68.379 84.783
```

The interval is quite wide. We can convert this to an interval for the age of the universe:

```
hubtoage(bci)
```

```
  2.5 % 97.5 %
x 14.32 11.549
```

The inversion needed to calculate the age changes the order of the limits to the interval. We see that the 95% confidence interval runs from 11.5 to 14.3 billion years.

1.1.3 Bayesian Analysis

We may be satisfied with the previous analysis. We have a point estimate and we have an expression of the uncertainty in this estimate. Yet if we consider the results more closely, we may feel some dissatisfaction and be willing to consider an alternative method of analysis.

It is tempting to view the confidence interval as claiming a 95% probability that the age of the universe lies in the interval 11.5 to 14.3 billion years. But according to the theory used to make confidence intervals this is not at all what is claimed. This theory, sometimes called *Frequentist* and sometimes called classical inference, views parameters as fixed, but usually unknown, quantities. So it makes no sense to assign probabilities to parameters according to this theory. The 95% confidence applies to the interval and not the parameter. It claims a 95% chance that the interval covers the true value of the parameter.

We might object that the data have now been observed and are known so the interval is no longer random. To answer this objection, we have to construct a rather elaborate argument. We envisage an unlimited number of alternative universes where galaxies are generated according to the linear model above. For each of these alternative universes we compute a confidence interval in the same way as above. In 95% of these alternate universes, the confidence interval will contain the true age of the universe.

Most users don't focus on this peculiar nature of the confidence interval and interpret it as being a probability statement about the age of the universe. Usually it doesn't matter too much if they do this from a practical perspective but sometimes it makes a difference.

In Bayesian statistics, we regard the parameters as random. We express our uncertainty about these parameters before fitting the model in terms of *prior probability*. We then use Bayes theorem to compute a *posterior probability* about the parameters by combining the prior with the likelihood of the data under the model. We explain later exactly how this can be done. Using this method, we can make straightforward probability statements about the age of the universe. We will also see other advantages.

1.1.4 INLA

Bayesian calculations are usually much more difficult than Frequentist calculations. This is the main reason why classical methods have been historically preferred to Bayesian methods. Now that computing is powerful and inexpensive, we can do the Bayesian calculation more easily. For a few simple models (like our Hubble example), Bayesian calculations can be done exactly. For anything more complicated, we must choose between two classes of computational methodology — simulation or approximation.

INLA, which stands for *Integrated Nested Laplace Approximation*, is, as you may guess from the name, an approximation method. Let's see how it works for the Hubble example. First we need to load the INLA R package — see Appendix A for details on the installation:

```
library(INLA)
```

To fit a Bayesian model, we must do more than specify the structural form of the model, which in this case is `y ~ x -1`. The `-1` indicates that the intercept should be omitted as otherwise it is included by default. We must also specify the prior distributions on the parameters. In this model, we have two — β and σ. Sometimes, we can rely on the default choice which strives to be *uninformative*. We shall do this for σ. We will discuss the defaults and the choice of prior in great detail later in the book. For now, we focus on β.

INLA applies to a wide class of models called *Latent Gaussian Models*. Membership of this class requires that some parameters in the model have priors with a Gaussian distribution. We will explain later what parameters have to be Gaussian but in our model, we need β to have a Gaussian prior. By default this prior has mean zero — we will stick with that for now. INLA likes to use the precision, which is the inverse of the variance, to express the spread of this prior. Large values of the precision indicate great certainty in the prior. In this case, we know very little, so we choose a very small value for the precision. The meaning of "very small" depends on the scale of the data. You may wish to standardize your data to avoid this scaling problem. In our example, that would require an extra step in the computation of the age of the universe to undo this standardization. For this reason, we go with the data as is and just make sure we pick a small enough precision. Let's fit our first model with INLA:

```
imod <- inla(y ~ x - 1, family = "gaussian",
    control.fixed = list(prec = 1e-09), data = hubble)
```

We specify the structural part of the model first using the standard R model notation. We need to set a distribution for the response. Here we choose Gaussian but this is not part of the Latent Gaussian requirement. We could pick binomial or Poisson or any of a quite large set of possibilities here. Next we need to specify the prior on the *fixed* parameter β. The term *fixed* is a little misleading in the Bayesian context because the parameters are usually not fixed at all. The terminology comes from mixed effect models, discussed in Chapter 5, which are defined with so-called fixed and random components. According to the same misleading terminology, σ^2 is the random component of the model but we rely on the default choice of prior for that. For the prior on β, we take the default mean of zero but specify a precision of 1×10^{-9}. Finally, we tell INLA where to find the `data`.

The model is simple and the dataset small so the computation is virtually instantaneous. Much is calculated but we focus on just the posterior distribution of β.

```
(ibci <- imod$summary.fixed)
```

```
    mean     sd 0.025quant 0.5quant 0.975quant   mode       kld
x 76.581 3.7638     69.152   76.581         84 76.581 1.406e-11
```

Various summary statistics are provided. The `kld` is a diagnostic that measures the accuracy of the INLA approximation. Small values, as seen here, are good. If you try this yourself, do not expect to get exactly the same numbers for reasons that are explained in Section 2.4. We can also plot the posterior density of β as seen in Figure 1.2. The R commands necessary to extract and plot the elements of INLA output can be rather complex so we postpone a description of these until later.

```
plot(imod$marginals.fixed$x, type = "l", xlab = "beta",
    ylab = "density", xlim = c(60, 100))
abline(v = ibci[c(3, 5)], lty = 2)
```

The posterior density for β expresses our uncertainty about Hubble's constant with the benefit of the data from the 24 galaxies. We can see it is most likely around 75 and unlikely to be less than 70 or more than 90. Here the mean, median and mode, as seen in summary output, are identical with the least squares estimate of 76.581.

We can express our uncertainty in terms of an interval containing 95% of the probability. One easy way to construct this interval is to take the 2.5% and 97.5% percentiles. This gives the interval as [69.2, 84.0]. This is called a *credible* interval to distinguish it from a confidence interval. Now we really do claim there is a 95% chance that Hubble's constant lies in this interval.

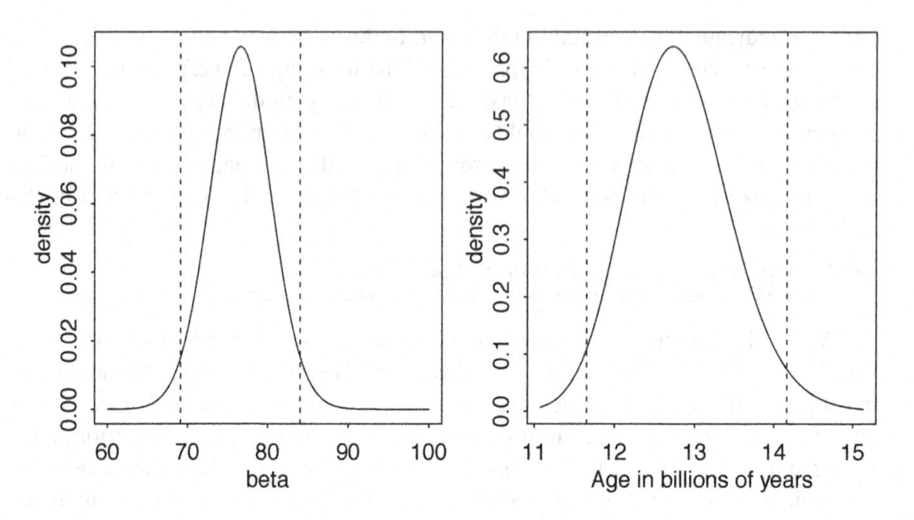

FIGURE 1.2
Posterior density for β on the left and for the age of the universe on the right. 95% credible intervals are shown as dashed lines.

The results for Hubble's constant can be transformed into the age of the universe scale:

```
hubtoage(ibci[c(1,3,4,5,6)])
```

```
    mean 0.025quant 0.5quant 0.975quant   mode
x 12.786      14.16   12.786     11.657 12.786
```

We have omitted the SE as getting the right transformation on this is more work. We can plot the posterior density with 95% credible interval as seen in Figure 1.2.

```
ageden <- inla.tmarginal(hubtoage, imod$marginals.fixed$x)
plot(ageden, type = "l", xlab = "Age in billions of years",
    ylab = "density")
abline(v = hubtoage(ibci[c(3, 5)]), lty = 2)
```

The 95% credible interval runs from 11.7 to 14.2 billion years which is slightly different from the previous confidence interval.

We used a non-informative prior in this analysis but we could use prior information. For one thing, we know the universe is expanding so we expect Hubble's constant to be positive. Before the launch of the Hubble Space Telescope, the age of the universe was thought to be between 10 and 20 billion years. Let's translate that to the scale of Hubble's constant, noting that our `hubtoage` works in reverse due to the inverse nature of the conversion.

```
hubtoage(c(10, 15, 20))
```

```
[1] 97.916 65.277 48.958
```

This suggests a prior mean of, say, 65. Suppose we assume that when people express their uncertainty with an interval of the form $[a, b]$, they are thinking, at least implicitly, of a 95% confidence/credible interval. That would imply the width of the interval is about four standard deviations. Since the width of the interval is 49, this suggests an SD of about 12. Of course, these people may not have been thinking in terms of 95% intervals, but, in the absence of further information, an SD of 12 is a reasonable place to start.

```
imod <- inla(y ~ x - 1, family = "gaussian",
    control.fixed = list(mean = 65, prec = 1/(12^2)), data = hubble)
(ibci <- imod$summary.fixed)
```

```
    mean     sd 0.025quant 0.5quant 0.975quant   mode        kld
x 75.479 3.7292     67.981   75.522     82.729 75.596 1.1216e-11
```

We convert that to the age of the universe scale.

```
hubtoage(ibci[c(1, 3, 4, 5, 6)])
```

```
    mean 0.025quant 0.5quant 0.975quant   mode
x 12.973     14.403   12.965     11.836 12.953
```

The predicted age is a bit greater than before since the prior mean of 15 billion years is somewhat higher than the LS estimate. Even so, the results are not very different. We prefer that our results not be very sensitive to the choice of prior. Although we accept the necessary subjectivity of the prior, it's nice if we can agree with other analysts who might have made different choices. If we have a reasonable amount of the data, the prior will not make much difference. The Bernstein–von Mises theorem provides some theoretical support to this empirical statement as explained in Le Cam (2012).

But what if the prior is entirely at odds with the data? According to the Ussher chronology, the universe was created on 22nd October, 4004BC. Other contemporary authors, including Isaac Newton, differed with Ussher, but the estimates were within a few hundred years of this date. Under this chronology, Hubble's constant would be:

```
(uhub <- hubtoage((2016 + 4004 - 1)/1e+09))
```

```
[1] 162678504
```

This results in a very large value of Hubble's constant. Let's use this value for the prior mean. Let's also assume a 5% variation in the estimates of the age of the universe and use this to set the precision:

```
imod <- inla(y ~ x - 1, family = "gaussian",
    control.fixed = list(mean = 65, prec = 1/(12^2)), data = hubble)
(ibci <- imod$summary.fixed)
```

```
    mean    sd 0.025quant 0.5quant 0.975quant   mode        kld
x 76.581 3.978     68.697   76.581     84.452 76.581 1.2644e-12
```

```
hubtoage(ibci[c(1, 3, 4, 5, 6)])
```

```
    mean 0.025quant 0.5quant 0.975quant   mode
x 12.786     14.233   12.786     11.607 12.786
```

We see that the results are quite similar to when the uninformative prior was used. The data is able to overcome the unusual prior. The normal distribution puts some weight on the whole real line. Although the probabilities assigned to values far from the mean are very small, they are not zero. This is sufficient to allow the data to overcome the prior.

In some cases, the prior can be so inconsistent with the data that difficulties will arise with computation and strange results may be observed. Such cases are called *data-prior conflicts* although one must also consider the possibility that the model itself is very wrong. See Evans et al. (2006) for discussion of this.

Before exploring the further application of INLA to statistical models, we need to build up some theory.

1.2 Bayes Theory

We make no attempt here at a full and thorough introduction to Bayesian theory. The reader is directed to other texts such as Gelman et al. (2014) or Carlin and Louis (2008). We provide merely a short tour through the ideas and methods we will need later in the book. Experienced Bayesians may skip to the next chapter.

We start with a parametric model $p(\mathbf{y}|\theta)$ for the data $\mathbf{y}$ with parameters θ. We propose this model given the structure of the data, information about how the data was collected and knowledge about the context from which it arises. There is uncertainty about the parameters which we hope the data will reduce but usually there is also uncertainty about the model itself. We rarely believe the model to be completely true but we hope that is an effective approximation to reality that will serve the purpose for which we intend it to be used. Given the uncertainty about the model, we may reasonably consider alternative models and we shall wish to check whether the model appears adequate for the data that are observed.

Instead of regarding the probability distribution, $p(\mathbf{y}|\theta)$ as a function of $\mathbf{y}$, we might view it as a function of θ. This *likelihood* $L(\theta)$ encompasses our uncertainty about θ. It is important to understand that $L(\theta)$ is not a probability density for θ.

In the frequentist approach, we wish to estimate the parameters θ. A natural and most common way to do this is to compute the most likely values achieved by maximizing $L(\theta)$. This is called the *maximum likelihood estimate* (MLE). This estimate has many favorable properties and one may readily derive useful associated quantities such as standard errors and confidence intervals. The theory behind such methods may be found in standard mathematical statistics books such as Bickel and Doksum (2015). In simpler cases, the MLE may be computed explicitly but usually numerical optimization methods are required. In many commonly used models, these methods are well understood and perform efficiently. For some more complex models, computing the MLE can become more problematic. Even so, the main reason we might prefer to use Bayesian methods is not for computational reasons since these methods are usually slower. The answers provided by a Bayesian analysis are conceptually different from the MLE approach.

In the frequentist view, the parameters θ are fixed but unknown. We use the data to estimate the parameters. In the Bayesian view, the parameters are random variables.

Before we observe the data, our uncertainty regarding the parameters is expressed through a prior density $\pi(\theta)$. We update our uncertainty about the parameters using the data and the model to produce a posterior density $p(\theta|\mathbf{y})$. Bayes theorem provides the means through which this transformation occurs:

$$p(\theta|\mathbf{y}) = \frac{p(\mathbf{y}|\theta)\pi(\theta)}{p(\mathbf{y})}.$$

We can write this equation in words as:

$$\text{Posterior} \quad \propto \quad \text{Likelihood} \quad \times \quad \text{Prior}$$

which makes it clear how the posterior is derived from combining the likelihood and the prior. Although this is conceptually simple, we need to compute the normalizing constant $p(\mathbf{y})$, to obtain the posterior density:

$$p(\mathbf{y}) = \int_{\theta} p(\mathbf{y}|\theta)\pi(\theta)d\theta.$$

This is the marginal probability of the data under the model. Computing integrals such as these is the reason why the conceptual simplicity of Bayes theorem requires rather more work than would first appear.

1.3 Prior and Posterior Distributions

Specifying the prior is an important part of a Bayesian analysis. It is perhaps the weakest part of Bayesian methodology since critics can reasonably question any choice of prior. When the conclusions of an analysis are contentious, it may be difficult to defend a particular choice of prior. One can defend against this criticism by pointing out that the choice of model is usually a subjective choice that cannot be strongly justified so a frequentist analysis cannot pretend to be entirely objective either. It is inevitable that almost any analysis will involve some amount of subjectivity and the Bayesian is just being honest about what choices have been made.

The best motivation for a choice of prior is substantive knowledge about θ. Sometimes this comes in the form of expert opinion which may require some mathematical formulation if the expert is not mathematically inclined. In other cases, we may have past experience regarding similar data which is helpful in the choice of prior. Indeed, our choice of prior may be the posterior from a previous experiment. Priors such as these are sometimes called *informative* or *substantive*. If this type of information is available, we should surely use it. But sometimes we are not so blessed and must find some other basis on which to make a decision.

A particular convenient choice is the *conjugate* prior. With this choice, the prior and posterior have the same distribution. Furthermore, there is an explicit formula for the posterior in terms of the prior parameters and the data. One simple example concerns data that consist of a univariate sample from a normal distribution. If one also chooses a normal density for the prior on the mean (with known variance), we obtain a normal posterior density. In the days before fast computation was available, such a choice of prior would be strongly advised. Unfortunately, there are some serious drawbacks. There are only a handful of simple models for which a conjugate choice of prior is available. Furthermore, although the conjugate choice of prior is convenient, there is no compelling reason why we should make this choice compared to other reasonable alternatives.

Sometimes the researcher has no strong information about what prior would be best and would like to make a safe choice that has no strong consequences for the conclusion. This is an understandable wish but it is surprisingly difficult to fulfill. Perhaps the most straightforward choice is the so-called *flat* prior:

$$\pi(\theta) \propto 1.$$

This apparently gives no preference to any particular choice of θ but there are some drawbacks, the first being that this is not a proper density as it does not (unless θ has finite range) integrate to a finite value. Nevertheless, we can often proceed, essentially normalizing the likelihood into a density. Sometimes the results are satisfactory, but other times, the outcome can be unreasonable. It is difficult to delineate exactly when these failures might occur so skepticism should be used when considering the posterior derived from a flat prior.

Furthermore, although the flat prior purports to give no preference to any particular choice of θ, this is, to some extent, unavoidable. For example, consider σ the SD and σ^2 the variance from a normal distribution. We can choose $\pi(\sigma) \propto 1$ or $\pi(\sigma^2) \propto 1$ but these choices are not equivalent. Being flat on one scale would become informative on the other. This lack of invariance bedevils other attempts at so-called *non-informative* priors. One possible solution is the use of the *Jeffreys* prior:

$$\pi(\theta) \propto |\mathbf{I}(\theta)|^{1/2}.$$

This prior is invariant to reparameterization. Unfortunately, the Jeffreys prior produces unreasonable outcomes under some models so it cannot be regarded as a universal solution to the prior choice problem.

Objective priors are an attempt to reduce the subjectivity in the choice of prior but it seems eliminating human judgment from the selection is not realistic. There are several other ideas for generating priors. Sometimes, for some classes of problems, researchers have grown into the habit of making certain prior choices. Following these same choices will insulate you from some criticism that your choice of prior is arbitrary. Nevertheless, one should understand that such choices can simply be customary and may lack any stronger justification.

INLA has a limited menu of preprogrammed prior choices (although you can define your own). Furthermore, we shall learn in the next chapter that we are constrained to a Gaussian prior for at least some of the parameters. For the parameters where we do have a choice within INLA, Simpson et al. (2017) state some general principles regarding prior construction which are implemented within INLA.

Although a very orthodox Bayesian view states that priors should be chosen once only in advance of seeing the data, there is a good reason to experiment with alternative priors. We perform the analyses under a range of reasonable prior choices. If the results are qualitatively similar, we can increase our confidence in the statement of the conclusions. On the other hand, if the results appear particularly sensitive to the choice of prior, we should be cautious in the strength of our claims. This idea is called "prior sensitivity" and we would like as little of it as possible.

Posterior distributions are rather less problematic than priors. After all, they are completely determined by the choice of prior (along with the likelihood). Nevertheless, understanding the posterior is not always straightforward. One common problem is that the posterior $p(\theta|\mathbf{y})$ is usually multivariate, sometimes of quite high dimension. This is difficult to visualize and so we often want to contemplate the marginal distribution for a single parameter, θ_i:

$$p(\theta_i|\mathbf{y}) = \int_{\theta_{-i}} p(\theta|\mathbf{y})d\theta_{-i},$$

where θ_{-i} means θ without the ith entry. Notice that this requires a multidimensional integration. For the multivariate normal, this is easily calculated but in other situations this can be quite demanding.

It is good practice to plot the marginal posterior distributions in order to fully understand the meaning of the fitted model. The posterior can also be summarized by familiar measures of center such as mean, median and mode. For relatively symmetric distributions there will not be much difference between these measures but for more skew distributions, it is worth understanding the distinctions and why a plot of the density may be particularly worthwhile. The spread of a density can be summarized by its standard deviation or selected quantiles.

We might select the 2.5% and 97.5% quantiles as this would form a 95% credible interval for the parameter. Although apparently similar, credible intervals differ from the confidence intervals used in frequentist statistics. A credible interval is a claim that there is specified probability that the parameter lies within the interval. Although many interpret a confidence interval in the same way, its real meaning is more convoluted. Credible intervals are not uniquely defined given the posterior. Quantiles are easiest to compute but we might instead ask for the shortest interval that contains the specified probability level. This is called the *highest posterior density* (HPD) interval. Somewhat more effort is required to compute these and is implemented using the `inla.hpdmarginal()` function.

1.4 Model Checking

The predictive distribution for a future observation z is

$$p(z|\mathbf{y}) = \int_{\theta} p(z|\theta)p(\theta|\mathbf{y})d\theta.$$

We will need these for making predictions but the idea is also useful for checking the model we have. The *conditional predictive ordinate* (CPO) introduced by Pettit (1990), is defined as:

$$\text{CPO}_i = p(y_i|\mathbf{y}_{-i}).$$

The notation $\mathbf{y}_{-i}$ means all the data except for the ith observation. The CPO statistic measures the probability (or density) of the observed value of y_i. Hence, particularly small values of this statistic would indicate unusual observations that do not fit the model well. Some observations can be so influential that they have a large effect on the fitted model. This is the reason for leaving out the observation itself in fitting the model used to compute the CPO statistic. It might seem that the computation of these statistics would be particularly onerous since we would need to refit the model for every case in the data but there are some effective shortcuts, as discussed in Held et al. (2010), which greatly reduce this burden.

The *probability integral transform* (PIT) is similar to the CPO statistic. It was introduced by Dawid (1984) and is defined as:

$$\text{PIT}_i = p(Y_i < y_i|\mathbf{y}_{-i}).$$

Notice that this makes most sense for continuous y. Across the whole data, we would expect the PIT statistics to be approximately uniformly distributed for a good model. Values of the PIT statistic close to zero or one would indicate observations which are much smaller or larger, respectively, than expected. As with the CPO, this would seem expensive to compute but INLA has ways to economize on this.

1.5 Model Selection

Sometimes we need to choose between several proposed models for the data. For this purpose, we need a criterion measuring how consistent the data are with a given model. The most well-known criterion is the *Akaike information criterion* (AIC) which is defined as:

$$AIC = -2\log p(\mathbf{y}|\hat{\theta}_{mle}) + 2k$$

where k is the number of parameters and $\hat{\theta}_{mle}$ is the MLE. Smaller values of AIC are preferred. The first part measures the fit of the model to the data. With only this part, larger models would always be preferred so we need the second, penalty part of the criterion, which prefers smaller models. There are two problems from a Bayesian perspective. Firstly, the criterion is based on the MLE so it is not Bayesian. Secondly, the number of parameters k is fine for simple models but problematic for hierarchical models where counting the effective number of parameters is not straightforward.

For Bayesian models, we might prefer the *deviance information criterion* (DIC) of Spiegelhalter et al. (2002) which is motivated by the Akaike information criterion. We define the deviance of the model as:

$$D(\theta) = -2\log(p(\mathbf{y}|\theta)).$$

In a Bayesian model, this is a random variable so we use the expected deviance $E(D(\theta))$ under the posterior distribution as a measure of fit. For counting the parameters, we introduce the idea of the *effective number of parameters*:

$$p_D = E(D(\theta)) - D(E(\theta)) = \bar{D} - D(\bar{\theta}),$$

and the DIC is then:

$$DIC = \bar{D} + p_D.$$

We compute the DIC for all the models of interest and choose the one with the smallest value. For further discussion of the DIC see Spiegelhalter et al. (2014).

An alternative is the *Watanabe Akaike information criterion* (WAIC) (Watanabe (2010)) which follows a more fully Bayesian approach to construct a criterion. Gelman et al. (2014) claims the WAIC is preferable to the DIC. They also explain why the so-called *Bayes information criterion* (BIC) is not comparable to the criteria we have already discussed.

Finally, we may well question why we need to select one model when we can use information from all the models we have considered. The idea is to assign a prior probability to each model which is then updated to a posterior probability for each model. We can then combine the information from all the models to make inferences or predictions, specifically. This idea is called *Bayesian model averaging* and is discussed in Draper (1995).

1.6 Hypothesis Testing

In the frequentist view of hypothesis testing, the null hypothesis H_0 is usually a single point while alternative hypothesis H_1 is everything else. We propose a test statistic suitable for choosing between these two hypotheses. We then compute the probability that, under the null hypothesis, a test statistic equal to or more extreme than the observed test statistic would be observed. This is called the p-value. Although it is frequently misinterpreted as such, the p-value is not the probability that the null hypothesis is true. The common practice is that, if the p-value is less than 0.05, we declare the result to be "statistically significant" and reject the null hypothesis. If p-value is not so small, we fail to reject the null hypothesis.

This procedure is sometimes called "null hypothesis significance testing" (NHST). There is a vast body of debate about this procedure that we will not reiterate here. In spite of the many criticisms, NHST is strongly embedded in scientific practice and there is no imminent sign of its demise. The practitioner needs to get his or her results published and/or accepted so it is impossible to ignore NHST. With this in mind, we should seek to develop comparable answers within the Bayesian framework that address the same questions that NHST attempts to answer. There is a wide acceptance of the idea of choosing between hypotheses. The Bayesian approach may reject the NHST formulation but it needs to provide its own answers that skeptics will find acceptable. This may mean providing answers which are similar to NHST answers even if the underlying motivation is different.

Suppose we have two hypotheses, H_0 and H_1, that we wish to compare. We assign prior probabilities, $p(H_0)$ and $p(H_1)$ to these hypotheses. We then find that:

$$\frac{p(H_0|\mathbf{y})}{p(H_1|\mathbf{y})} = \frac{p(\mathbf{y}|H_0)}{p(\mathbf{y}|H_1)} \times \frac{p(H_0)}{p(H_1)}.$$

The term $p(\mathbf{y}|H_0)/p(\mathbf{y}|H_1)$ is called the *Bayes factor* and $p(H_0)/p(H_1)$ is called the *prior odds*. Thus the resulting *posterior odds* is obtained by multiplying the prior odds by the Bayes factor. Admittedly, we may find it difficult to specify the prior odds which determines the relative belief we give to the two hypotheses but the Bayes factor tells us how much that belief is changed by the data. We can simply report the Bayes factor and let the reader decide based on their own prior odds. To compute the Bayes factor, we need the marginal probabilities of each model, $p(\mathbf{y})$, but this is just the normalizing constant we needed earlier in the computation of the posterior.

The Bayes factor is attractive since it avoids the need to commit about priors on the hypotheses and yet priors are still required to fit the two models under consideration. So one does not avoid the specification of some priors. Indeed, one must be quite careful in setting the priors for the two models so as not to favor one model over the other. Although the Bayes factor has some philosophical appeal, it has never really caught on in scientific publications.

Very often, H_0 represents a restriction of the form $\theta_k = 0$. Sometimes k is a single index and sometimes it is a set of indices. This suggests we might want to find the posterior probability $P(\theta_k = 0)$. We could compute a Bayesian model and find this probability. The difficulty is that, assuming we have assigned a continuous prior to θ_k, we will have a continuous posterior and $P(\theta_k = 0) = 0$. One possible solution is to use a prior that assigns positive probability to the event $\theta_k = 0$, but this is not easy to do. First we must ask what positive probability should be chosen and secondly, priors which have combinations of discrete and continuous elements like this result in very difficult computations. A wider philosophical objection is that we would not reasonably believe any continuous parameter to be exactly any specific value in many applications. In other words, the point null hypothesis might be clearly untrue, even without seeing the data.

One possible solution is to compute the posterior probability $P(|\theta_k| < c)$. The c is chosen as large as possible but with the proviso that we can still claim that θ_k is negligibly small. Provided we have a way of setting c that others will find convincing, we have a solution. Unfortunately, choosing c may not be easy. We can report the marginal posterior so that the reader can choose their own c and make their own calculation but scientific publications usually require us to be more assertive.

We can borrow one idea from frequentist hypothesis testing — in many situations, we can perform the hypothesis test $H_0 : \theta_k = 0$ by checking whether zero falls within a 95% confidence interval. If it does, we do not reject the null, but if it falls outside, we reject the null. Now some people like to have more than just a binary outcome — they want a measure of how reasonable or unreasonable the null hypothesis is. Although, the p-value is not really what they were asking for, it is the kind of answer they want. One way to compute the p-value via confidence intervals is to find the largest p such that the $100(1-p)\%$ confidence interval does not contain zero. This is the p-value. We can reproduce the same calculations using the marginal posterior distribution. We can make the 95% credible interval and check whether it contains zero. We can also see how large we can make the credible interval without containing zero. For relatively symmetric posteriors, suppose the posterior mean is positive. We can then compute our so-called p-value as

$$p = 2P(\theta < 0).$$

We would use the other tail if the posterior mean is negative. In truth, this is just a measure of where zero falls on the posterior distribution but it has the advantage of being like a p-value and it would not be unreasonable to use it in the same manner.

1.7 Bayesian Computation

Bayesian theory is clear but the computation of the posterior and the other quantities of interest is not always easy. There are three main ways to compute the posterior which we discuss in this section.

1.7.1 Exact

A conjugate prior results in a posterior density of the same family of distributions. Usually, this means we can explicitly state the parameters of the posterior in terms of the parameters of the prior and some function of the data. There is a relatively small number of conjugate priors that apply to quite simple models, usually without covariates. More complex models have parameters of different types and so more than one prior, often from different families, is required. In such circumstances, conjugacy cannot be achieved.

Historically, models where conjugate priors could be used provided the only full kind of Bayesian analysis possible. This seriously limited the uptake of Bayesian statistics, no matter how philosophically appealing it may have been. Now since we have powerful computation, we can consider a wide range of models for Bayesian analysis and conjugacy has become a side issue, convenient for undergraduate examinations but only useful in limited circumstances.

The posterior distribution is proportional to the product of the likelihood and the prior. To compute the posterior mode, we merely need to find the value of θ that maximizes this product. We do not need to know the normalization to compute the mode. This is called the *maximum a posteriori* (MAP) estimator. Notice that if we use a flat prior, this is also the MLE. Since this method avoids the need for normalization, it only requires some numerical optimization at worst. The drawback is that it is only a point estimate with no assessment of uncertainty. This is only a partial solution at best.

1.7.2 Sampling

The goal of sampling-based methods is to generate samples from the posterior distribution. Although various schemes have been developed for simpler models, the main contenders in this category generate a Markov chain whose stationary distribution is the posterior distribution. The idea dates back to the work of Metropolis et al. (1953) and Hastings (1970), but became more widely used in statistics after Gelfand and Smith (1990). Collectively, these methods are described as Markov chain Monte Carlo (MCMC). Since this book is about a competitor to MCMC, we do not propose to explain how MCMC works. We will just give an overview of the steps of an MCMC analysis.

Let's suppose we have decided on our model and prior and have our data readily available. We could derive an MCMC-based method and implement it in code, but more likely, we would want to use a package in which our choice of prior and model can be fit. Let's also suppose that we restrict ourselves to software that is an R package or can be called from R. We have a wide choice.

The Bayesian task view on CRAN lists a large selection of R packages which implement MCMC for a specific class of models. These packages are typically called in R but written in a compiled language, usually C. One advantage of these packages is that they can be called from R without needing to understand an additional language. The other advantage is that they run faster because the most expensive parts of the computation are written in a compiled language that can be optimized especially for the chosen model class. The disadvantage is that they are quite inflexible in that they only work for a narrow class of models and limited choice of priors. Perhaps the most general example is the `MCMCglmm` package of Hadfield (2010) which will handle a good subset of the models considered in this book.

Moving from the specific to the general, there are several general software packages that allow the user to fit a wide range of Bayesian models. These are written in a specialist language that one must understand in order to specify the model and prior. The programs run separately from R but can be called from R and can send back the results for processing within R.

The first of these general software packages for Bayesian computing was BUGS (Bayesian inference Using Gibbs Sampling) introduced in 1989. The book by Lunn et al. (2012) provides a general introduction. BUGS developed into WinBUGS and more recently OpenBUGS. Just another Gibbs sampler (JAGS) is another variant of BUGS. It was introduced by Plummer et al. (2003). More recently, STAN became available and is described in Stan Development Team (2016). BayesX, described in Umlauf et al. (2015), is a somewhat more specific Bayesian computing package that partly uses MCMC.

There are several steps to an MCMC-based analysis of a Bayesian model:

1. You need to write code specifying your chosen model and prior and how the data can be passed to the program. For users who are more familiar with fitting a model with a single line command in R, this requires significantly more learning and expertise.
2. As mentioned earlier, an MCMC method generates a Markov chain whose

stationary distribution we desire. But the chain has to start somewhere so initial values need to be specified. Ideally, it should not matter too much where we start but it is quite possible to specify implausible, or just unfortunate, starting values and the chain may never converge. Some experimentation with different starting values is advisable and some wisdom is necessary to recognize when the chain has gone awry. STAN uses four sets of random starting values — this is a sensible precaution no matter what MCMC package you choose.

3. We allow the chain to run for some time. After some point, we need to recognize when the chain has reached a stationary state. All the observations before this point are discarded as "burn-in" or "warm-up" values. Diagnostic methods are available to help us determine when and if a stationary distribution has been achieved. Even so, it is possible for the chain never to venture into some plausible reasons of the parameters space but the diagnostics may appear entirely satisfactory.
4. Observations from the chain are positively correlated so the sample from the posterior generated by the chain is not as valuable as an independent sample. This means that the sample needs to be larger. Sometimes, users "thin" the sample by keeping only every n^{th} observation but this is an economy measure, saving the cost of storage, rather than a necessity.
5. The end product is a sample from the posterior and not the posterior itself. In principle, you can estimate any functional of the posterior from this information. In practice, you'll need to trade off the cost of generating the samples against the accuracy of the estimation. For estimating a posterior mean, you need fewer samples, but for a more extreme quantile, you'll need substantially more.

As computing speeds have increased and MCMC methods have improved over time, the range of models and sizes of dataset that these models can reasonably tackle has increased. Even so, there remain some combinations of model and data for which MCMC methods either fail or would take an unreasonable amount of time to run. Even for the models which can be fit with a little patience, the cost of experimenting with multiple models is prohibitive. This is an obstacle to the modern style of data analysis which considers many possible models.

We must recognize the tremendous success of MCMC methods but there are two main drawbacks: they are difficult to use and they are slow.

1.7.3 Approximation

Exact solutions to Bayes modeling problems are mostly restricted to the narrow and simple class of models where conjugacy can be used. Approximate solutions require the use of numerical integration. As we saw earlier, the main difficulty in computing the posterior is that this requires high dimensional approximation. In the next chapter we will describe how methods of quadrature and the Laplace approximation in particular can be used to produce accurate approximations. In statistics, these methods started from the seminal paper of Tierney and Kadane (1986). In addition to INLA, there are other methods based on approximation such as variational Bayes. More about these other methods can be found within the Bayesian task view at:

```
cran.r-project.org/web/views/Bayesian.html
```

The advantage of INLA over MCMC-based methods is that it is much faster. In larger cases, INLA finds a solution where MCMC methods would take far too long. For smaller problems, the speed of computation allows us to take a more exploratory and interactive approach to model construction and testing. In frequentist analyses, the MLE can usually be computed quickly which means we can devote much more attention to model diagnostics and exploring the model space. This exploration and checking is crucial to executing an effective analysis. Getting the model right is more important than the method of inference used. If the cost of fitting a model is high enough, we will have to economize on the exploration. This is the main drawback of slow MCMC methods. In contrast, INLA is fast enough that we can explore with freedom.

Furthermore, INLA is easier to use. There is no separate programming language we need to learn and we do not need to navigate the difficult waters of MCMC diagnostics. INLA is, for the most part, non-random, meaning that analyses are more reproducible. For MCMC, the outcome is random. We can make it reproducible by specifying the random seed but a truly independent replication would come out differently. We can mitigate this by taking a large enough sample but the problem cannot be entirely erased.

Nevertheless, we do not want to exaggerate the case in favor of INLA. The most general MCMC-based packages cover a wider range of models than INLA ever will. As we shall see in the next chapter, there is a broad, but limited, class to which INLA applies. Furthermore, if you really care about your data, it is wise to make sure your analysis is not sensitive to some peculiarity of the software you have used. Once you have selected the model using an INLA-based analysis, it is well worthwhile to repeat the fitting of the model using MCMC. If the results agree, you will have greater qualitative confidence in your conclusions. If they disagree, it will be interesting to find out why.

2

Theory of INLA

The integrated Laplace approximation (INLA) methodology was first introduced by Rue et al. (2009), followed by developments in Martins et al. (2013), and is most recently reviewed in Rue et al. (2017). It is a deterministic approach to approximate Bayesian inference for latent Gaussian models (LGMs). In most cases INLA is both faster and more accurate than MCMC alternatives for LGMs. The INLA `R` package (see `www.r-inla.org`) can be used for quick and reliable Bayesian inference in practical applications. A list of recent applications of INLA can be found in Rue et al. (2017).

In this chapter we reveal the "secrets" that make INLA successful. There are three key components required by INLA: the LGM framework, a Gaussian Markov random field (GMRF) and the Laplace approximation. We introduce these components using a top-down approach, starting with LGMs and the type of statistical models that may be viewed as LGMs (Section 2.1). We then discuss the concept of a GMRF, a class of Gaussian processes that are computationally efficient within this formulation (Section 2.2). Finally, we illustrate how INLA makes use of the Laplace approximation, an old technique for approximating integrals, to perform accurate and fast Bayesian inference on LGMs (Section 2.3).

The theory behind the INLA method is not easy and some readers may wish to skip Section 2.3 at the first attempt. Even so, it is worth making the effort to understand LGMs and GMRFs as this will allow you to distinguish which models can be attempted with INLA from those for which INLA is impossible or just impractical.

2.1 Latent Gaussian Models (LGMs)

The INLA approach is restricted to a specific class of models, so called latent Gaussian models (LGMs). LGMs have a wide-ranging list of applications, and most structured Bayesian models are in fact of this form (see e.g., Fahrmeir and Tutz, 2001). In this book we focus on regression models, the most extensively used subset of LGMs. Other common LGMs include dynamic models, spatial models and spatial-temporal models (see e.g., Rue et al., 2009; Blangiardo and Cameletti, 2015).

A simple example of an LGM is the Bayesian generalized linear model (GLM) (described in detail in Chapter 4). It corresponds to the linear predictor

$$\eta_i = \beta_0 + \beta_1 x_{i1}, \quad i = 1, \ldots, n,$$

where β_0 is the intercept, x_{i1} is the covariate and the slope (or linear effect) is β_1. The response y_i is assumed to follow a distribution from an exponential family, and its (conditional) mean μ_i is associated with η_i via a link function $g()$ such that $\eta_i = g(\mu_i)$. There is a variety of likelihood models available in INLA package and the Gaussian model is the default choice. One may view the list by

```
library(INLA)
names(inla.models()$likelihood)
```

For each model a detailed description and an example of usage are provided on INLA's website (www.r-inla.org/models/likelihoods). Although the likelihood itself does not have to be Gaussian, each *latent* η_i must follow a normal distribution given its hyperparameter(s) in a LGM. It means that we must use Gaussian priors on β_0 and β_1, i.e., $\beta_0 \sim N(\mu_0, \sigma_0^2)$ and $\beta_1 \sim N(\mu_1, \sigma_1^2)$, which results in $\eta_i \sim N(\mu_0 + \mu_1 x_{i1}, \sigma_0^2 + \sigma_1^2 x_{i1}^2)$. With some linear algebra we can show that $\boldsymbol{\eta} = (\eta_1, \ldots, \eta_n)'$ is a Gaussian process with mean vector $\boldsymbol{\mu}$ and covariance matrix Σ. The hyperparameters σ_0^2 and σ_1^2 are to be either fixed or estimated by taking hyperpriors on them.

We can also take a more general additive form on η_i:

$$\eta_i = \beta_0 + \sum_{j=1}^{J} \beta_j x_{ij} + \sum_{k=1}^{K} f_k(z_{ik}), \tag{2.1}$$

by adding more covariates and *model components* $f_k()$, which can be used to relax the linear relationship of the covariate, or introduce random effects, or both. For modeling (smooth) nonlinear effects, one may use parametric nonlinear (e.g., quadratic) terms, or nonparametric models such as random walk models (Fahrmeir and Tutz, 2001; Rue and Held, 2005), P-spline models (Lang and Brezger, 2004), and Gaussian processes (Besag et al., 1995). To account for overdispersion caused by unobserved heterogeneity or for correlation in longitudinal data, we may consider using random effects in the model, which can be introduced by letting f_k follow independent zero-mean normal distributions (Fahrmeir and Lang, 2001). In many applications, the linear predictor is a sum of various model components, such as random effects, and both linear and smooth effects of some covariates, as shown in model (2.1). Such models can be termed generalized additive models (GAMs) (see details in Chapter 9). For each linear effect and each model component we must take a Gaussian prior that has either a univariate or multivariate normal density to make this additive $\boldsymbol{\eta}$ be Gaussian as required by the LGM. As we will see in the next section and following chapters in the book, there exists a class of Gaussian models, called Gaussian Markov random field (GMRF) models, which are quite flexible and efficient with regard to modeling various possible effects used in an LGM.

Now let's write a generic three-stage hierarchical model formulation for each LGM. Letting $\boldsymbol{y} = (y_1, \ldots, y_n)'$, we assume in the first stage each variable in $\boldsymbol{y}$ is conditionally independent with a certain exponential family distribution

$$\boldsymbol{y} \mid \boldsymbol{\eta}, \boldsymbol{\theta}_1 \sim \prod_{i=1}^{n} p(y_i \mid \eta_i, \boldsymbol{\theta}_1),$$

given η and hyperparameters θ_1. In the second stage we specify η to be a latent Gaussian random field with density function given by

$$p(\eta \mid \theta_2) \propto |\mathbf{Q}_{\theta_2}|_+^{1/2} \exp\Big(-\frac{1}{2}\eta' \mathbf{Q}_{\theta_2}\eta\Big),$$

where Q_{θ_2} is a semi-positive definite matrix that depends on hyperparameters θ_2, and $|Q_{\theta_2}|_+$ denotes the product of its non-zero eigenvalues. The matrix Q_{θ_2} is called the *precision matrix* that describes the underlying dependence structure of the data, and its inverse (if it exists) is a covariance matrix (see Section 2.2).

In the final stage we assume $\theta = (\theta_1, \theta_2)$ follow a prior distribution $\pi(\theta)$. It could be a joint distribution or a product of several distributions. As a result, the joint posterior distribution of η and θ reads

$$\begin{aligned} p(\eta, \theta \mid y) &\propto \pi(\theta)\, \pi(\eta \mid \theta_2) \prod_i p(y_i \mid \eta_i, \theta_1) \\ &\propto \pi(\theta) |Q_{\theta_2}|^{1/2} \exp\Big(-\frac{1}{2}\eta' Q_{\theta_2}\eta + \sum_i \log \pi(y_i \mid \eta_i, \theta_1)\Big). \end{aligned} \quad (2.2)$$

Bayesian inference on this LGM is derived from this expression. In INLA, the (nested) Laplace approximation method, as will be described in Section 2.3, is applied to (2.2) to obtain (approximate) posterior distributions for every unknown parameter.

Unfortunately, not every LGM can be fitted efficiently by INLA. We (in general) need the following additional assumptions:

1. The number of hyperparameters θ should be small, typically 2 to 5, but not exceeding 20.
2. When n is big (10^4 to 10^5) η must be a Gaussian Markov random field (GMRF) (see Section 2.2).
3. Each y_i only depends on one component of η, e.g., η_i.

These assumptions are crucial both for computational reasons and to ensure that the Laplace approximations are accurate. Fortunately, many commonly used in LGMs literature satisfy these assumptions. Note that the first two assumptions must be strictly satisfied, while the last one can be somehow relaxed to a certain extent (see Martins et al., 2013, Section 4.5)

2.2 Gaussian Markov Random Fields (GMRFs)

The latent field η should not only be Gaussian, but also be a Gaussian Markov random field (GMRF), in order for INLA to work efficiently. To explain the concept, let's consider a simple GMRF case where we let η_i follow a *first-order autoregressive process*, AR(1). Assuming the current value is based on the immediately preceding value, the AR(1) model can be defined as:

$$\begin{cases} \eta_1 \sim N\left(0, \sigma_\eta^2/(1-\rho^2)\right), \\ \eta_i \mid \eta_{i-1}, \ldots, \eta_1 \sim N(\rho\eta_{i-1}, \sigma_\eta^2), \quad i = 2, \ldots, n, \end{cases}$$

where given other variables each η_i (except η_1) follows a normal distribution with mean $\rho\eta_{i-1}$ and constant variance σ_η^2. The parameter ρ ($|\rho| < 1$) is the correlation between η_i and η_{i-1}. We can prove that the marginal distribution of each η_i is Gaussian with mean 0 and variance $\sigma_\eta^2/(1-\rho^2)$. The covariance between η_i and η_j is $\sigma_\eta^2\rho^{|i-j|}/(1-\rho^2)$, which decays as the distance $|i-j|$ increases. As a result, the AR(1) η is a Gaussian process with mean vector of zeroes and covariance matrix $\mathbf{\Sigma}$, i.e., $\boldsymbol{\eta} \sim N(\mathbf{0}, \mathbf{\Sigma})$. $\mathbf{\Sigma}$ is an $n \times n$ dense matrix. This is troublesome because the calculations involving such a matrix are generally expensive, making it apparently inconvenient to use an AR(1) model for large data, especially under Bayesian framework.

Fortunately, the AR(1) is a special Gaussian process which has a sparse precision matrix. This can be shown by computing the joint distribution of $\boldsymbol{\eta} = (\eta_1, \ldots, \eta_n)$ using

$$p(\boldsymbol{\eta}) = p(\eta_1)p(\eta_2 \mid \eta_1)p(\eta_3 \mid \eta_1, \eta_2) \cdots p(\eta_n \mid \eta_{n-1}, \ldots, \eta_1).$$

This turns out to be multivariate normal with precision matrix

$$\boldsymbol{Q} = \sigma_\eta^{-2} \begin{pmatrix} 1 & -\rho & & & \\ -\rho & 1+\rho^2 & -\rho & & \\ & \ddots & \ddots & \ddots & \\ & & -\rho & 1+\rho^2 & -\rho \\ & & & -\rho & 1 \end{pmatrix},$$

a *banded matrix* that has nonzero elements only on the main diagonal, the first diagonal below this, and the first diagonal above the main diagonal. The reason $\boldsymbol{Q}$ has this sparse structure is that given other variables, η_i only depends on the immediately preceding η_{i-1}. In other words, η_i and η_j are *conditionally independent* for all $|i-j| > 1$. For example, η_2 and η_4 are conditionally independent because

$$\begin{aligned} p(\eta_2, \eta_4 \mid \eta_1, \eta_3) &= p(\eta_2 \mid \eta_1)p(\eta_4 \mid \eta_1, \eta_2, \eta_3) \\ &= p(\eta_2 \mid \eta_1)p(\eta_4 \mid \eta_3). \end{aligned}$$

The conditional density of η_2 does not depend on η_4 and vice versa. Therefore their joint conditional density can be written as the product of two unrelated conditional densities. This property of an AR(1) process is reflected in $\boldsymbol{Q}$ by having the element in the ith row and j^{th} column be $Q_{ij} = 0$ for $|i-j| > 1$, knowing that the *conditional correlation* between η_i and η_j is given by $-Q_{ij}/\sqrt{Q_{ii}Q_{jj}}$. The higher order the AR process is, the less sparse the corresponding precision matrix $\boldsymbol{Q}$ is, because η_i conditionally depends on more preceding variables. In summary, the AR process is a

Gaussian process with a *conditional independence* property, making the corresponding precision matrix have a particular sparse structure.

We are now ready for a general definition of GMRF. We say η is a GMRF if it has a multivariate normal density with additional conditional independence (also called the "Markov property"). There exists a variety of GMRFs and they have been extensively used in various fields (see Rue and Held, 2005). In the following chapters we will present a class of GMRFs that are particularly useful in Bayesian regression framework. They have different conditional independence structures to reflect our belief in how the random variables for each field locally depend on each other. However, there is one thing in common between different GMRFs: they all have a sparse precision matrix. This provides a huge computational benefit when making Bayesian inference, as calculating with a sparse $n \times n$ matrix costs much less than calculating with a dense matrix. Let's take the AR process as an example. Recall that the covariance matrix of AR(1) is an $n \times n$ dense matrix and computing its inverse costs $O(n^3)$ time. However, the corresponding precision matrix is tridiagonal and can be factorized in $O(n)$ time (see Rue and Held, 2005, Section 2.4). The memory requirement is also reduced from $O(n^2)$ to $O(n)$, which makes it much easier to run larger models.

When using GMRFs to construct the additive models like the one in (2.1), the following fact provides some of the "magic" used in INLA: The joint distribution of η in (2.1) is also a GMRF and its precision matrix consists of sums of the precision matrices of the covariates and the model components. We will see in the next section that this joint distribution needs to be formed many times in INLA, as it depends on the hyperparameters $\boldsymbol{\theta}$. Fortunately, we can treat it as a GMRF with a precision matrix that is easy to compute. It is one of the key reasons that the INLA approach is so efficient. Also, the sparse structure of the precision matrix boosts computational efficiency. See a detailed demonstration in Rue et al. (2017).

2.3 Laplace Approximation and INLA

Laplace approximation is used to approximate integral $I_n = \int_x \exp(nf(x))dx$ as $n \to \infty$. Letting x_0 be the mode of $f(x)$, we do a Taylor expansion on $f(x)$ at x_0:

$$\begin{aligned} I_n &\approx \int_x \exp\left(n\left(f(x_0) + (x-x_0)f'(x_0) + \frac{1}{2}(x-x_0)^2 f''(x_0)\right)\right) dx \\ &= \exp(nf(x_0)) \int \exp\left(\frac{n}{2}(x-x_0)^2 f''(x_0)\right) dx \\ &= \exp(nf(x_0)) \sqrt{\frac{2\pi}{-nf''(x_0)}} = \tilde{I}_n, \end{aligned}$$

where $f'(x_0)$ and $f''(x_0)$ denote the first-order and second-order derivatives at x_0, respectively. Note that I_n is the Gaussian integral and $f'(x_0) = 0$ since x_0 is the mode. If $nf(x)$ is interpreted as the sum of log-likelihoods and x as the unknown parameter, the Laplace approximation will be exact as $n \to \infty$, if the central limit theorem holds. It can be immediately extended to higher dimensional integrals with a good error rate (Rue et al., 2017).

Let's see how the Laplace approximation is the foundation stone of INLA with a very small dataset. Suppose we observe two counts, i.e., $y_1 = 1$ and $y_2 = 2$, from a Poisson distribution with mean λ_i $(i = 1,2)$. We use a log link to associate λ_i with linear predictor η_i, and model η_i with an AR(1) process. The resulting posterior distribution is given by

$$p(\eta \mid \boldsymbol{y}) \propto \exp\left(-\frac{1}{2}\boldsymbol{\eta}' \boldsymbol{Q_\theta} \boldsymbol{\eta}\right) \exp\left(\eta_1 + 2\eta_2 - e^{\eta_1} - e^{\eta_2}\right), \quad \boldsymbol{Q_\theta} = \tau \begin{pmatrix} 1 & -\rho \\ -\rho & 1 \end{pmatrix},$$

where $\boldsymbol{\theta} = (\tau, \rho)$, $\tau > 0$ is the scale parameter and $-1 < \rho < 1$ is the autocorrelation parameter. For easy demonstration let's assume both τ and ρ to be constants first. The task now is to approximate posterior marginals for η_1 and η_2. One may apply the two-dimensional Laplace approximation directly to $p(\boldsymbol{\eta} \mid \boldsymbol{\theta}, \boldsymbol{y})$, and then obtain the Gaussian approximation for each marginal. However, such approximations fail to correctly capture both location and skewness in the marginals. The performance can be much improved by applying a sequence of Gaussian approximations using the Laplace approximation. More specifically, we approximate the marginal of η_1 using the formula:

$$p(\eta_1 \mid \boldsymbol{y}) = \frac{p(\boldsymbol{\eta} \mid \boldsymbol{y})}{p(\eta_2 \mid \eta_1)} \approx \frac{p(\boldsymbol{\eta} \mid \boldsymbol{y})}{\tilde{p}(\eta_2 \mid \eta_1)},$$

where $\tilde{p}(\eta_2 \mid \eta_1)$ is the Gaussian approximation of the full conditional of η_2 for each value of η_1. The marginal of η_2 can be approximated in a similar way. As shown in (Rue et al., 2017), the resulting approximated marginal is quite accurate and only runs into slight trouble where the likelihood changes abruptly. This approach provides more accuracy because each conditional distribution is much closer to Gaussian than their joint distribution. This is the key idea of the INLA method: Laplace approximations are only applied to densities that are near-Gaussian, replacing complex dependencies with conditioning.

We now extend the example shown above to a more general LGM framework. In the first stage, we observe n counts y_i that follow independent Poisson distributions with mean λ_i for $i = 1, \ldots, n$. In the second stage, we model λ_i using a log link and first-order autoregressive process. The first two stages of this LGM can be written as

$$\begin{aligned} p(\boldsymbol{y} \mid \boldsymbol{\eta}, \boldsymbol{\theta}) &\propto \prod_{i=1}^{n} \exp\left(\eta_i y_i - e^{\eta_i}\right)/y_i!, \\ \pi(\boldsymbol{\eta} \mid \boldsymbol{\theta}) &\propto |\boldsymbol{Q_\theta}|^{1/2} \exp\left(-\frac{1}{2}\boldsymbol{\eta}' \boldsymbol{Q_\theta} \boldsymbol{\eta}\right), \end{aligned}$$

where $\boldsymbol{\theta} = (\tau, \rho)$ denotes the hyperparameters, and their prior $\pi(\boldsymbol{\theta})$ needs to be specified in the third stage. Then, the joint posterior distribution of the unknowns reads:

$$p(\boldsymbol{\eta}, \boldsymbol{\theta} \mid \boldsymbol{y}) \propto \pi(\boldsymbol{\theta}) \pi(\boldsymbol{\eta} \mid \boldsymbol{\theta}) p(\boldsymbol{y} \mid \boldsymbol{\eta}, \boldsymbol{\theta}).$$

Our goal is to accurately approximate the posterior marginals $p(\eta_i \mid \boldsymbol{y})$ for $i = 1, \ldots, n$ and $p(\theta_j \mid \boldsymbol{y})$ for $j = 1, 2$. The direct implementation of Laplace approximations is problematic here because we have a product of a Gaussian and a (very) non-Gaussian. The strategy used in INLA is to reformulate the problem as a series of subproblems and only apply the Laplace approximation to the densities that are almost Gaussian. More specifically, the method can be divided into three main tasks: firstly propose an approximation $\tilde{p}(\boldsymbol{\theta} \mid \boldsymbol{y})$ to the joint posterior of $p(\boldsymbol{\theta} \mid \boldsymbol{y})$, secondly propose an approximation $\tilde{p}(\eta_i \mid \boldsymbol{\theta}, \boldsymbol{y})$ to the marginals of the conditional distribution of η_i given the data and the hyperparameters $p(\eta_i \mid \boldsymbol{\theta}, \boldsymbol{y})$, and finally explore $\tilde{p}(\boldsymbol{\theta} \mid \boldsymbol{y})$ and use it for numerical integration. As a result, the approximated posterior marginals of interest returned by INLA have the following form:

$$\tilde{p}(\eta_i \mid \boldsymbol{y}) = \sum_k \tilde{p}(\eta_i \mid \boldsymbol{\theta}^{(k)}, \boldsymbol{y}) \tilde{p}(\boldsymbol{\theta}^{(k)} \mid \boldsymbol{y}) \, \Delta\theta^{(k)}, \tag{2.3a}$$

$$\tilde{p}(\theta_j \mid \boldsymbol{y}) = \int \tilde{p}(\boldsymbol{\theta} \mid \boldsymbol{y}) d\boldsymbol{\theta}_{-j}, \tag{2.3b}$$

where $\boldsymbol{\theta}_{-j}$ denotes the vector of $\boldsymbol{\theta}$ with its jth element excluded, and $\tilde{p}(\boldsymbol{\theta}^{(k)} \mid \boldsymbol{y})$ are the density values computed during an exploration on $\tilde{p}(\boldsymbol{\theta} \mid \boldsymbol{y})$. Below we describe how each task is completed in detail.

Approximating $p(\boldsymbol{\theta} \mid \boldsymbol{y})$

The Laplace approximation of a joint posterior of the hyperparameters is given by

$$\tilde{p}(\boldsymbol{\theta} \mid \boldsymbol{y}) \propto \left. \frac{p(\boldsymbol{\eta}, \boldsymbol{\theta} \mid \boldsymbol{y})}{\tilde{p}(\boldsymbol{\eta} \mid \boldsymbol{\theta}, \boldsymbol{y})} \right|_{\boldsymbol{\eta} = \boldsymbol{\eta}^*(\boldsymbol{\theta})}, \tag{2.4}$$

where $\tilde{p}(\boldsymbol{\eta} \mid \boldsymbol{\theta}, \boldsymbol{y})$ is a Gaussian approximation to the full conditional of $\boldsymbol{\eta}$ obtained by matching the modal configuration and the curvature at the mode, and $\boldsymbol{\eta}^*(\theta)$ is the mode of the full conditional for $\boldsymbol{\eta}$ for a given value of $\boldsymbol{\theta}$. This Laplace approximation will be exact if $p(\boldsymbol{\eta} \mid \boldsymbol{\theta}, \boldsymbol{y})$ is Gaussian. It is noteworthy that the approximation $\tilde{p}(\boldsymbol{\eta} \mid \boldsymbol{\theta}, \boldsymbol{y})$ has the following form

$$\tilde{p}(\boldsymbol{\eta} \mid \boldsymbol{\theta}, \boldsymbol{y}) \propto |\boldsymbol{R}_{\boldsymbol{\theta}}|^{1/2} \exp\left(-\frac{1}{2}(\boldsymbol{\eta} - \boldsymbol{\mu}_{\boldsymbol{\theta}})' \boldsymbol{R}_{\boldsymbol{\theta}} (\boldsymbol{\eta} - \boldsymbol{\mu}_{\boldsymbol{\theta}})\right), \tag{2.5}$$

where $\boldsymbol{\mu}_{\boldsymbol{\theta}}$ is the location of the mode, $\boldsymbol{R}_{\boldsymbol{\theta}} = \boldsymbol{Q}_{\boldsymbol{\theta}} + \text{diag}(\boldsymbol{c}_{\boldsymbol{\theta}})$ and the vector $\boldsymbol{c}_{\boldsymbol{\theta}}$ on the diagonal contains the negative second derivatives of the log-likelihood at the mode with respect to η_i (see Rue and Held, 2005, Section 4.4.1). There are two big advantages of using (2.5). First, it is a GMRF with the same dependence structure as from $\boldsymbol{Q}_{\boldsymbol{\theta}}$ because the impact from accounting for the observations is only a shift in the mean and the diagonal of the precision matrix. Therefore, it is easy to evaluate (2.5) for every value of $\boldsymbol{\theta}$. Second, the Gaussian approximation is likely to be quite accurate since the impact of conditioning on the observations is only on the diagonal, which shifts the mean, reduces the variance and might introduce some skewness into the marginals. More importantly, it does not change the Gaussian dependence structure.

We can improve approximation (2.4) using variance-stabilizing transformations of $\boldsymbol{\theta}$, such as log transformation, the Fisher transform of correlations, etc. For example, INLA uses the following transformations for $\boldsymbol{\theta} = (\theta_1, \theta_2)$ of the AR(1) model:

$$\begin{cases} \theta_1 = \log(\kappa), \quad \kappa = \tau(1-\rho^2), \\ \theta_2 = \log\left(\frac{1+\rho}{1-\rho}\right), \end{cases} \tag{2.6}$$

where κ is the marginal precision. Additionally, we can use the Hessian matrix at the mode to construct almost independent linear combinations (or transformations) of $\boldsymbol{\theta}$ (see Rue et al., 2009). These transformations tend to diminish long tails and reduce skewness, which gives much simpler and better-behaved posterior densities.

Approximating $p(\eta_i \mid \boldsymbol{\theta}, \boldsymbol{y})$

For approximating $p(\eta_i \mid \boldsymbol{\theta}, \boldsymbol{y})$, there are three options available and they vary in terms of speed and accuracy. The fastest option is to use the marginals of the Gaussian approximation $\tilde{p}(\boldsymbol{\eta} \mid \boldsymbol{\theta}, \boldsymbol{y})$ already computed by (2.5). The only extra cost to obtain $\tilde{p}_G(\eta_i \mid \boldsymbol{\theta}, \boldsymbol{y})$ is the computation of the marginal variances from the sparse precision matrix $\boldsymbol{R_\theta}$ (see Rue et al., 2009, for details). Although it often gives reasonable results, the Gaussian approximation can cause errors in the location and/or the skewness (Rue and Martino, 2007).

A more accurate approach is to perform a Laplace approximation one more time

$$\tilde{p}_{LA}(\eta_i \mid \boldsymbol{\theta}, \boldsymbol{y}) \propto \left. \frac{p(\boldsymbol{\eta}, \boldsymbol{\theta} \mid \boldsymbol{y})}{\tilde{p}(\boldsymbol{\eta}_{-i} \mid \eta_i, \boldsymbol{\theta}, \boldsymbol{y})} \right|_{\boldsymbol{\eta}_{-i} = \boldsymbol{\eta}^*_{-i}(\eta_i, \boldsymbol{\theta})}, \tag{2.7}$$

for $i = 1, \ldots, n$, where $\tilde{p}(\boldsymbol{\eta}_{-i} \mid \eta_i, \boldsymbol{\theta}, \boldsymbol{y})$ is the Gaussian approximation with the modal configuration $\boldsymbol{\eta}^*_{-i}(\eta_i, \boldsymbol{\theta})$. However, this method involves the location of the mode and the factorization of a $(n-1) \times (n-1)$ matrix many times for each i, which is simply too demanding, especially when n is large.

A third option denoted by $\tilde{p}_{SLA}(\eta_i \mid \boldsymbol{\theta}, \boldsymbol{y})$ is called the *simplified Laplace approximation*, which is obtained by doing a Taylor expansion on the numerator and denominator of the expression (2.7) up to third order, thus correcting the Gaussian approximation for location and skewness with a much lower cost when compared to $\tilde{p}_{LA}(\eta_i \mid \boldsymbol{\theta}, \boldsymbol{y})$.

We refer to Rue et al. (2009) for a detailed description of the Gaussian, Laplace and simplified Laplace approximations. These three approximation strategies can be specified via "`strategy=`" in `control.inla` argument in `inla()` function. They are termed as "gaussian", "simplified.laplace" and "laplace", respectively. For example, when the Laplace approximation strategy is needed we may specify the argument as follows:

```
inla(..., control.inla = list(strategy = 'laplace'), ...)
```

The default strategy is the simplified Laplace approximation.

Exploring $\tilde{p}(\boldsymbol{\theta} \mid \boldsymbol{y})$

When approximating $p(\eta_i \mid \boldsymbol{y})$ we only need $\tilde{p}(\boldsymbol{\theta} \mid \boldsymbol{y})$ to integrate out uncertainty with respect to $\boldsymbol{\theta}$. Hence there is no need of a detailed exploration of $\tilde{p}(\boldsymbol{\theta} \mid \boldsymbol{y})$ as long as we are able to select good evaluation points for the numerical integration in (2.3a). Rue et al. (2009) propose three different exploration schemes depending on the number of hyperparameters. All schemes, however, require a reparameterization of $\boldsymbol{\theta}$-space to make the density more regular. Without loss of generality we assume $\boldsymbol{\theta} = (\theta_1, \ldots, \theta_m) \in \mathbf{R}^m$ and proceed as follows. First, find the mode $\boldsymbol{\theta}^*$ of $\tilde{p}(\boldsymbol{\theta} \mid \boldsymbol{y})$ and compute the negative Hessian matrix $\boldsymbol{H}$ at the modal configuration. Then, we standardize $\boldsymbol{\theta}$ to obtain a new variable

$$\boldsymbol{z} = (\boldsymbol{V}\boldsymbol{\Lambda}^{1/2})^{-1}(\boldsymbol{\theta} - \boldsymbol{\theta}^*),$$

with $\boldsymbol{H}^{-1} = \boldsymbol{V}\boldsymbol{\Lambda}\boldsymbol{V}'$ being the eigen-decomposition.

If the dimension of $\boldsymbol{\theta}$ is small, say $m \leq 2$, the $\boldsymbol{z}$-parameterization is used to build a grid covering the bulk of the probability mass of $\tilde{p}(\boldsymbol{\theta} \mid \boldsymbol{y})$ (see the left panel of Figure 2.1). It turns out that a rough grid is enough to give accurate results if our purpose is on $p(\eta_i \mid \boldsymbol{y})$ only. Unfortunately, the computational cost of such a grid search method grows exponentially with m. For instance, if we only use three evaluation points in each dimension the computational cost is $O(3^m)$, making it inapplicable to the case where we have even a moderate number of hyperparameters. One alternative approach, although it is a little extreme, is *empirical Bayes*: we only use the model configuration to integrate over $p(\boldsymbol{\theta}|\boldsymbol{y})$. This "plug-in" method will obviously underestimate uncertainty, but it will provide reasonable results given the fact that the variability in the latent field is not dominated by the variability in the hyperparameters.

An intermediate approach between full numerical integration and the plug-in method is described in Section 6.5 of Rue et al. (2009). Guided by mode $\boldsymbol{\theta}^*$ and the negative Hessian matrix $\boldsymbol{H}$ some "points" in the m-dimensional space are found to approximate the unknown function with a second-order surface and a classical quadratic design like the central composite design (CCD) is used. A CCD contains an embedded factorial or fractional factorial design with center points augmented with a group of points that allow for estimating the curvature. These integration points are approximately located on an appropriate level set for the joint posterior of $\boldsymbol{\theta}$ (see the right panel of Figure 2.1). This CCD integration requires much less computation compared to the grid search, and is still able to capture variability in $\boldsymbol{\theta}$-space when it is too wide to be explored via the grid search.

The three integration strategies: grid search, empirical Bayes and CCD, can be specified via the "`int.strategy=`" in the `control.inla` argument. They are termed as `grid`, `eb` and `ccd`, respectively. For example, when the grid search is needed we may specify the argument as follows:

```
inla(..., control.inla = list(int.strategy = 'grid'), ...)
```

There is another "auto" option, which is the default choice in the argument. It chooses between the three integration strategies depending on m, the number of hyperparam-

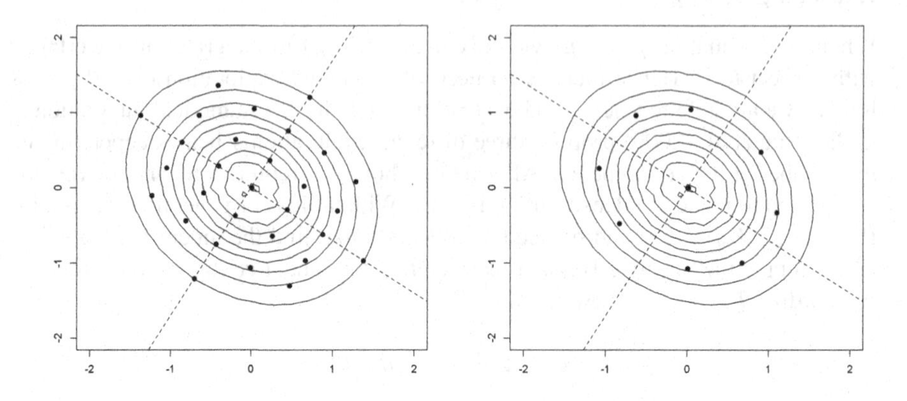

FIGURE 2.1
Location of the integration points in a two-dimensional $\boldsymbol{\theta}$-space using the grid (left) and the CCD (right) strategy.

eters. If $m \leq 2$ the grid search method is used. If $m > 2$ the `ccd` method is used. The empirical Bayes method is never used unless it is specified in the argument.

Approximating $p(\theta_j \mid \boldsymbol{y})$

If the dimension of $\boldsymbol{\theta}$ is not too high, it is possible to derive marginals for θ_j directly from $\tilde{p}(\boldsymbol{\theta}|\boldsymbol{y})$ by summing out the variables $\boldsymbol{\theta}_{-j}$ in (2.3b). It is, however, a pretty costly method because every value of $\boldsymbol{\theta}$ would require an evaluation of (2.5) and the computation of numerical integration grows exponentially with the dimension of $\boldsymbol{\theta}$. Another intuitive approach is to apply the Laplace approximation as in (2.7), where the numerator is $\tilde{p}(\boldsymbol{\theta}|\boldsymbol{y})$ obtained in (2.4) and the denominator is the Gaussian approximation to $\tilde{p}(\boldsymbol{\theta}_{-j}|\theta_j, \boldsymbol{y})$ built by matching the mode and the curvature at the mode. However, this approach suffers from expensive computation as well as some other issues, making it infeasible for most LGMs of interest (Martins et al., 2013). Therefore it is desirable to build algorithms that use the density points already evaluated in the grid exploration of $\tilde{p}(\boldsymbol{\theta}|\boldsymbol{y})$ as described in the above section. Those grid points have already been computed in order to integrate out the uncertainty of $\boldsymbol{\theta}$ using (2.3a), so the algorithms to compute $\tilde{p}(\theta_j|\boldsymbol{y})$ using these points will yield little extra cost. It is quite a challenge to find a quick and reliable approach to deriving all the marginal distributions from (2.4), while keeping the number of evaluation points low. We provide below a brief description of the current remedy used in INLA, and refer to Martins et al. (2013) for technical details.

The joint distribution $p(\boldsymbol{\theta}|\boldsymbol{y})$ can be approximated by a multivariate normal distribution by matching the mode and the curvature at the mode of $\tilde{p}(\boldsymbol{\theta}|\boldsymbol{y})$. This Gaussian approximation for $p(\theta_j|\boldsymbol{y})$ comes without extra computational effort since the mode and the negative Hessian matrix of $\tilde{p}(\boldsymbol{\theta}|\boldsymbol{y})$ have already been computed to approximate (2.3a). However, the true marginals can be rather skewed so that we have to correct the Gaussian approximations for the lack of skewness with minimal additional costs. Such a correction is done by approximating the joint distribution as a sum of the mixture of normal distributions with scaling parameters allowed to vary according to different axes and their directions. Then, the marginals can be computed via numerical integration of the approximated joint distribution. This algorithm was successfully used in the INLA package for a long time, giving accurate results with short computational time. However, the multi-dimensional numerical integration becomes unstable when fitting models with a higher number of hyperparameters, resulting in approximated posterior marginals densities with undesirable spikes. To fix this problem, Martins et al. (2013) proposed an integration-free algorithm, where the posterior marginal of each θ_j is directly approximated by a mixture of normal distributions

$$\tilde{p}(\theta_j \mid \boldsymbol{y}) = \begin{cases} N(0, \sigma^2_{j+}), & \theta_j > 0 \\ N(0, \sigma^2_{j-}), & \theta_j \leq 0 \end{cases}$$

and the scaling parameters σ^2_{j+} and σ^2_{j-} are estimated without using numerical integration. As shown by Martins et al., this algorithm gives sensible results with almost no extra computation time, although it loses some accuracy compared to the grid exploration method which is much more computationally intensive. It has become the default method to compute the posterior marginals for the hyperparameters in INLA. In order to get more accurate results via the grid search method we may use `inla.hyperpar(result)`, where `result` is the output of the `inla()` function.

Example: Simulated Data

Now let's use INLA to fit a simple LGM given by

$$y_i \mid \eta_i \sim \text{Poisson}(E_i \lambda_i), \quad \lambda_i = \exp(\eta_i),$$

where E_i are offset terms and $\boldsymbol{\eta} = (\eta_1, \ldots, \eta_{50})$ follows an AR(1) process with correlation $\rho = 0.8$ and precision $\tau = 10$. We then simulate data as follows:

```
set.seed(1)
n <- 50
rho <- 0.8
prec <- 10
E <- sample(c(5, 4, 10, 12), size = n, replace = TRUE)
eta <- arima.sim(list(order=c(1,0,0), ar=rho), n=n, sd=sqrt(1/prec))
y <- rpois(n, E*exp(eta))
```

To estimate τ and ρ we need to assign them hyperpriors. In INLA the two parameters are first reparameterized to θ_1 and θ_2 as in (2.6), and a log-gamma prior is then taken on θ_1 and a normal prior on θ_2. We then fit this LGM by

```
data <- list(y = y, x = 1:n, E = E)
formula <- y ~ f(x, model = "ar1")
result.sla <- inla(formula, family = "poisson", data = data, E = E)
```

The `formula` defines how the response depends on the covariate with the AR(1)

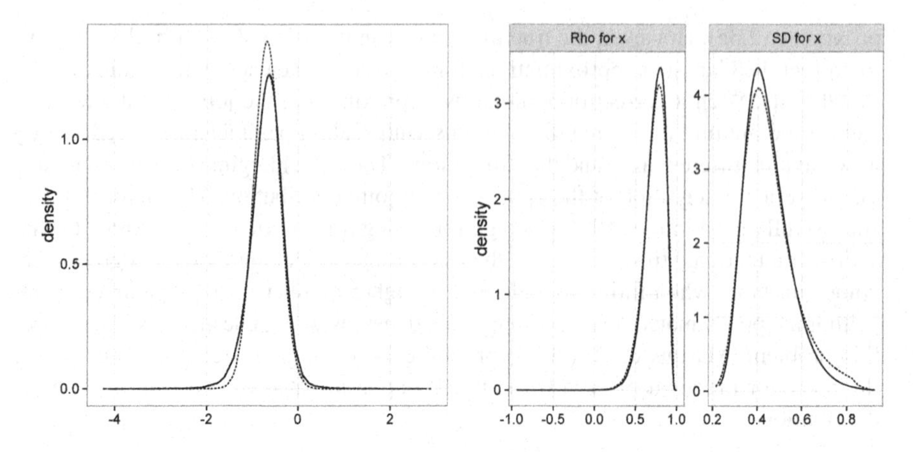

FIGURE 2.2
Posterior densities of η_{50} (left panel) using simplified Laplace approximation (solid), the Gaussian approximation (dotted) and the best possible Laplace approximation (dashed). Posterior densities of the hyperparameters using the integration-free method (solid) and the grid method (dotted).

model specified in `f()` function. The `inla()` function is where we specify likelihood, data, offset, etc., and implement INLA methodology.

In the left panel of Figure 2.2 we show three estimates for the posterior marginal of η_{50}. The solid line is the default estimate using the simplified Laplace approximation as outlined in Section 2.3, given by the R commands above. The dotted line is yielded by the Gaussian approximation that avoids integration over $\boldsymbol{\theta}$ (empirical Bayes), which is obtained by

```
result.gau <-  inla(formula, family = "poisson", data = data, E = E,
   ↪ control.inla = list(strategy='gaussian', int.strategy="eb"))
```

The dashed line represents the Laplace approximation and accurate integration over $\boldsymbol{\theta}$. It is the best approximation provided by the current software, although it takes a much longer time. We fit the model as follows:

```
result.la <-  inla(formula, family = "poisson", data = data, E = E,
   ↪ control.inla = list(strategy = 'laplace', int.strategy = "grid"
   ↪ , dz = 0.1, diff.logdens = 20))
```

Here `dz` is the step-length in the standardized scale for the integration of the hyperparameters, and `diff.logdens` is the difference of the log density for the hyperparameters to stop numerical integration. It is hard to see the dashed line as it is almost entirely covered by the solid line, indicating the simplified Laplace approximation is very close to being exact in this example. We also note that the uncertainty increases by integrating out $\boldsymbol{\theta}$ (as it should), and the skewness is successfully accounted for in the estimate. The left panel of Figure 2.2 is made by the following code:

```
i <- 50
```

```
marg.lst <- list(result.sla$marginals.random$x[[i]], result.gau$
    ↪ marginals.random$x[[i]], result.la$marginals.random$x[[i]])
names(marg.lst) <- as.factor(c(1,2,3))
pp <- data.frame(do.call(rbind, marg.lst))
pp$method <- rep(names(marg.lst), times = sapply(marg.lst, nrow))
library(ggplot2)
ggplot(pp, aes(x = x, y = y, linetype = method)) + geom_line(show.
    ↪ legend = FALSE) + ylab("density") + xlab("") + theme_bw(base_
    ↪ size = 20)
```

Let's look at how well the hyperparameters are estimated. In the right panel of Figure 2.2 we present the solid lines which are posterior marginals of ρ and the marginal SD estimated by the default integration-free method. Note that $\rho = 0.8$ and the marginal SD is

```
marg.prec <- prec*(1 - rho^2)
(marg.sd <- 1/sqrt(marg.prec))
```

```
[1] 0.5270463
```

As we can see, the estimated density curves are centered around the true parameters. We also tried the grid method for more accurate estimates

```
res.hyper <- inla.hyperpar(result.sla)
```

and the results are dotted curves in the right panel of Figure 2.2. Compared to the solid curves, the uncertainties slightly increase while the overall shapes remain unchanged. The figure is made by the following code:

```
library(brinla)
p1 <- bri.hyperpar.plot(result.la)
p2 <- bri.hyperpar.plot(res.hyper)
pp <- rbind(p1, p2)
pp$method <- as.factor(c(rep(1, dim(p1)[1]), rep(2, dim(p2)[1])))
ggplot(pp, aes(x = x, y = y, linetype = method)) + geom_line(show.
    ↪ legend = FALSE) + facet_wrap(~parameter, scales = "free") +
    ↪ ylab("density") + xlab("") + theme_bw(base_size = 20)
```

2.4 INLA Problems

Reproducibility: INLA is a computationally intensive method and so we want to get the most of our hardware. Most modern computers have multiple cores. In standard operation, INLA makes use of the OpenMP multiple processing interface. In theory, this could reduce computation time by a factor of the number of cores. In practice, the improvement will be somewhat less than this but still welcome for extensive computations. Unfortunately, this increase in speed comes with an unwanted side effect. Due to the intrinsic nature of OpenMP, small variations in the computation will be observed in repeated runs. One would not be too concerned about variations in, say, the seventh decimal place of a statistic but sometimes the variation can be more substantial than this. Some of the computations are sensitive to perturbations in starting values and so the effects can be magnified to a more noticeable level. Even so, we suggest that the user not be too concerned about this as the variation will be much smaller than parametric and model uncertainty that is unavoidably present. Sometimes, statistical results are presented with many significant digits which creates the false illusion of accuracy. In truth, estimates are rarely accurate beyond the second or third digit and so worrying about detail beyond this is pointless.

This variation is of a different nature to that generated by simulation methods such as MCMC. Variation in simulation studies is substantial and can only be reduced with the substantial cost of increasing the simulation size. The variation caused by OpenMP is much smaller. In simulation studies, one can ensure reproducibility by setting the seed of the random number generator. This seed has no relevance to our problem.

Sometimes exact reproducibility is required. Other people may insist on it or you may wish to check your computation for other changes at a later date for debugging or software version purposes. You can achieve this by using the option `num.threads = 1` in the `inla()` command call. This forces the use of a single processor and ensures bit reproducibility. Note that repeating the computation on different hardware will not necessarily give exactly the same result. Also understand that the single and multiple threaded answers are equally good and there is no advantage to the single threaded answer other than the reproducibility.

In this book, we use the default multiprocessing method. This means that if you run the same commands, you may get slightly different results. The INLA method is an approximation that could be made more accurate at the expense of longer computation. By using it, we accept this tradeoff as reasonable. Using similar reasoning, we are willing to trade bit reproducibility for faster computation.

Approximation accuracy: Numerical integration requires many choices regarding the location and number of function evaluations. Optimization also requires many choices about starting values, step lengths used in iterations, termination criteria and more. One can improve the accuracy of the computation by adjusting the parameters that control these integrations and optimizations but only at the cost of greater compute time. You can see many of these parameters with:

```
inla.set.control.inla.default()
```

If you are concerned about the accuracy of the approximation, you might try the following options to your `inla()` call:

```
inla(..., control.inla = list(strategy = "laplace",
    int.strategy = "grid", dz=0.1, diff.logdens=20), num.threads=1)
```

This will use a more accurate integration strategy with a finer grid with a bit reproducible result. If the `inla()` call on the default settings computes quickly, you may well afford the extra time for this deluxe calculation. In the rest of this book, we will use the default settings unless there are special reasons to deviate.

Failure: Any statistical procedure can fail. For simpler and well-established methods, the modes and signs of failure are more familiar. For more complex methods, such as INLA, there are more things that can go wrong and diagnosing the problem is more difficult.

The most gentle form of failure is due to a syntactic error in specifying the INLA commands. As is sometimes the case in R, the error message is not so helpful because it reports the immediate rather than the root difficulty. In such circumstances, the generic advice applies. Search the internet with key words from the error message. You are probably not the first person to make such a mistake. Check the INLA help pages and INLA FAQ on the INLA website. An INLA mailing list is archived on the INLA website — a search of this can be rewarding.

The next level of failure occurs when the INLA commands are correct but the program terminates with an error. Again, the error message reports what went wrong immediately before the termination but the real source of the problem may lie with the specification of the model or the data available. You can obtain more information about the progress of the INLA procedure by adding the argument `verbose=TRUE` to the `inla()` call. This may suggest the source of the problem. In some cases, the problem is due to insufficient data to estimate the chosen model. For example, it is rather difficult to estimate a variance with only one observation. Consider the parameters of your model and whether the data is sufficiently rich to estimate these parameters. In some cases, we can succeed in fitting a model by specifying more informative priors. If the data are inadequate, we need to make stronger assumptions to progress.

Simplification is a debugging strategy that can be revealing. Reduce the complexity of your model by removing elements such as covariates or random effects. If you can fit a simpler model, you can narrow down which feature of the model is the source of the problem. For very large problems, you should try a small subsample of the data to see whether the problem is due to the size of the dataset. Indeed, for large datasets and complex models, we may start INLA and wait impatiently for an answer that never seems to come. This is a good reason to start with a smaller dataset and/or model so you have a realistic expectation of how long the larger problem will take. Your personal computer may not be adequate to the task but a larger machine may do the job.

A more insidious form of failure occurs when INLA returns a result without complaint but this answer is wrong. Here is an example. The INLA methodology needs the full conditional density for the latent field to be "near" Gaussian. This is usually achieved by replications, or borrowing strength/smoothing from the GMRF priors. We present here a simple example which does not meet this requirement. Suppose y_i follows a Bernoulli distribution with success probability $p_i = e^{\eta_i}/(1+e^{\eta_i})$, and assume $\eta_i \sim N(\mu, \sigma^2)$ independently with $\mu = \sigma^2 = 1$. Such data can be simulated as follows:

```
set.seed(1)
n <- 100
u <- rnorm(n)
eta <- 1 + u
p <- exp(eta)/(1+exp(eta))
y <- rbinom(n, size = 1,  prob = p)
```

For each binary observation there is an i.i.d. Gaussian random effect with unknown mean and variance. Since there is no smoothing, borrowing strength, or replications, the full conditional density of η_i is not near Gaussian. In this situation INLA has a tendency to underestimate the variance σ^2:

```
data <- data.frame(y = y, x = 1:n)
result <- inla(y ~ 1 + f(x, model = "iid"), data = data, family = "
    ↪ binomial", Ntrials = 1)
round(bri.hyperpar.summary(result), 4)
```

```
          mean     sd q0.025   q0.5 q0.975   mode
SD for x 0.012 0.0124 0.0036 0.0084 0.0431 0.0057
```

We see the posterior mean of σ is 0.012 is much smaller than its true value 1. The mean of η_i, however, is reasonably well estimated:

```
round(result$summary.fixed, 4)
```

```
              mean     sd 0.025quant 0.5quant 0.975quant   mode kld
(Intercept) 0.9947 0.2253     0.5649   0.9903     1.4499 0.9813   0
```

The main reason that INLA fails to estimate the variance in the example above is that the usual assumptions ensuring asymptotic validity of the Laplace approximation do not hold here (see Section 4 of Rue et al., 2009, for details on asymptotic results). The independence of the random effects make the effective number of parameters (Spiegelhalter et al., 2002) on the order of the number of data points, and there is a lack of strong asymptotic results for the models with a large effective number of parameters, like what we have here. Moreover, the data simulated here provide little information about the parameters, with the shape of the likelihood function adding to the problem. For a single $y_i = 1$, the log likelihood is an increasing function of increasing η_i, and gets very flat for high values of η_i, making inference very difficult (Ferkingstad and Rue, 2015).

In this simulated example, we have the advantage of divine knowledge of the source of the problem, because we are able to compare the INLA result with the known true value. In practical examples, we cannot compare our results to the truth for this is unknown and perhaps unknowable. However, there are a number of measures we can take to increase our confidence in the validity of our results.

Where possible, we recommend that you compute the maximum likelihood estimates (MLE). For the models considered in this text, there is usually an R command or package that will produce these without significant additional effort. We have done this in several of our examples. You should compare the MLE with the posterior modes from the INLA output. The MLE is based on a maximum so this is most analogous to the mode rather than the median or mean. There is no requirement that these agree exactly but with flat priors, the maximum a posteriori (MAP) estimate (which is the posterior mode) will be very similar to the MLE. With more informative priors, one can expect more of a difference. If the results are similar, we will feel more confident about the INLA result. If there is some difference, we can try to understand why this has occurred. INLA allows Bayesian inference while maximum likelihood provides frequentist conclusions. Computing the MLE is just for validation of the INLA result — we plan to stick with INLA for the inference.

We can also implement the model using MCMC-based methods such as BUGS, JAGS or STAN. We are reluctant to do this because this will require greater programming effort to reproduce the modeling in these packages. Furthermore, it will take significantly longer to compute (or fail entirely in more difficult cases). Avoiding this trouble may have motivated us to use INLA over MCMC-based methods in the first place. Even so, if the results are sufficiently important and we need to be assured of their validity, this may be a price we must pay. We can also give the same advice to users who started with MCMC — check your results with INLA.

Simulation is another strategy for validating our results. We can simulate data from our model with known parameters and check whether INLA is able to match these known values. We may need to simplify the model or reduce the data size to do this effectively since some replication will be necessary. Our example above is a simple demonstration of how this might be done. You will also find that the examples on the INLA help pages use simulated data to demonstrate the methods. Simulation can provide additional assurance to our results but it is not foolproof. Unless you can afford large numbers of replications, there will be some sampling variability. The true parameters may be quite different from your simulation.

As with any statistical model, the data may be generated from a quite different model from the one we are using. In such circumstances, INLA cannot be said to fail but the results will, nevertheless, be misleading. Diagnostic methods can be helpful in detecting such model failures and in suggesting improved models. We discuss diagnostic methods for many of the models in this book. Unfortunately, diagnostic procedures for Bayesian models are generally less well-developed compared to the corresponding frequentist models. This is partly due to a conceptual reluctance by some Bayesian to engage in model diagnostics. We, however, have no reticence in recommending that you check your models thoroughly by any means available.

We hope this section has not discouraged you from using INLA. The advice we have given here applies to any complex statistical procedure. This is simply the price you must pay.

2.5 Extensions

Approximating Marginal Likelihood

The marginal likelihood $p(\boldsymbol{y})$ is a useful quantity for model comparison. For example, Bayes factors are defined as ratios of marginal likelihoods of two competing models; the computation of deviance information criterion (DIC) also involves this likelihood (see Section 1.6). Based on expression (2.4) an intuitive approximation to $p(\boldsymbol{y})$ is given by

$$\tilde{p}(\boldsymbol{y}) = \int \left. \frac{p(\boldsymbol{\eta}, \boldsymbol{\theta} \mid \boldsymbol{y})}{\tilde{p}(\boldsymbol{\eta} \mid \boldsymbol{\theta}, \boldsymbol{y})} \right|_{\boldsymbol{\eta}=\boldsymbol{\eta}^*(\boldsymbol{\theta})} d\boldsymbol{\theta}.$$

It is actually the normalizing constant of $\tilde{p}(\boldsymbol{\theta}|\boldsymbol{y})$. This approximation allows for the departure from being Gaussian because $\tilde{p}(\boldsymbol{\theta}|\boldsymbol{y})$ is treated in a "non-parametric" way.

However, this method could fail if the posterior marginal of $\boldsymbol{\theta}$ is multimodal. This is not specific to the evaluation of the marginal likelihood but applies generally to the INLA approach. Fortunately, the latent Gaussian models almost always generate unimodal posterior distributions (Rue et al., 2009).

Model Selection

To compare different models two criteria are available in INLA: the deviance information criterion (DIC) (Spiegelhalter et al., 2002) and the Watanabe Akaike information criterion (WAIC) (Gelman et al., 2014) (see Section 1.5 for details).

The DIC depends on the deviance defined as $-2\log(p(\boldsymbol{y}))$. We need to compute its posterior expectation and evaluate it at the posterior expectation. Although it may seem odd, we evaluate the deviance at the posterior *mean* of $\boldsymbol{\eta}$ and the posterior *mode* of $\boldsymbol{\theta}$ instead of doing it at the posterior mean of all parameters. This is because the posterior marginals of $\boldsymbol{\theta}$ can be highly skewed, making the posterior expectation a bad representation of location. The WAIC criterion is computed in a similar way.

To compute these two criteria, we need to add the following argument:

```
result <- inla(..., control.compute = list(dic = TRUE, waic = TRUE))
```

The estimates become available as `result$dic` and `result$waic`.

Model Checking

For model checking, INLA provides two "leave-one-out" predictive measures — the conditional predictive ordinate (CPO) and the probability integral transform (PIT) (see also Section 1.4). These two measures are both computed on the basis of the predictive distribution for the observed y_i given all the other observations, that is $p(y_i|\boldsymbol{y}_{-i})$. Instead of reffiting the model n times, we may approximate this quantity simply as follows. The marginals of η_i and $\boldsymbol{\theta}$ change when we remove y_i from the data:

$$\begin{aligned} p(\eta_i \mid \boldsymbol{y}_{-i}, \boldsymbol{\theta}) &\propto p(\eta_i \mid \boldsymbol{y}, \boldsymbol{\theta}) / p(y_i \mid \eta_i, \boldsymbol{\theta}), \\ p(\boldsymbol{\theta} \mid \boldsymbol{y}_{-i}) &\propto p(\boldsymbol{\theta} \mid \boldsymbol{y}) / p(y_i \mid \boldsymbol{y}_{-i}, \boldsymbol{\theta}), \end{aligned}$$

where a one-dimensional integral is required to compute:

$$p(y_i \mid \boldsymbol{y}_{-i}, \boldsymbol{\theta}) = \int p(y_i \mid \eta_i, \boldsymbol{\theta}) p(\eta_i \mid \boldsymbol{y}_{-i}, \boldsymbol{\theta}) d\eta_i.$$

We then integrate out the effect of $\boldsymbol{\theta}$ from $p(y_i \mid \boldsymbol{y}_{-i}, \boldsymbol{\theta})$ the same way as in equation (2.3a). Given the approximated marginal $\tilde{p}(y_i|\boldsymbol{y}_{-i})$, we can compute CPO and PIT statistics for each observation. To enable the computation of these quantities, we need to add an argument to `inla()` as follows:

```
result <- inla(..., control.compute = list(cpo = TRUE), ...)
```

The predictive measures are available as `result$cpo$cpo` for CPO and `result$cpo$pit` for PIT.

The CPO/PIT measures are computed with the fixed integration-points, and therefore we may consider improving the grid integration by adding the argument

```
inla(..., control.inla = list(int.strategy="grid", diff.logdens=4))
```

In some cases, the tail behavior of the marginals is important for the CPO/PIT calculations. Hence we'd better increase the accuracy of the tails using the Laplace approximation approach, i.e.,

```
inla(..., control.inla = list(strategy = "laplace", npoints = 21))
```

where we add more evaluation points `npoints = 21` instead of 9 by default .

One should be also aware that a few implicit assumptions are made in the INLA program, and there are internal checks about these assumptions when computing CPO/PIT. These checks are reflected in `result$cpo$failure`, where some assumption is violated for the ith observation if `result$cpo$failure[i]` > 0, and the higher the value (maximum 1) the more serious the violation. We therefore should recompute the CPO/PIT values for the observations in violation, which must be done manually by removing "y[i]" from the dataset, fitting the model and predicting "y[i]". Fortunately, an efficient implementation of this is available in INLA using `inla.cpo(result)`, which returns the improved estimates of the measures.

In practice one should always pay attention to the cases with unusually small CPO/PIT values no matter if they violate the assumptions or not. This is because the validation with these measures depends on the tail behavior, and it is difficult for INLA to estimate it. As a result, we suggest you should always verify that the smallest measures are estimated correctly by setting their `res$cpo$failure` values to be 1, and re-estimate them with `inla.cpo()`, especially if you plan to call the case an outlier.

Linear Combinations of the Latent Field

Sometimes, we might be interested in more than just the posterior marginals of the elements in the latent field. We might also want linear combinations of those elements. Such linear combinations can be written as $\boldsymbol{v} = \boldsymbol{A}\boldsymbol{\eta}$, where $\boldsymbol{A}$ is a $k \times n$ matrix with k being the number of linear combinations and n is the size of the latent field $\boldsymbol{\eta}$. For example, one may be interested in the joint distribution of $\boldsymbol{v} = (\eta_1, \eta_2)^T$, and the corresponding $\boldsymbol{A}$ is a $2 \times n$ matrix that only has nonzero entries $\boldsymbol{A}[1,1] = \boldsymbol{A}[2,2] = 1$. The functions `inla.make.lincomb()` and `inla.make.lincombs()` are used to define a linear combination and many linear combinations at once, respectively. The resulting linear combination(s) is then added to `lincomb` argument in `inla()`.

INLA provides two approaches for dealing with $\boldsymbol{v}$. In the first approach, we create an enlarged latent field $\tilde{\boldsymbol{\eta}} = (\boldsymbol{\eta}, \boldsymbol{v})$ and then fit the enlarged model as usual, using the Gaussian, Laplace or simplified Laplace approximations discussed in Section 2.3. We therefore have posterior marginals for each element of $\tilde{\boldsymbol{\eta}}$, including the linear combinations $\boldsymbol{v}$. The drawback of this approach is that the addition of many linear combinations yields more dense precision matrices, which consequently slows down the computations. To implement this approach, we need the following argument:

```
inla(..., control.inla = list(lincomb.derived.only = FALSE), ...)
```

In the second approach $\boldsymbol{v}$ is not included in the latent field, but a post-processing of the INLA output is performed and the conditional distribution $\boldsymbol{v}|\boldsymbol{\theta},\boldsymbol{y}$ is approximated by a Gaussian with mean $\boldsymbol{A}\tilde{\boldsymbol{\mu}}$ and covariance matrix $\boldsymbol{A}\tilde{\boldsymbol{Q}}^{-1}\boldsymbol{A}^T$, where $\tilde{\boldsymbol{\mu}}$ is the mean of the best marginal approximation used for $p(\eta_i|\boldsymbol{\theta},\boldsymbol{y})$ (i.e., Gaussian, simplified Laplace or Laplace approximation) and $\tilde{\boldsymbol{Q}}$ is the precision matrix of the Gaussian approximation $\tilde{p}(\boldsymbol{\eta}|\boldsymbol{\theta},\boldsymbol{y})$ as presented in (2.5). We then integrate out $\boldsymbol{\theta}$ from the approximation of $p(\boldsymbol{v}|\boldsymbol{\theta},\boldsymbol{y})$ in a process similar to equation (2.3a). The advantage of this approach is that the computation does not enlarge the latent field, leading to a much faster approximation. Consequently, this is the default method in INLA, but more accurate approximations can be obtained by switching to the first approach.

When using the faster approach, there is an option to compute the posterior correlation matrix between all the linear combinations using the following argument:

```
inla(..., control.inla = list(lincomb.derived.correlation.matrix
        = TRUE), ...)
```

This correlation matrix could be used, for example, to build a Gaussian copula to approximate the joint density of some components of the latent field, as discussed in Section 6.1 of Rue et al. (2009).

3

Bayesian Linear Regression

Linear regression is one of the most common statistical approaches for modeling the relationship between a scalar dependent variable (or response) and one or more explanatory variables (or independent variables). It is the study of linear, additive relationships between variables. The methodology was the first type of regression analysis to be studied rigorously, and has been a topic of innumerable textbooks (Chatterjee and Hadi, 2015). Although it may seem to be too simple compared to some of the more modern statistical regression techniques described in later chapters of this book, linear regression is still considered as one of the most useful and powerful tools in practical applications. This chapter can serve as a good starting point for newer and more complex modeling approaches that we will discuss in the later chapters. Having a deep understanding of standard linear regression is of importance, since many fancy regression techniques can be viewed as generalizations or extensions of it.

3.1 Introduction

Let

$$\{x_{i1}, \dots, x_{ip}, y_i\}, \qquad i = 1, \dots, n,$$

represent n observation units, each of which consists of a measurement of the p-vector of predictors $(x_1, \dots, x_p)$ and a measurement of the response variable y. The multiple linear regression model takes the form

$$y_i = \beta_0 + \beta_1 x_{i1} + \dots + \beta_p x_{ip} + \varepsilon_i. \tag{3.1}$$

In this linear model (3.1), the relationship between y and $(x_1, \dots, x_p)$ is modeled using the linear predictor function $\mu = \beta_0 + \beta_1 x_1 + \dots + \beta_p x_p$, and a disturbance term or error variable ε. The unknown model parameters $(\beta_0, \beta_1, \dots, \beta_p)$ are estimated from the data. The model becomes a simple linear regression when $p = 1$. Linearity here is with respect to the unknown parameters. Sometimes the predictor function contains a nonlinear function of a predictor. For example, a polynomial regression of degree 3 is expressed as $y_i = \beta_0 + \beta_1 x_i + \beta_2 x_i^2 + \beta_3 x_i^3 + \varepsilon_i$. This model remains linear since it is linear in the parameter vector.

The linear model (3.1) can be written in a matrix form,

$$\mathbf{y} = \mathbf{X}\boldsymbol{\beta} + \varepsilon, \tag{3.2}$$

where

$$\mathbf{y} = \begin{bmatrix} y_1 \\ y_2 \\ \vdots \\ y_n \end{bmatrix}, \quad \mathbf{X} = \begin{bmatrix} 1 & x_{11} & \cdots & x_{1p} \\ 1 & x_{21} & \cdots & x_{2p} \\ \vdots & \vdots & \ddots & \vdots \\ 1 & x_{n1} & \cdots & x_{np} \end{bmatrix}, \quad \boldsymbol{\beta} = \begin{bmatrix} \beta_0 \\ \beta_1 \\ \vdots \\ \beta_p \end{bmatrix}, \quad \varepsilon = \begin{bmatrix} \varepsilon_1 \\ \varepsilon_2 \\ \vdots \\ \varepsilon_n \end{bmatrix}.$$

We begin our discussion by assuming that the errors are independent and normally distributed with mean zero and constant variance, i.e., the error term ε in (3.2) is assumed to be distributed as $N(0, \sigma^2\mathbf{I})$ with an unknown variance parameter σ^2.

In frequentist statistics, the parameters can be estimated using the maximum likelihood estimation (MLE) or the least squares method. Specifically, the likelihood function for the model (3.2) is,

$$L(\boldsymbol{\beta}, \sigma^2|\mathbf{X}, \mathbf{y}) = \left(\frac{1}{\sqrt{2\pi}\sigma}\right)^n \exp\left[-\frac{1}{2\sigma^2}(\mathbf{y} - \mathbf{X}\boldsymbol{\beta})^T(\mathbf{y} - \mathbf{X}\boldsymbol{\beta})\right], \tag{3.3}$$

which yields the score equations

$$S_1(\boldsymbol{\beta}, \sigma^2) = \frac{\partial \log L}{\partial \boldsymbol{\beta}} = -\frac{1}{2\sigma^2}\mathbf{X}^T(\mathbf{y} - \mathbf{X}\boldsymbol{\beta}), \tag{3.4}$$

$$S_2(\boldsymbol{\beta}, \sigma^2) = \frac{\partial \log L}{\partial \sigma} = -\frac{n}{\sigma} + \frac{1}{2\sigma^3}(\mathbf{y} - \mathbf{X}\boldsymbol{\beta})^T(\mathbf{y} - \mathbf{X}\boldsymbol{\beta}). \tag{3.5}$$

Assuming $\mathbf{X}^T\mathbf{X}$ is of full rank and setting (3.4) and (3.5) to zero, we obtain the maximum likelihood estimators

$$\hat{\boldsymbol{\beta}} = (\mathbf{X}^T\mathbf{X})^{-1}\mathbf{X}^T\mathbf{y}, \tag{3.6}$$

and

$$\hat{\sigma}^2 = \frac{1}{n}(\mathbf{y} - \mathbf{X}\hat{\boldsymbol{\beta}})^T(\mathbf{y} - \mathbf{X}\hat{\boldsymbol{\beta}}). \tag{3.7}$$

Note that $\hat{\boldsymbol{\beta}}$ is an unbiased estimator of $\boldsymbol{\beta}$, and the least squares estimator of $\boldsymbol{\beta}$ is also $\hat{\boldsymbol{\beta}}$ (Chatterjee and Hadi, 2015). However, $\hat{\sigma}^2$ is not an unbiased estimator of σ^2. The more commonly used estimator of σ^2, which is unbiased, is

$$S^2 = \frac{1}{n-p-1}(\mathbf{y} - \mathbf{X}\hat{\boldsymbol{\beta}})^T(\mathbf{y} - \mathbf{X}\hat{\boldsymbol{\beta}}).$$

3.2 Bayesian Inference for Linear Regression

We now consider Bayesian inference for the model (3.2). In Bayesian analysis, the inverse of the variance parameter plays an important role and is called the *precision*,

$\tau = \sigma^{-2}$. We shall use the precision τ in manipulating the distributions. Based on the assumption of the model, we have

$$\mathbf{y}|\boldsymbol{\beta},\tau \sim N(\mathbf{X}\boldsymbol{\beta},\tau^{-1}\mathbf{I}).$$

We further assume $\boldsymbol{\beta}$ and τ are independent. Therefore, the joint posterior distribution of the unknown parameters, thus, is

$$\pi(\boldsymbol{\beta},\tau|\mathbf{X},\mathbf{y}) \propto L(\boldsymbol{\beta},\tau|\mathbf{X},\mathbf{y})p(\boldsymbol{\beta})p(\tau),$$

where $p(\boldsymbol{\beta})$ and $p(\tau)$ are the priors for the parameters $\boldsymbol{\beta}$ and τ. Closed form of the posterior distributions of $\boldsymbol{\beta}$ and τ are only available under certain restricted prior distributions.

An important problem in Bayesian analysis is how to define the prior distribution. If prior information about the parameters is available, it should be incorporated in the prior distribution. If we have no prior information, we want that a prior distribution can be guaranteed to have a minimal influence on the inference. Noninformative prior distribution, for example $p(\boldsymbol{\beta}) \propto 1$, has always been appealing, since many real applications lack information on the parameters. However, the major drawback of noninformative prior is that it is not invariant for transformation of the parameters. See more discussions in Appendix B. We also refer to Gelman et al. (2014) for a comprehensive discussion on prior selection.

In INLA, we assume that the model is a latent Gaussian model, that is, we have to assign $\boldsymbol{\beta}$ a Gaussian prior. For the hyperparameter τ, we often assume a diffuse prior, a probability distribution with an extremely large variance. A typical prior choice for $\boldsymbol{\beta}$ and τ is

$$\boldsymbol{\beta} \sim N_{p+1}(\mathbf{c}_0,\mathbf{V}_0), \quad \tau \sim Gamma(a_0,b_0).$$

Here the prior of $\boldsymbol{\beta}$ is $p+1$-dimensional multivariate normal with known $\mathbf{c}_0$ and $\mathbf{V}_0$. We often assume that $\mathbf{V}_0$ is diagonal, which is equivalent to specifying separate univariate normal priors on the regression coefficients. The precision τ follows a dispersed gamma distribution with a known shape parameter a_0 and a known rate parameter b_0 (that is, we have mean a_0/b_0 and variance a_0/b_0^2). In linear regression, the gamma prior is conditionally conjugate for τ since the conditional posterior distribution, $p(\tau|\mathbf{X},\mathbf{y})$, is also in that class.

Although the posterior is intractable under these priors, it is straightforward to construct a blocked Gibbs sampling algorithm and be suitable for MCMC implementation (Gelman et al., 2014). Specifically, the algorithm iterates between the pair of conditional distributions:

$$\begin{cases} \pi(\boldsymbol{\beta}|\mathbf{X},\mathbf{y},\tau) \propto L(\boldsymbol{\beta},\tau|\mathbf{X},\mathbf{y})p(\boldsymbol{\beta}), \\ \pi(\tau|\mathbf{X},\mathbf{y},\boldsymbol{\beta}) \propto L(\boldsymbol{\beta},\tau|\mathbf{X},\mathbf{y})p(\tau). \end{cases}$$

Instead of MCMC simulations, the INLA approach provides approximations to the posterior marginals of the parameters which are both very accurate and extremely fast to compute (Rue et al., 2009). The marginal posterior $\pi(\tau|\mathbf{X},\mathbf{y})$ is approximated using

$$\tilde{\pi}(\tau|\mathbf{X},\mathbf{y}) \propto \left.\frac{\pi(\boldsymbol{\beta},\tau,\mathbf{X},\mathbf{y})}{\tilde{\pi}(\boldsymbol{\beta}|\tau,\mathbf{X},\mathbf{y})}\right|_{\boldsymbol{\beta}=\boldsymbol{\beta}^*(\tau)}, \tag{3.8}$$

which is the Gaussian approximation to the full conditional distribution of β evaluated in the mode $\beta^*(\tau)$ for a given τ. Expression (3.8) is equivalent to the Laplace approximation of a marginal posterior distribution (Tierney and Kadane, 1986), and it is exact when $\pi(\beta|\tau,\mathbf{X},\mathbf{y})$ is Gaussian.

Posterior marginals for the model parameters, $\tilde{\pi}(\beta_j|\tau,\mathbf{X},\mathbf{y})$, $j = 0, 1, ..., p$, are then approximated via numerical integration as:

$$\begin{aligned}\tilde{\pi}(\beta_j|\mathbf{X},\mathbf{y}) &= \int \tilde{\pi}(\beta_j|\tau,\mathbf{X},\mathbf{y})\tilde{\pi}(\tau|\mathbf{X},\mathbf{y})d\tau \\ &\approx \sum_k \tilde{\pi}(\beta_j|\tau_k,\mathbf{X},\mathbf{y})\tilde{\pi}(\tau_k|\mathbf{X},\mathbf{y})\Delta_k,\end{aligned}$$

where the sum is over values of τ with area weights Δ_k. For more technical details, we refer back to Chapter 2.

The approximate posterior marginals obtained from the INLA procedure can then be used to compute summary statistics of interest, such as posterior means, variances and quantiles. As a by-product of the main computations, INLA also computes other quantities of interest like *deviance information criterion* (DIC), marginal likelihoods, etc., which are useful to compare and validate models.

Let us look at an example of multiple linear regression of analyzing the air pollution data. The dataset has been discussed in Everitt (2006). The data were collected to investigate the determinants of pollution for 41 cities in the United States. Table 3.1 displays the variables being recorded and their descriptions. In this study, `SO2` level is considered as the dependent variable and the other six variables are considered as potential explanatory variables. Among these potential predictors, two of them are related to human ecology (`pop`, `manuf`) and four others are related to climate (`negtemp`, `wind`, `precip`, `days`). Note that the variable, `negtemp`, represents the negative value of average annual temperature. Using the negative values here is because all variables are such that high values represent a less attractive environment.

TABLE 3.1
Description of variables in the air pollution data.

Variable Name	Description	Codes/Values
SO2	sulfur dioxide content of air	micrograms per cubic meter
negtemp	negative value of average annual temperature	fahrenheit
manuf	number of manufacturing enterprises employing 20 or more workers	integers
pop	population size in thousands (1970 census)	numbers
wind	average annual wind speed	miles per hour
precip	average annual precipitation	inches
days	average number of days with precipitation per year	integers

Prior to performing a regression analysis on the data, it is useful to graph the data in a certain way so that we can have an insight into the overall structure of them. Figure 3.1 presents a matrix of scatterplots for all variables on the upper triangular, and the correlation coefficients displayed on the lower triangle, with nonparametric kernel density plots for each variable on the main diagonal. The figure is created using the following R code:

```
library(ggplot2, GGally)
data(usair, package = "brinla")
pairs.chart <- ggpairs(usair[,-1], lower = list(continuous = "cor"),
    ↪ upper = list(continuous = "points", combo = "dot")) + ggplot2::
    ↪ theme(axis.text = element_text(size = 6))
pairs.chart
```

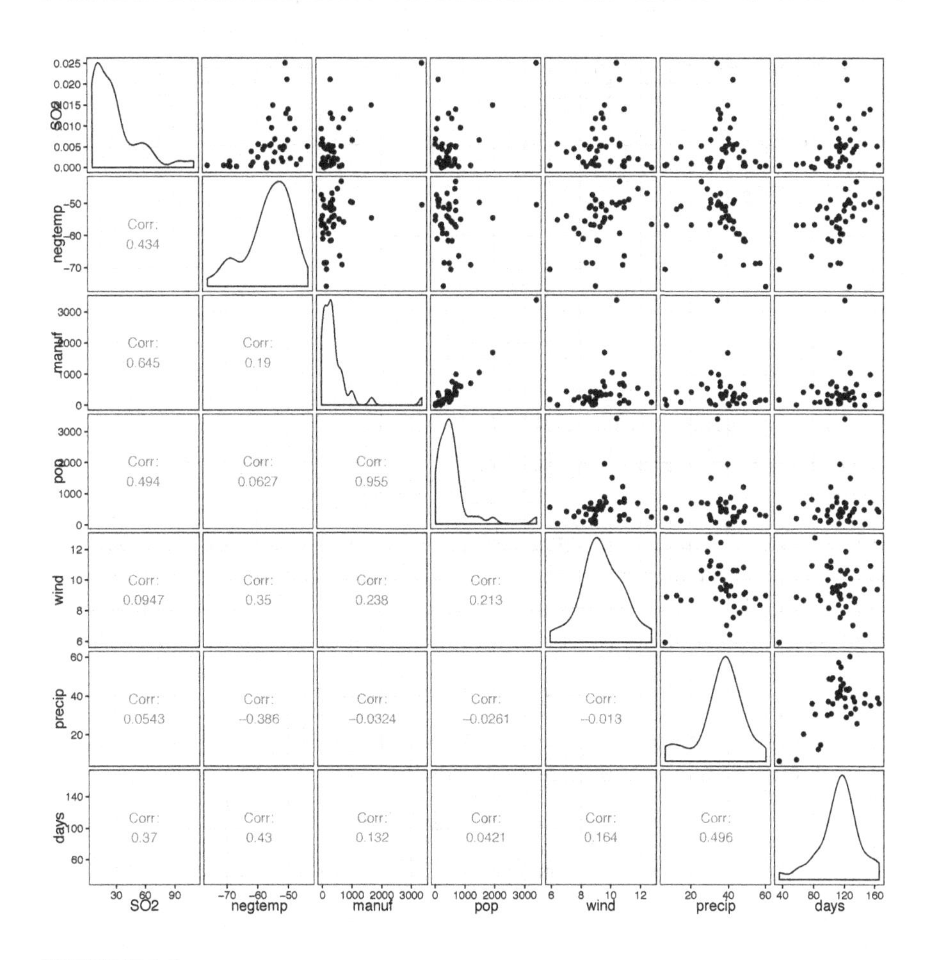

FIGURE 3.1

Scatterplot matrix of the variables in the air pollution data: The upper triangular displays the paired scatterplots; the lower triangle shows the paired correlation coefficients; and the main diagonal presents nonparametric kernel density plots for each variable.

From Figure 3.1 we notice that manuf and pop are highly correlated. Checking their sample correlation, we find that it is as high as 0.955. This phenomenon is known as multicollinearity, in which two or more predictors in a multiple regression model are highly correlated. In this situation the coefficient estimates of the multiple regression are very sensitive to slight changes in the data and to the addition or deletion of variables in the equation. The estimated regression coefficients often have large sampling errors, which affect both inference and prediction that is based on the model (Chatterjee and Hadi, 2015).

To avoid the multicollinearity issue, we use a simple approach by keeping only one of the two variables, manuf, in our regression analysis. A more sophisticated method to deal with multicollinearity is addressed in Section 3.7.

For comparison purposes, we begin the analysis with the conventional maximum likelihood method. We first fit a regression model with the five predictors, negtemp, mauf, wind, precip, and days:

```
usair.formula1 <-  SO2 ~ negtemp + manuf + wind + precip + days
usair.lm1 <- lm(usair.formula1, data = usair)
round(coef(summary(usair.lm1)), 4)
```

```
            Estimate Std. Error t value Pr(>|t|)
(Intercept) 135.7714    50.0610  2.7121   0.0103
negtemp       1.7714     0.6366  2.7824   0.0086
manuf         0.0256     0.0046  5.5544   0.0000
wind         -3.7379     1.9444 -1.9224   0.0627
precip        0.6259     0.3885  1.6111   0.1161
days         -0.0571     0.1748 -0.3265   0.7460
```

The estimated residual standard error, $\hat{\sigma}$, is given by:

```
round(summary(usair.lm1)$sigma, 4)
```

```
[1] 15.79
```

The results for this model show that the predictors negtemp and manuf are significant predictors while the predictors wind, precip, and days are not. We now fit Bayesian models using INLA with the default priors. In the INLA package, the default choice of priors for $\beta_j, j = 0, ..., p$

$$\beta_j \sim N(0, 10^6), \quad j = 0, ..., p,$$

and the prior for the variance parameter is defined internally in terms of logged precision, $\log(\tau)$. It follows a log gamma distribution,

$$\log(\tau) \sim \text{logGamma}(1, 10^{-5}).$$

We fit a Bayesian linear regression using INLA with the following code:

```
library(INLA)
usair.inla1 <- inla(usair.formula1, data = usair, control.compute =
    ↪ list(dic = TRUE, cpo = TRUE))
```

The inla function returns an object, here named usair.inla1, which has a class attribute, inla. This is a list containing a lot of objects which can be explored with names(usair.inla1). For example, the summary of the fixed effects can be obtained by the command:

```
round(usair.inla1$summary.fixed, 4)
```

```
                mean      sd 0.025quant 0.5quant 0.975quant     mode kld
(Intercept) 135.4892 50.0629    36.6158 135.4955   234.1778 135.5116   0
negtemp       1.7690  0.6370     0.5111   1.7690     3.0247   1.7692   0
manuf         0.0256  0.0046     0.0165   0.0256     0.0347   0.0256   0
wind         -3.7229  1.9424    -7.5570  -3.7234     0.1080  -3.7241   0
precip        0.6249  0.3888    -0.1429   0.6249     1.3913   0.6249   0
days         -0.0567  0.1749    -0.4020  -0.0567     0.2882  -0.0567   0
```

The summary of the hyperparameter is obtained by

```
round(usair.inla1$summary.hyperpar, 4)
```

```
                                  mean    sd 0.025quant 0.5quant 0.975quant  mode
Precision
for the Gaussian observations 0.0042 9e-04     0.0026   0.0042     0.0063 0.004
```

Summaries of these posterior distributions include posterior means and 95% credible intervals, which can be used as Bayesian alternatives to the maximum likelihood estimates and 95% confidence intervals, respectively. For example, the posterior mean of the coefficient fcr `negtemp` is 1.7690, and the 95% credible interval is (0.5111, 3.0247). These indicate that, with very high probability, `negtemp` is positively associated with the response, `SO2`. Unlike confidence intervals, which are calculated by assuming large sample approximations, Bayesian interval estimates are typically appropriate in small samples. More importantly, the Bayesian 95% credible interval estimates have an intuitively appealing interpretation as the interval containing the true parameter with 95% probability. This interpretation is often preferable to that of the 95% confidence interval, which is the range of values containing the true parameter 95% of the time in repeated sampling.

Another simple way to look at the result from an `inla` object is to use `summary` function, which produces default summaries of the results of the `inla` fitting function:

```
summary(usair.inla1)
```

The result summaries, which we do not display here, output the summary statistics of posteriors distributions of the fixed effects and the hyperparameters for the model including posterior means, standard deviations, the quartiles and others. Some model statistics, such as marginal log-likelihood, and the model fitting index, DIC (when we specify `DIC = TRUE` in `inla`), are printed. Users can also selectively print certain model results by their needs by making use of the `$` sign, as we have shown above. Some other useful information, for example, posterior marginal distributions of the fixed effects parameters and the hyperparameters can be obtained by the commands, `usair.inla1$marginals.fixed` and `usair.inla1$marginals.hyperpar`.

The `INLA` library includes a set of functions to operate on marginal distributions. The commonly used functions include `inla.dmarginal`, `inla.pmarginal`, `inla.qmarginal`, `inla.mmarginal`, and `inla.emarginal` to compute the density, distribution, quantile function, mode, and expected values of marginals, respectively. The function `inla.rmarginal` is used to generate random numbers, and the function `inla.tmarginal` can be used to transform a given marginal distribution. Here we show an example of how to make use of the functions.

By default, the posterior summaries of the precision τ is outputted from an `inla` object. However, we are often interested in the posterior mean of σ. The estimate can be obtained by using the `inla.emarginal` function:

```
inla.emarginal(fun = function(x) 1/sqrt(x), marg = usair.inla1$
    ↪ marginals.hyperpar$`Precision for the Gaussian observations`)
```

```
[1] 15.66331
```

Comparing the estimated residual standard error from the conventional maximum likelihood method, we obtain very similar results. For the user's convenience, we have written an R function `bri.hyperpar.summary` for producing the summary statistics of hyperparameters in terms of σ in our `brinla` package:

```
library(brinla)
round(bri.hyperpar.summary(usair.inla1),4)
```

```
                                      mean     sd  q0.025    q0.5 q0.975    mode
SD for the Gaussian observations 15.6617 1.8379 12.5321 15.4871 19.759 15.1513
```

We want to further look at the plot of the posterior distribution of σ. The function `bri.hyperpar.plot` in `brinla` can be applied to produce this posterior density directly:

```
bri.hyperpar.plot(usair.inla1)
```

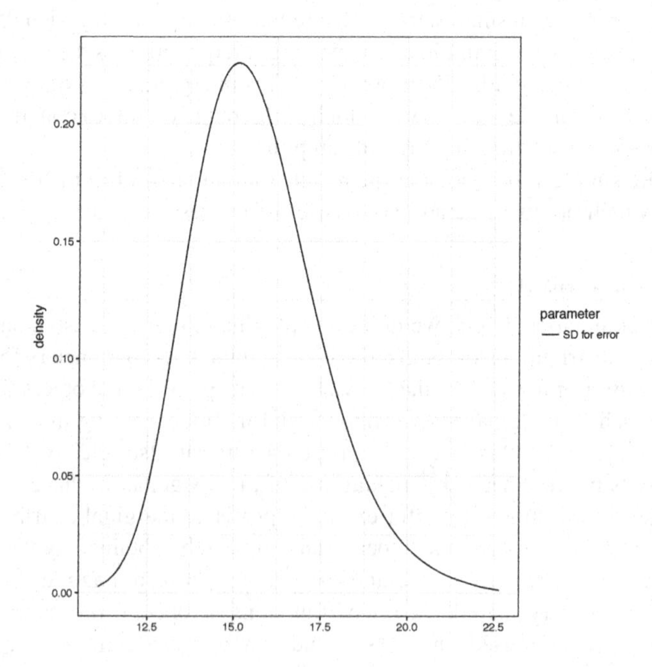

FIGURE 3.2
Posterior density for the parameter σ in the US air pollution study.

In Figure 3.2, we see a slightly right-skewed posterior distribution for σ.

The INLA program allows the user to change the prior for the regression parameters. Suppose that we have certain prior information for the intercept β_0 and the coefficients of negtemp and wind. For example, we assume that $\beta_0 \sim N(100, 100)$, $\beta_{negtemp} \sim N(2, 1)$, and $\beta_{wind} \sim N(-3, 1)$. The prior specification can be achieved using the option control.fixed in inla function. By default, a diffuse gamma prior is assumed on the precision parameter τ. If we want to specify, for instance, a log-normal prior to τ (equivalent to assuming a normal prior on the logarithm of τ), this can be specified using the option control.family:

```
usair.inla2 <- inla(usair.formula1, data = usair, control.compute =
    ↪ list(dic = TRUE, cpo =TRUE), control.fixed = list(mean.
    ↪ intercept = 100, prec.intercept = 10^(-2), mean = list(negtemp
    ↪ = 2, wind = -3, default =0), prec = 1), control.family = list(
    ↪ hyper = list(prec = list(prior="gaussian", param =c(0,1))))
```

Note that here we change the priors for the intercept as well as two fixed parameters for negtemp and wind. The statement mean = list(negtemp = 2, wind = -3, default = 0) assigns prior mean equal to 2 for negtemp, -3 for wind, and zero means for all other parameters of the remaining predictors, using list. A list has to be also specified for prec if we have different precision assumptions. Certainly, users need to be careful in changing priors. The model selection and checking methods discussed in Section 3.4 can be used for comparing models with different priors.

3.3 Prediction

Suppose we apply the regression model to a new set of data, for which we have observed the vector of explanatory variables $\tilde{x} = (\tilde{x}_1, \tilde{x}_2, ..., \tilde{x}_p)^T$, and wish to predict the outcome, $\tilde{y}$. Typically, the Bayesian prediction is made based on the *posterior predictive distribution*, $p(\tilde{y}|\mathbf{y})$. Here "posterior" means that it is conditional on the observed $\mathbf{y}$, and "predictive" means that it is a prediction for $\tilde{y}$. Let $\boldsymbol{\theta} = (\boldsymbol{\beta}, \tau)$. We have

$$
\begin{aligned}
p(\tilde{y}|\mathbf{y}) = \frac{p(\tilde{y}, \mathbf{y})}{p(\mathbf{y})} &= p(\mathbf{y})^{-1} \int p(\tilde{y}|\boldsymbol{\theta}) p(\mathbf{y}|\boldsymbol{\theta}) p(\boldsymbol{\theta}) \mathrm{d}\boldsymbol{\theta} \\
&= p(\mathbf{y})^{-1} \int p(\tilde{y}|\boldsymbol{\theta}) p(\boldsymbol{\theta}|\mathbf{y}) p(\mathbf{y}) \mathrm{d}\boldsymbol{\theta} \\
&= \int p(\tilde{y}|\boldsymbol{\theta}) p(\boldsymbol{\theta}|\mathbf{y}) \mathrm{d}\boldsymbol{\theta}.
\end{aligned}
$$

The analytic form of the posterior predictive distribution in most regression models is not available. In conventional Bayesian analysis, the prediction can be done by posterior predictive simulation, i.e., drawing random samples from $p(\tilde{y}|\mathbf{y})$.

Going back to the air pollution example, suppose that we have the following new observations:

```
new.data <- data.frame(negtemp = c(-50, -60, -40), manuf = c(150, 100,
```

```
    ↪ 400), pop = c(200, 100, 300), wind = c(6, 7, 8), precip = c
    ↪ (10, 30, 20), days = c(20, 100, 40))
```

To predict SO2 from the MLE fit, `usair.lm1`, in R, we run:

```
predict(usair.lm1, new.data, se.fit = TRUE)
```

```
$fit
       1        2        3
33.72743 18.94993 55.47696

$se.fit
        1         2         3
14.936928  5.329492 17.639438

$df
[1] 35

$residual.scale
[1] 15.78998
```

The R output includes a vector of predictions (`$fit`), a vector of standard error of predicted means (`$se.fit`), the degrees of freedom for residuals (`$df`), and the residual standard deviation (`$residual.scale`).

In the `INLA` library, there is no function "predict" as for `lm` in R. However, we do not need a posterior predictive simulation like in MCMC approaches. Predictions can be done as a part of the model fitting itself in INLA. As prediction is the same as fitting a model with some missing data, we need to set the response variables "`y[i] = NA`" for those "observations" we want to predict. The prediction in INLA is implemented through the following R code:

```
usair.combined <- rbind(usair, data.frame(SO2 = c(NA, NA, NA), new.
    ↪ data))
usair.link <- c(rep(NA, nrow(usair)), rep(1, nrow(new.data)))
usair.inla1.pred <- inla(usair.formula1, data = usair.combined,
    ↪ control.predictor = list(link = usair.link))
usair.inla1.pred$summary.fitted.values[(nrow(usair)+1):nrow(usair.
    ↪ combined),]
```

```
                         mean        sd 0.025quant 0.5quant 0.975quant     mode
fitted.predictor.42 33.65338 14.938107   4.191079 33.65547   63.09751 33.65951
fitted.predictor.43 18.92744  5.329609   8.416159 18.92807   29.43290 18.92930
fitted.predictor.44 55.40423 17.644354  20.605306 55.40629   90.18455 55.41030
```

Note that we set the `control.predictor` option as `control.predictor = list(link = usair.link)`, where the object `usair.link` is set to be a vector of `NA` if the corresponding response is observed in the original dataset and `1` if the corresponding response is missing (and is to be predicted). The summary statistics of the predicted responses from INLA show concordance with the results from the MLE method.

3.4 Model Selection and Checking

In Chapter 1, we briefly discussed model selection and checking for a Bayesian model. Here we show the details on how to implement the analysis in INLA using the air pollution data example.

3.4.1 Model Selection by DIC

In regression analysis, we often want to find a reduced model with the best subset of the variables from the full model. The model selection in frequentist analysis is commonly based on Akaike information criterion (AIC), a MLE-based criterion. Back to the air pollution data example, a stepwise model selection procedure using AIC can be implemented by the function `stepAIC` in R library `MASS`:

```
library(MASS)
usair.step <- stepAIC(usair.lm1, trace = FALSE)
usair.step$anova
```

```
Stepwise Model Path
Analysis of Deviance Table

Initial Model:
SO2 ~ negtemp + manuf + wind + precip + days

Final Model:
SO2 ~ negtemp + manuf + wind + precip


    Step Df Deviance Resid. Df Resid. Dev      AIC
1                           35   8726.322 231.7816
2 - days  1 26.57448        36   8752.897 229.9063
```

It turns out that the variable, `days`, is dropped from the full model. The final reduced model includes the four predictors, `negtemp`, `manuf`, `wind`, and `precip`. Let us fit the final reduced model:

```
usair.formula2 <- SO2 ~ negtemp + manuf + wind + precip
usair.lm2 <- lm(usair.formula2, data = usair)
round(coef(summary(usair.lm2)), 4)
```

```
            Estimate Std. Error t value Pr(>|t|)
(Intercept) 123.1183    31.2907  3.9347   0.0004
negtemp       1.6114     0.4014  4.0148   0.0003
manuf         0.0255     0.0045  5.6150   0.0000
wind         -3.6302     1.8923 -1.9184   0.0630
precip        0.5242     0.2294  2.2852   0.0283
```

In Bayesian analysis, DIC, a generalization of AIC, is one of the most popular measures for Bayesian model comparison, which is defined as the sum of a measure of goodness of fit plus a measure of model complexity (Spiegelhalter et al., 2002). In the `INLA` library, the `dic=TRUE` flag makes the `inla()` function compute the model's DIC. The model with the lower DIC provides the better trade off between fit and model complexity. Unfortunately, there is no function available for stepwise model selection by DIC in the INLA library. In this example, there are only five predictors in the full model. We may perform a backward elimination procedure using DIC (i.e., manually eliminating variables based on DIC). In this air pollution study, it turns out that the final model with four predictors, negtemp, mauf, wind, and precip, has the lowest DIC (=348.57), which is concordant with the above model selection result using the frequentist approach:

```
usair.inla3 <- inla(usair.formula2, data = usair, control.compute =
    ↪ list(dic = TRUE, cpo = TRUE))
round(usair.inla3$summary.fixed, 4)
```

```
                mean      sd 0.025quant 0.5quant 0.975quant     mode kld
(Intercept) 122.9366 31.2837    61.1617 122.9406   184.5970 122.9507   0
negtemp       1.6102  0.4016     0.8172   1.6102     2.4019   1.6103   0
manuf         0.0255  0.0045     0.0165   0.0255     0.0344   0.0255   0
wind         -3.6168  1.8905    -7.3480  -3.6172     0.1111  -3.6179   0
precip        0.5239  0.2296     0.0706   0.5239     0.9766   0.5240   0
```

The cpo=TRUE flag will be discussed in a later section. Comparing the DICs between two Bayesian models, we have a smaller DIC for the last model:

```
c(usair.inla1$dic$dic, usair.inla3$dic$dic)
```

```
[1] 350.6494 348.5703
```

From the output, regression coefficients of the variables negtemp, mauf, and precip are significantly different from zero (in the Bayesian sense). That is, the 95% credible intervals of these coefficients do not contain zero. They have high posterior probabilities of being positively associated with the response SO2. The variable wind is negatively associated with SO2, but is not significant. Model coefficients can be interpreted as follows. For example, the posterior mean of precip, 0.5239, means that for every additional inch of average annual precipitation we expect the sulfur dioxide content of air to increase 0.5239 micrograms per cubic meter, when other covariates are fixed. The 95% credible interval for precip, is (0.0706, 0.9766), which contains the true parameter of precip with 95% probability.

3.4.2 Posterior Predictive Model Checking

Checking the model fit is critical in statistical analysis. In Bayesian analysis, model assessment is often based on posterior predictive checks or leave-one out cross-validation predictive checks. Held et al. (2010) compared two approaches for estimating Bayesian models using MCMC and INLA. Bayesian model posterior predictive check was originally proposed by Gelman et al. (1996). The key concept of such a check is the posterior predictive distribution of a replicate observation y_i^* which has density

$$p(y_i^*|\mathbf{y}) = \int p(y_i^*|\theta)p(\theta|\mathbf{y})d\theta.$$

The corresponding *posterior predictive p-value*,

$$p(y_i^* \leq y_i|\mathbf{y}),$$

is used as a measure of model fit (Meng, 1994). Extreme posterior predictive p-values ("extreme" means that p-value is very close to 0 or 1 here) can be used to identify observations that diverge from the assumed model.

In the INLA package, the posterior predictive p-value can be obtained by the R function inla.pmarginal, which returns the distribution function of marginals obtained by inla. The following R code generates the histogram of posterior predictive p-values for the reduced final model in the US air pollution study.

```
usair.inla3.pred <- inla(usair.formula2, data = usair, control.
    ↪ predictor = list(link = 1, compute = TRUE))
post.predicted.pval <- vector(mode = "numeric", length = nrow(usair))
for(i in (1:nrow(usair))) {
  post.predicted.pval[i] <- inla.pmarginal(q=usair$SO2[i], marginal =
      ↪ usair.inla3.pred$marginals.fitted.values[[i]])
}
hist(post.predicted.pval, main="", breaks = 10, xlab="Posterior
    ↪ predictive p-value")
```

Figure 3.3 shows that many posterior predictive p-values are close to 0 or 1. However, one drawback about interpreting posterior predictive p-values is that they could not have a uniform distribution even if the data come from the true model. See Hjort et al. (2006); Marshall and Spiegelhalter (2007) for the details. From the scatterplot matrix of the air pollution data (Figure 3.1), the response, SO2, is right-skewed and the predictor, manuf, has a very large variance and contains some outliers. The posterior predictive p-values could be affected by the nature of the data. So, although the plot of the posterior predictive p-values is not satisfied, we want to further check the model using other model assessment methods.

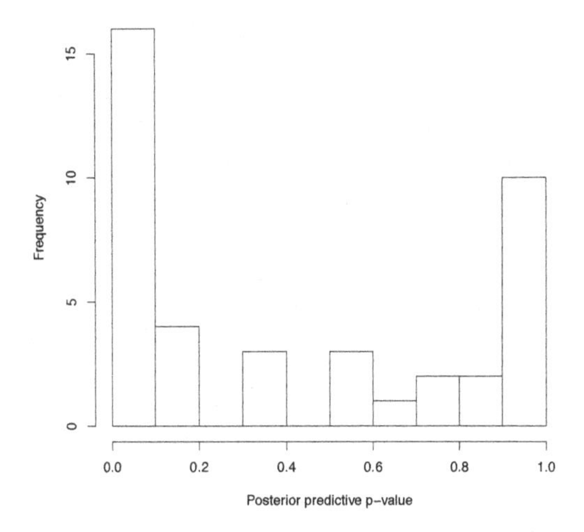

FIGURE 3.3
Histogram of the posterior predictive p-values for the reduced final model in the US air pollution study.

3.4.3 Cross-Validation Model Checking

The other methods based on the predictive distribution are the leave-one-out cross-validation. Two quantities, *conditional predictive ordinate* (CPO) and *probability integral transform* (PIT), are used for evaluating the goodness of the model:

$$\begin{aligned}\text{CPO}_i &= p(y_i|\mathbf{y}_{-i}),\\ \text{PIT}_i &= p(y_i^* \le y_i|\mathbf{y}_{-i}).\end{aligned}$$

Here $\mathbf{y}_{-i}$ denotes the observations $\mathbf{y}$ with the ith observation omitted. Note that the only difference between PIT and the posterior predictive p-value is that PIT is computed based on $\mathbf{y}_{-i}$ rather than $\mathbf{y}$.

In INLA, these quantities are computed without rerunning the model for each observation in turn (Held et al., 2010). To obtain CPOs and PITs, we need to simply add the argument `control.compute = list(cpo = TRUE)` into `inla` function. For example, in our resulting object `usair.inla3` for the final reduced model using INLA, we can find the predictive CPOs and PITs using the commands `usair.inla3$cpo$cpo` and `usair.inla3$cpo$pit`. Held et al. (2010) showed numerical problems may occur when CPOs and PITs are computed using INLA. There are internal checks in the INLA program for the potential problems, which appears as `usair.inla3$cpo$failure`. It is a vector containing 0 or 1 for each observation. A value equal to 1 indicates that the estimate of CPO or PIT is not reliable for the corresponding observation. In our example, we can check if there are any failures by:

```
sum(usair.inla3$cpo$failure)
```

```
[1] 0
```

So, there is no issue of the computation of CPOs and PITs in the fit `usair.inla3`. The uniformity of the PIT values indicates that the predictive distributions match the observations from the data and thus it is an indication of a well-fitted model (Diebold et al., 1998; Gneiting et al., 2007). We now plot the histogram and the uniform Q-Q plot of PITs.

```
hist(usair.inla3$cpo$pit, main="", breaks = 10, xlab = "PIT")
qqplot(qunif(ppoints(length(usair.inla3$cpo$pit))), usair.inla3$cpo$
    ↪ pit, main = "Q-Q plot for Unif(0,1)", xlab = "Theoretical
    ↪ Quantiles", ylab = "Sample Quantiles")
qqline(usair.inla3$cpo$pit, distribution = function(p) qunif(p), prob
    ↪ = c(0.1, 0.9))
```

Figure 3.4 shows that the distribution of the PITs is close to a uniform distribution, suggesting that the model reasonably fits the data. Note that the PIT histogram is much closer to a uniform distribution than the corresponding posterior predictive histogram shown in Figure 3.3.

If we think of the product of all of the CPO values as a "pseudo marginal likelihood," this gives a cross-validatory summary measure of fit. The *log pseudo marginal likelihood* (LPML), proposed by Geisser and Eddy (1979), is simply the log of this

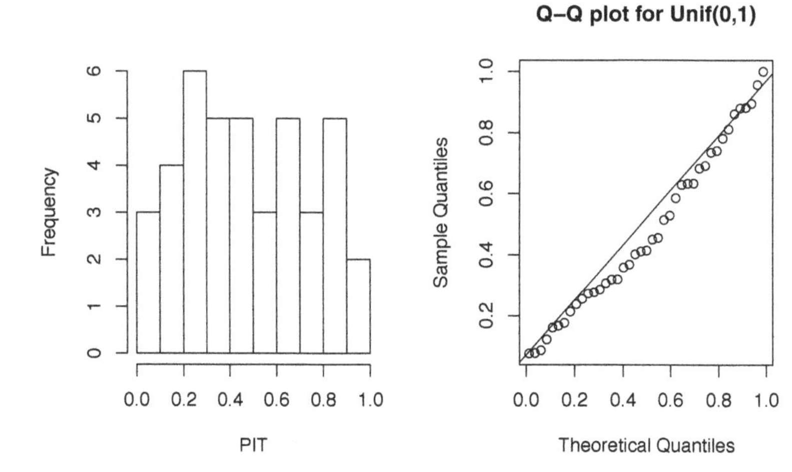

FIGURE 3.4
Histogram and uniform Q-Q plot of the cross-validated PIT for the reduced final model in the US air pollution study.

measure,

$$LPML = \log\left\{\prod_{i=1}^{n} p(y_i|\mathbf{y}_{-i})\right\} = \sum_{i=1}^{n} \log p(y_i|\mathbf{y}_{-i}) = \sum_{i=1}^{n} \log \text{CPO}_i,$$

which is often used as an alternative measure for DIC. Draper and Krnjajic (2007, Sec. 4.1) have shown that DIC approximates the LPML for approximately Gaussian posteriors. LPML remains computationally stable (Carlin and Louis, 2008). Unlike DIC, a model with a larger LPML is better supported by the data. Let us compute the LPMLs for the full model and reduced model in the air pollution study:

```
LPML1 <- sum(log(usair.inla1$cpo$cpo))
LPML3 <- sum(log(usair.inla3$cpo$cpo))
c(LPML1, LPML3)
```

```
[1] -176.9495 -175.6892
```

LPML for the reduced model is larger than that for the full model, indicating that the reduced model is preferred. Oftentimes, we can also perform a graphical analysis of a point-wise comparison of CPOs to choose a model.

```
plot(usair.inla1$cpo$cpo, usair.inla3$cpo$cpo, xlab="CPO for the full
    ↪ model", ylab="CPO for the reduced model")
abline(0,1)
```

Figure 3.5 shows a scatterplot of the pointwise comparison of CPOs between the full model and the reduced model, along with a reference line marking where the values are equal for the two models, in the US air pollution study. Since larger

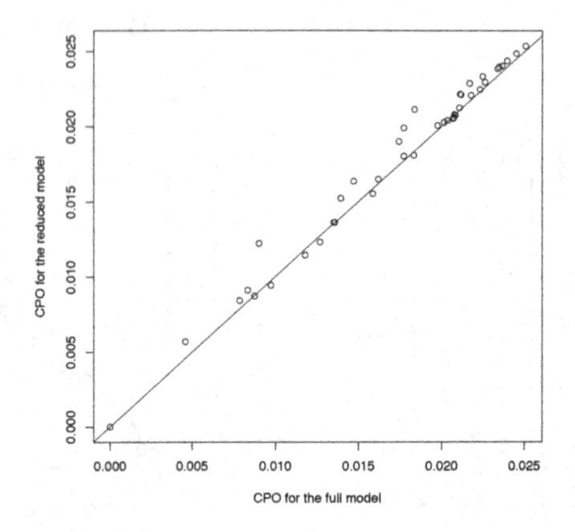

FIGURE 3.5
Bayesian model choice based on the pointwise comparison of CPOs between the two models in the US air pollution study.

CPO values are indicative of better model fit, the predominance of points above the reference line implies a preference for the reduced model. This is in agreement with our previous findings using DIC and LPML criteria.

3.4.4 Bayesian Residual Analysis

In the linear regression setting, the model is given as $y_i = \mathbf{x}_i^T \beta + \varepsilon_i, i = 1..., n$. Given the unknown parameters β and the predictors $\mathbf{x}_i$ for a data point y_i, *Bayesian residuals* are defined as

$$r_i = \varepsilon_i(\beta) = y_i - x_i^T \beta, \quad i = 1, ..., n. \tag{3.9}$$

These are sometimes called "realized residuals" in contrast to the classical or estimated residuals, $y_i - x_i^T \hat{\beta}$, which is based on a point estimate $\hat{\beta}$ of the unknown parameters. A priori of these residuals are distributed as $N(0, \sigma^2)$, and the posterior distribution of them can be obtained from the posterior distributions of β and σ^2.

Checking residual plots is a common way for model diagnostics in regression analysis. In order to obtain a Bayesian residual plot, we generate samples from the posterior distribution of β and σ^2 and then substitute these samples into (3.9) to produce samples from the posterior distribution of the residuals. The posterior mean or median of the r_i can be calculated and examined. Plotting these residuals versus index or the fitted values might detect outliers as well as reveal failure in model assumptions, such as the normality or the homogeneity of variance assumption. Summing their squares or absolute values could provide an overall measure of fit. An early application of Bayesian residuals analysis can be found in Chaloner and Brant (1988). See Gelman et al. (2014) for more discussions on this topic.

Back to our air pollution example, we have written a convenience function

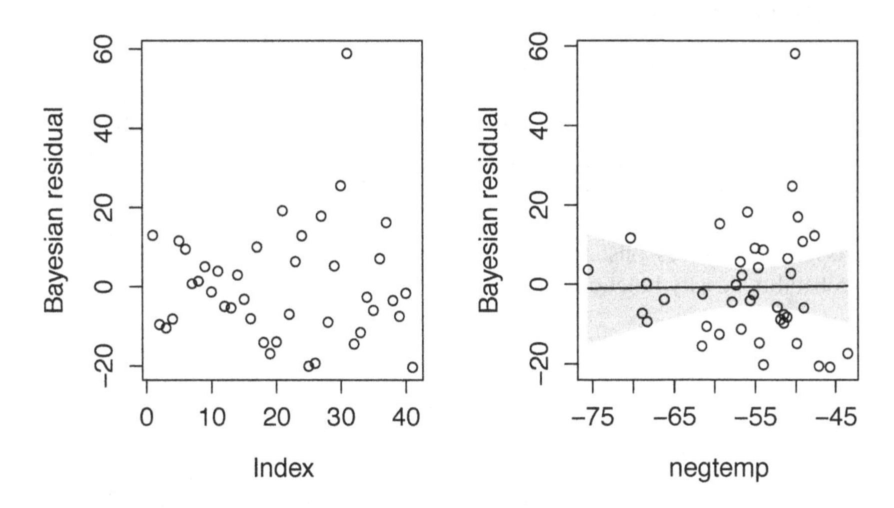

FIGURE 3.6
Bayesian residual plots in the US air pollution study: the left panel is the Bayesian residual index plot; the right panel is the plot of the residual versus the predictor, `negtemp`.

`bri.lmresid.plot` for producing the Bayesian residual plot in our `brinla` package. The following code generates a Bayesian residual index plot as well as a plot of the residual versus the predictor, `negtemp`:

```
bri.lmresid.plot(usair.inla2)
bri.lmresid.plot(usair.inla2, usairl$negtemp, xlab = "negtemp", smooth
    ↪ =TRUE)
```

Figure 3.6 shows that the Bayesian residuals generally show a random pattern around zero in both plots. We find that the observation number 31 seems to be an outlier, where its Bayesian residual is as high as 59.6927. The observation is from the city of Providence. Its SO2 level, 94, is much higher than all other cities. More investigations could be conducted in terms of this city. The argument `smooth = TRUE` in the `bri.mresid.plot` function is to add the smooth curve with its 95% credible intervals on the residual plot. Here the smooth curves are estimated using the random walk model of order 2 with INLA, which is discussed in detail in Chapter 7. The curves do not show a special trend, which indicates that the model assumptions are valid.

3.5 Robust Regression

Statistical inference based on the normal distribution is often known to be vulnerable to outliers. Lange et al. (1989) introduced t distributed errors regression models as a robust extension of the traditional normal models. The t regression models provide a useful extension of the normal regression models for the studies involving residual errors with longer-than-normal tails, as well as for the cases that extreme points exist in a dataset.

Assume that the response variable Y follows a t distribution, denoted by $t(\mu,\sigma,\nu)$. Its probability density function is given by,

$$f(y|\mu,\sigma,\nu) = \frac{\nu^{\nu/2}\Gamma((\nu+1)/2)}{\sigma\sqrt{\pi}\Gamma(\nu/2)}\left\{\nu+\left(\frac{y-\mu}{\sigma}\right)^2\right\}^{-(\nu+1)/2},$$

where μ is location parameter, σ is dispersion parameter, ν is degree of freedom, and $\Gamma(\cdot)$ is the gamma function. Note that the distribution $t(\mu_i,\sigma,\nu)$ can be viewed as a mixture of normal and gamma distributions (Lange et al., 1989; Liu and Rubin, 1995), i.e., if a latent variable $\eta_i \sim \Gamma(\nu/2,\nu/2)$, and $y_i|\eta_i \sim N(\mu_i,\sigma^2/\eta_i)$, then $y_i \sim t(\mu_i,\sigma,\nu)$.

The t regression model can be expressed as

$$\begin{cases} y_i \sim t(\mu_i,\sigma,\nu), \\ \mu_i = \beta_0+\beta_1 x_{i1}+\ldots+\beta_p x_{ip}, \quad i=1,\ldots,n. \end{cases}$$

In the model, the parameter σ is the unknown dispersion parameter which models the variance of y_i's. In INLA, we model the precision parameter $\tau = \sigma^{-2}$.

Let us go back to our air pollution example. In our Bayesian residual analysis, we find that the observation number 31 is an outlier, which is from the city of Providence. To avoid this data point influencing the regression equation too strongly, let us fit a t regression model. In INLA, this analysis is implemented through specifying the argument `family = "T"`:

```
usair.inla4 <- inla(usair.formula2, family = "T", data = usair,
  ↪ control.compute = list(dic = TRUE, cpo = TRUE))
round(usair.inla4$summary.fixed, 4)
```

```
                mean      sd 0.025quant 0.5quant 0.975quant     mode kld
(Intercept) 119.2344 36.4249    47.5236 119.2752   190.6313 119.3441   0
negtemp       1.4588  0.4752     0.5319   1.4564     2.3978   1.4509   0
manuf         0.0266  0.0051     0.0163   0.0267     0.0363   0.0269   0
wind         -3.9300  2.2315    -8.3388  -3.9220     0.4322  -3.9039   0
precip        0.4356  0.2653    -0.0722   0.4307     0.9705   0.4198   0
```

```
round(usair.inla4$summary.hyperpar, 4)
```

```
                            mean     sd 0.025quant 0.5quant 0.975quant   mode
precision for the
student-t observations    0.0053 0.0016     0.0030   0.0051     0.0092 0.0046
degrees of freedom for
student-t                10.9839 8.7846     3.5675   8.3886    34.1122 5.6104
```

Comparing the results of the t regression model with those of the normal regression model, the estimates of intercept, `negtemp`, and `precip` are decreased noticeably, and the standard deviation of the estimated parameters are increased. These are due to the fact of the heavier tails in the t distribution. The presence of extreme observations, therefore, has a smaller influence on the regression parameters.

3.6 Analysis of Variance

Analysis of variance (ANOVA) is a collection of statistical models used to determine the degree of difference or similarity between two or more groups of data. It has been heavily used in the analysis of experimental data. The experimenter adjusts factors and measures responses in an attempt to determine an effect. ANOVA and the design of experiments have been extensively discussed in a number of books (Montgomery, 2013; Chatterjee and Hadi, 2015). Here, we give a brief discussion of the main concepts and show an example on solving the ANOVA models using INLA.

In fact, an ANOVA is just a special case of regression with categorical variables. Let us consider the case of a single factor A with levels $i = 1, 2, ..., a$. Assume that the response variable Y is normally distributed and the effect of the factor variable influences the mean of Y. Thus, the model can be summarized by

$$y_{ij} = \mu_i + \varepsilon_{ij}, \tag{3.10}$$

where y_{ij} is the jth ($j = 1, ..., n_i$) observed value for the ith factor level, and the error $\varepsilon_{ij} \sim N(0, \sigma^2)$. Model (3.10) is often rewritten as

$$y_{ij} = \mu_0 + \alpha_i + \varepsilon_{ij}, \tag{3.11}$$

In (3.11), there are $a+1$ parameters to be estimated. To make the model identifiable, we need a constraint on the parameters. A commonly used constraint in statistical literature is the *corner constraint*. The effect of one level $r \in \{1, 2, ..., a\}$ is set equal to zero: $\alpha_r = 0$. This level r is referred to as the baseline or reference level of factor A. Typically, we use the first level as the reference level, i.e.,

$$\alpha_1 = 0.$$

Therefore, μ_0 becomes the mean for the first level. The corner constraint is the constraint that is used in the `INLA` library.

The one-way ANOVA model can be easily extended to accommodate additional categorical variables. Assume that we have a two-by-two factorial design: each level of the factor A (with a levels) is crossed with each level of the factor B (with b levels). Suppose that there are n replicates with each of the ab cells. The observation y_{ijk} is modeled as:

$$y_{ijk} = \mu_0 + \alpha_i + \beta_j + \gamma_{ij} + \varepsilon_{ijk}, \tag{3.12}$$

for $i = 1,...,a$, $j = 1,...,b$, and $k = 1,...,n$. In this model, μ_0 is the grand mean, α_i is the main effect of level i from the factor A, β_j is the main effect of level j from the factor B, γ_{ij} is the interaction effect of the ith level of A and the jth level of B, and the error $\varepsilon_{ijk} \sim N(0, \sigma^2)$.

Note that model (3.12) contains $1+a+b+ab$ parameters, however the data supply only ab sample means. Hence, we need the constraints on the parameters. In the corner-point parameterization we impose the following constraints:

$$\alpha_1 = \beta_1 = \gamma_{11} = ... = \gamma_{1b} = \gamma_{21} = ... = \gamma_{a1} = 0.$$

In frequentist literature, model (3.11) or model (3.12) could be treated as either a fixed-effects or random-effects model. For example, in (3.11), if α_i's in the one-way classification are viewed as non-random, (3.11) is called a *fixed-effects ANOVA*. If α_i's are treated as a random sample from a probability distribution, for instance, $\alpha_i \sim N(0, \sigma_\alpha^2)$, (3.11) is called a *random-effects ANOVA*. Choosing the fixed-effects or random-effects models is often based on how the levels are defined or selected. For example, treatment effects in a clinical trial are considered as fixed effects, while the effects of patients may be regarded as random effects.

From a Bayesian perspective, the distinction between fixed and random effects is less distinct because all parameters are viewed as random variables. The fixed-effects or random-effects can be reflected through the selection of priors. In (3.11), the "fixed effects" corresponding to the factor A may be assigned independent diffuse Gaussian prior distributions with known precision, i.e., $\alpha_i \sim N(0, \sigma_0^{-1})$ where σ_0 is fixed. The "random effects" corresponding to the factor A may be assigned the prior $\alpha_i|\sigma_\alpha^2 \sim N(0, \sigma_\alpha^2)$ with σ_α^2 assigned a hyperprior.

An ANOVA model can be easily fitted using the INLA approach. We will only focus fixed-effects ANOVA models in the following example. For random-effects models we refer to Chapter 5.

The `painrelief` dataset was from an experiment examining the effects of codeine and acupuncture on post-operative dental pain in male subjects (Kutner et al., 2004). The study used a randomized block design with two treatment factors occurring in a factorial structure. Both treatment factors have two levels. The variable `Codeine` has two groups: take a codeine capsule or a sugar capsule. The variable `Acupuncture` has two groups: apply to two inactive acupuncture points or two active acupuncture points. There are four distinct treatment combinations due to this factorial treatment structure. The variable `PainLevel` is the blocking variable. The response variable `Relief` is the pain relief score (the higher the score, the more relief the patient has). Totally, 32 subjects are assigned to eight blocks of four subjects each based on an assessment of pain tolerance. The description of variables for the data set is displayed in Table 3.2.

In INLA, the specification of an ANOVA model is analogous to a linear regression analysis. The only difference is that the explanatory variable needs to be a `factor`, rather than a numeric vector in R. The following statement invokes `inla` function, where the blocking variable and the two treatment factors with their interaction appear in the model.

TABLE 3.2
Description of variables in the painrelief dataset.

Variable Name	Description	Codes/Values
Relief	pain relief score	numeric number
PainLevel	pain level	level from 1 to 8
Codeine	use codeine or other	1: a codeine capsule; 2: a sugar capsule.
Acupuncture	acupuncture method	1: apply to 2 inactive acupuncture points; 2: apply to 2 active acupuncture points.

```
data(painrelief, , package = "brinla")
painrelief$PainLevel <- as.factor(painrelief$PainLevel)
painrelief$Codeine <- as.factor(painrelief$Codeine)
painrelief$Acupuncture <- as.factor(painrelief$Acupuncture)
painrelief.inla <- inla(Relief ~ PainLevel + Codeine*Acupuncture, data
    ↪ = painrelief)
round(painrelief.inla$summary.fixed, 4)
```

```
                       mean     sd 0.025quant 0.5quant 0.975quant   mode kld
(Intercept)          0.0188 0.0706    -0.1210   0.0188     0.1583 0.0188   0
PainLevel2           0.1500 0.0851    -0.0186   0.1500     0.3183 0.1500   0
PainLevel3           0.3250 0.0851     0.1564   0.3250     0.4933 0.3250   0
PainLevel4           0.3000 0.0851     0.1314   0.3000     0.4683 0.3000   0
PainLevel5           0.6750 0.0851     0.5064   0.6750     0.8433 0.6750   0
PainLevel6           0.9750 0.0851     0.8064   0.9750     1.1433 0.9750   0
PainLevel7           1.0750 0.0851     0.9064   1.0750     1.2433 1.0750   0
PainLevel8           1.1500 0.0851     0.9814   1.1500     1.3183 1.1500   0
Codeine2             0.4625 0.0602     0.3433   0.4625     0.5815 0.4625   0
Acupuncture2         0.5750 0.0602     0.4558   0.5750     0.6940 0.5750   0
Codeine2:Acupuncture2 0.1500 0.0851    -0.0186   0.1500     0.3183 0.1500   0
```

The main effects of both treatment factors, codeine and acupuncture, are highly significant in the Bayesian sense. However, the interaction between the two factors is not significant at the 95% credible level, indicating that there is no interaction effect between them. The blocking effect, painlevel, has multiple levels. To better understand the significance of the effect, we generate a plot of posterior mean estimates and 95% credible levels for the different pain levels (Figure 3.7; the R code is not displayed here). The horizontal line is the reference line for pain level 1, which is set to zero. For the higher pain level, subjects receive larger pain relief scores (except pain level 4). PainLevel is clearly a significant confounding factor that needs to be considered into the model.

3.7 Ridge Regression for Multicollinearity

In regression modeling, multicollinearity is a phenomenon of high intercorrelations or inter-associations among the independent variables. Statistical inferences using

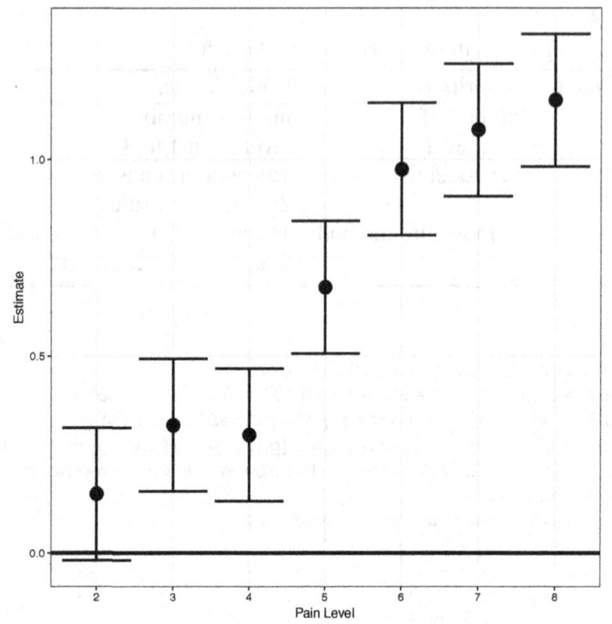

FIGURE 3.7
Posterior estimates and the error bars for the `PainLevel` effect in the `PainRelief` study. The horizontal line is the reference line for pain level 1, which is set to zero.

ordinary least squares may not be reliable if multicollinearity is present in the data. Multicollinearity can result in many problems in the regression analysis. For instance, the partial regression coefficient due to multicollinearity may not be estimated precisely. The standard errors are typically very high. Multicollinearity makes it tedious to assess the relative importance of the predictors in explaining the variation caused by the response variable.

There are many possible ways to deal with multicollinearity. Principal component regression is one of the traditional methods. One performs the multivariate reduction of the set of correlated predictors to a smaller set of uncorrelated predictors through principal component analysis. Instead of regressing the dependent variable on the correlated explanatory variables directly, the principal components of the explanatory variables are used as regressors (Jolliffe, 1982).

Ridge regression, another popular solution to the multicollinearity problem, addresses the issues of ordinary least squares by imposing a penalty on the size of coefficients (Hoerl and Kennard, 1970). The ridge coefficients minimize a penalized residual sum of squares,

$$\min_{\beta}\{(\mathbf{y}-\mathbf{X}\beta)^T(\mathbf{y}-\mathbf{X}\beta)\}+\lambda\beta^T\beta,$$

where $\lambda \geq 0$ is a complexity parameter that controls the amount of shrinkage: the larger the value of λ, the greater the amount of shrinkage.

The resulting ridge regression estimator using the least squares method is

$$\hat{\boldsymbol{\beta}}_{ridge}(\lambda) = (\mathbf{X}^T\mathbf{X} + \lambda\mathbf{I})^{-1}\mathbf{X}^T\mathbf{y}.$$

This method will induce bias (which increases with λ) however yield more precise parameter estimates (i.e., smaller variances for the parameters).

Indeed, the ridge regression is closely related to a version of the standard posterior Bayesian regression estimate, but using a specific prior distribution on the elements of the regression parameters. Assume β_i has the prior distribution $\beta_j \sim N(0, \sigma_0^2/\lambda), j = 1, ..., p$ independently, where σ_0 is known. A large value of λ corresponds to a prior that is more tightly concentrated around zero, and hence leads to greater shrinkage towards zero.

The mean of the posterior distribution of $\boldsymbol{\beta}$ given the data $(\mathbf{X}, \mathbf{y})$ is then

$$(\mathbf{X}^T\mathbf{X} + \lambda\mathbf{I})^{-1}\mathbf{X}^T\mathbf{y},$$

which is identical to the form of the ridge estimator (Hsiang, 1975). Note that since λ is applied to the squared norm of the parameter vector $\boldsymbol{\beta}$, one wants to standardize all of the predictors to make them have a similar scale. To implement the Bayesian ridge regression with INLA, we need some programming tricks, which we will illustrate in the following case study.

We use an example that concerns import activity in the French economy. The data have been analyzed in Chatterjee and Hadi (2015). The dependent variable is import (`IMPORT`), domestic production (`DOPROD`), stock formation (`STOCK`), and domestic consumption (`CONSUM`). All variables are measured in billions of French francs for the years 1949 to 1966. We consider a linear regression model for the data. Let us first examine the sample correlations among the predictors:

```
data(frencheconomy, package = "brinla")
round(cor(frencheconomy[,-1]),4)
```

```
       IMPORT DOPROD  STOCK CONSUM
IMPORT 1.0000 0.9842 0.2659 0.9848
DOPROD 0.9842 1.0000 0.2154 0.9989
STOCK  0.2659 0.2154 1.0000 0.2137
CONSUM 0.9848 0.9989 0.2137 1.0000
```

We notice that the correlation between `DOPROD` and `CONSUM` is very high, which is equal to 0.9989. Collinearity is an issue to fit a standard linear regression model for the data.

Bayesian ridge regression assumes that the elements of the coefficients of all predictors, $(\beta_1, ..., \beta_p)$ are drawn from a common normal density. Note that a preliminary standardization of the predictors is often necessary to make this prior assumpation more plausible. So, we first standardize the predictors:

```
fe.scaled <- cbind(frencheconomy[, 1:2], scale(frencheconomy[, c
   ↪ (-1,-2)]))
```

To implement ridge regression with INLA, we need to use the "`copy`" feature in the `INLA` library, and change the dataset. The following `R` commands fit a Bayesian ridge regression with INLA:

```
n <- nrow(fe.scaled)
fe.scaled$beta1 <- rep(1,n)
fe.scaled$beta2 <- rep(2,n)
fe.scaled$beta3 <- rep(3,n)
param.beta = list(prec = list(param = c(1.0e-3, 1.0e-3)))
formula.ridge  = IMPORT ~ f(beta1, DOPROD, model="iid", values = c
    ↪ (1,2,3), hyper = param.beta) + f(beta2, STOCK, copy="beta1",
    ↪ fixed=T) + f(beta3, CONSUM, copy="beta1", fixed=T)
frencheconomy.ridge <- inla(formula.ridge, data = fe.scaled)
ridge.est <- rbind(frencheconomy.ridge$summary.fixed, frencheconomy.
    ↪ ridge$summary.random$beta1[,-1])
round(ridge.est,4)
```

```
               mean     sd 0.025quant 0.5quant 0.975quant    mode kld
(Intercept) 30.0778 0.5176    29.0498  30.0778    31.1042 30.0778   0
1            5.1608 5.2530    -6.3008   5.3693    15.3948  5.6275   0
2            0.7205 0.5427    -0.3564   0.7203     1.7972  0.7199   0
3            6.9279 5.2557    -3.2703   6.7037    18.4260  6.4066   0
```

We want to compare the results with the standard Bayesian linear regression analysis using INLA:

```
formula <- IMPORT ~  DOPROD + STOCK + CONSUM
frencheconomy.inla <- inla(formula, data = fe.scaled, control.fixed =
    ↪ list(prec = 1.0e-3), control.family = list(hyper = param.beta))
round(frencheconomy.inla$summary.fixed, 4)
```

```
               mean      sd 0.025quant 0.5quant 0.975quant    mode kld
(Intercept) 30.0778  0.5688    28.9456  30.0778    31.2086 30.0778   0
DOPROD       2.9951 10.8776   -18.4346   2.9553    24.6253  2.8820   0
STOCK        0.7197  0.5995    -0.4737   0.7198     1.9113  0.7200   0
CONSUM       9.1424 10.8734   -12.5047   9.1816    30.5379  9.2557   0
```

The results show that the estimates are quite different using ridge regression vs. the standard approach. In our preliminary data exploratory analysis, we note that not only the correlation DOPROD and CONSUM is high but also both of them are highly correlated with the response variable, IMPORT. However, in the regression equation with standard priors, the contribution of DOPROD is much smaller than that of CONSUM (2.9951 vs. 9.1424). In the ridge estimation, the coefficient of DOPROD increases to 5.1608 while the coefficient of CONSUM decreases to 6.9279. These estimates provide a different and more plausible representation of the IMPORT relationship than is obtained from the results with standard priors. Noticeably, the standard deviations of the coefficients of DOPROD and CONSUM are decreased using the ridge method rather than using the standard approach.

Lastly, we want to compare the results using traditional frequentist ridge regression analysis and INLA. The R function lm.ridge in MASS library allows us to fit the model:

```
library(MASS)
ridge2 <- lm.ridge(IMPORT ~  DOPROD + STOCK + CONSUM, data = fe.scaled
    ↪ , lambda = seq(0, 1, length=100))
ridge2.final <- lm.ridge(IMPORT ~  DOPROD + STOCK + CONSUM, data = fe.
    ↪ scaled, lambda = reg2$kHKB)
ridge2.final
```

```
                DOPROD      STOCK     CONSUM
30.0777778  4.9559793  0.7176308  7.1669364
```

In the above commands, we first fit the model with a sequence of ridge constants. Then the final model is determined by the Hoerl–Kennard–Baldwin (HKB) criterion (Hoerl et al., 1975). The results show similar estimates as those using Bayesian ridge regression with INLA.

3.8 Regression with Autoregressive Errors

In the standard linear regression, we assume that the error term ε in (3.2) follows $N(0, \sigma^2 \mathbf{I}_n)$. However, it can happen that the errors have non-constant variance or are correlated in many applications. A more general assumption on the error term is that

$$\varepsilon \sim N(0, \mathbf{\Sigma}),$$

where the error covariance matrix $\mathbf{\Sigma}$ is symmetric and positive definite. Different diagonal entries in $\mathbf{\Sigma}$ correspond to non-constant error variances, while non-zero off-diagonal entries in $\mathbf{\Sigma}$ correspond to correlated errors. Suppose that $\mathbf{\Sigma}$ is known. In frequentist analysis, generalized least squares technique minimizes

$$(\mathbf{y} - \mathbf{X}\beta)^T \mathbf{\Sigma}^{-1} (\mathbf{y} - \mathbf{X}\beta),$$

which is solved by

$$\hat{\beta} = (\mathbf{X}^T \mathbf{\Sigma}^{-1} \mathbf{X})^T \mathbf{X}^T \mathbf{\Sigma}^{-1} \mathbf{y}$$

with covariance matrix

$$\text{Cov}(\hat{\beta}) = (\mathbf{X}^{\mathrm{T}} \Sigma^{-1} \mathbf{X})^{\mathrm{T}}.$$

Certainly, the covariance matrix $\mathbf{\Sigma}$ is generally unknown in applications, and has to be estimated from the data. However, there are $n(n+1)/2$ distinct elements in the matrix. It is not possible to estimate the model without any further restrictions. We often model $\mathbf{\Sigma}$ using a small number of parameters.

A common case in which the errors from a regression model are correlated is in time series data, where a collection of observations is obtained through repeated measurements over time. When the observations have a natural sequential order, adjacent observations tend to be similar. This correlation among the observations is referred to as autocorrelation. Ignoring the presence of autocorrelation could have several effects on the regression analysis. For instance, the estimate of σ^2 and the standard errors of the regression coefficients are often seriously underestimated and thus the statistical inferences commonly employed would no longer be strictly valid.

Consider a linear regression model with autoregressive errors:

$$y_t = \beta_0 + \beta_1 x_{t1} + \ldots + \beta_p x_{tp} + \varepsilon_t, \quad t =, 1 \ldots, n, \tag{3.13}$$

where the regression errors, ε_t's, are *stationary*. That is, $\mathrm{E}(\varepsilon_t) = 0$, $\mathrm{Var}(\varepsilon_t) = \sigma^2$, and the correlation between two errors depends only upon their separation s in time: $\text{Cor}(\varepsilon_t, \varepsilon_{t+s}) = \text{Cor}(\varepsilon_t, \varepsilon_{t-s}) = \rho_s$, where ρ_s is called the error *autocorrelation* at lag s.

Many models have been proposed and investigated for stationary time-series (Shumway and Stoffer, 2011). The most common one for autocorrelated regression errors is the *first-order autoregressive process*, AR(1):

$$\varepsilon_t = \rho\varepsilon_{t-1} + \eta_t, \quad t = 1, ..., n,$$

where $\eta_t \sim N(0, \sigma_\eta^2)$ independently. In this situation, the error-covariance matrix has the following structure:

$$\Sigma = \sigma^2 \begin{bmatrix} 1 & \rho & \rho^2 & \dots & \rho^{n-1} \\ \rho & 1 & \rho & \dots & \rho^{n-2} \\ \rho^2 & \rho & 1 & \dots & \rho^{n-3} \\ & \dots & & \dots & \\ \rho^{n-1} & \rho^{n-2} & \rho^{n-3} & \dots & 1 \end{bmatrix},$$

where $\sigma^2 = \sigma_\eta^2/(1-\rho^2)$. In this model, the error autocorrelation $\rho_s = \rho^s, s = 1, 2, ...$ decay exponentially to 0 as s increases.

Higher-order autoregressive models can be directly generalized from the AR(1) model. For instance, the second-order autoregressive model is

$$\varepsilon_t = \rho\varepsilon_{t-1} + \tau\varepsilon_{t-2} + \eta_t, \quad t = 1, ..., n.$$

Other popular time-series models include moving-average (MA) process and ARMA process. See Shumway and Stoffer (2011) for details.

Let $\hat{\varepsilon}_1, \hat{\varepsilon}_2, ..., \hat{\varepsilon}_n$ be the residuals from a regression model. The sample autocorrelation function of the residuals is defined as

$$\hat{\rho}(s) = \hat{\gamma}(s)/\hat{\gamma}(0),$$

where $\hat{\gamma}(s) = n^{-1}\sum_{t=1}^{n-s}(\hat{\varepsilon}_{t+s} - \sum_{t=1}^{n}\hat{\varepsilon}_t/n)(\varepsilon_t - \sum_{t=1}^{n}\hat{\varepsilon}_t/n)$ is the sample autocovariance function.

The partial autocorrelation at lag s for $s \geq 2$, $\omega(s)$ is defined as the direct correlation between ε_t and ε_{t-s} with the linear dependence between the intermediate variables ε_h with $t-s < h < t$ removed. Consider the standard regression model: $\varepsilon_t = \alpha_0 + \alpha_1\varepsilon_{t-1} + ... + \alpha_s\varepsilon_{t-s} + u_t$. The estimate of the partial autocorrelation is equal to the estimate of the coefficient, $\hat{\omega}(s) = \hat{\alpha}_s$.

An AR(s) process has an exponentially decaying autocorrelation function, and a partial autocorrelation function with s non-zero spike at the first s lags. So, in regression analysis, examining the residual autocorrelations and partial autocorrelations from a standard linear regression (assuming independent error) can help us to identify a suitable form for the error-generating process.

Several frequentist tests for autocorrelation have been available in the literature. One traditional method is the Dubin–Watson test, which is based on the statistics,

$$d_s = \frac{\sum_{t=s+1}^{n}(\hat{\varepsilon}_t - \hat{\varepsilon}_{t-s})^2}{\sum_{t=s+1}^{n}\hat{\varepsilon}_t^2}.$$

In Bayesian statistics, Dreze and Mouchart (1990) suggested using the classical Durbin–Watson statistic and examining the autocorrelation plot of the residuals as a quick check. Bauwens and Rasquero (1993) proposed two Bayesian tests of residual autocorrelation, which check if an approximate highest posterior density region of the parameters of the autoregressive process of the errors contains the null hypothesis.

In frequentist analysis, the regression with autoregressive errors (3.13) can be fit using the iteratively reweighted least squares fitting algorithm (Carroll and Ruppert, 1988). This model can be also fit using INLA. We now look at an example of time series data. The New Zealand unemployment data include the quarterly unemployment rates for both youth (15 – 19 years old) and adult (greater than 19 years old) from March 1986 to June 2011. Since June 2008, the New Zealand government has abolished the act of the differential youth minimum wage. Here we would like to study the relationship of the unemployment rates between adult and youth before and after the abolition of the act. Table 3.3 displays the variables in the dataset.

TABLE 3.3
Description of variables in the New Zealand unemployment data.

Variable Name	Description	Codes/Values
quarter	quarters from March 1986 to June 2011	characters
adult	unemployment rate	percent
youth	unemployment rate	percent
policy	type of minimum wage law	“Different”; "Equal"

Let us first load the dataset and create a variable for the index of the time series.

```
data(nzunemploy, package = "brinla")
nzunemploy$time <- 1:nrow(nzunemploy)
```

We begin our analysis by examining the time series for the unemployment rate of youth and adult.

```
qplot(time, value, data = gather(nzunemploy[,c(2,3,5)], variable,
    ↪ value, -time), geom = "line") + geom_vline(xintercept = 90) +
    ↪ facet_grid(variable ~ ., scale = "free") + ylab("Unemployment
    ↪ rate") + theme_bw()
```

Figure 3.8 shows the time series for the unemployment rate by youth and adult. The vertical lines in both subplots indicate the time of the abolition of the act of the differential youth minimum wage. We can see that the unemployment rates fluctuated substantially but gradually during this historical period. The patterns of two time series are similar, indicating the rate for youth is correlated with the rate for adult.

In order to make the coefficients easier to understand, we center adult unemployment rate on its mean over the time series.

```
nzunemploy$centeredadult = with(nzunemploy, adult - mean(adult))
```

We shall estimate the regression of `youth` on `centeredadult`, `policy` and their interaction. A preliminary standard regression with INLA produces the following fit to the data:

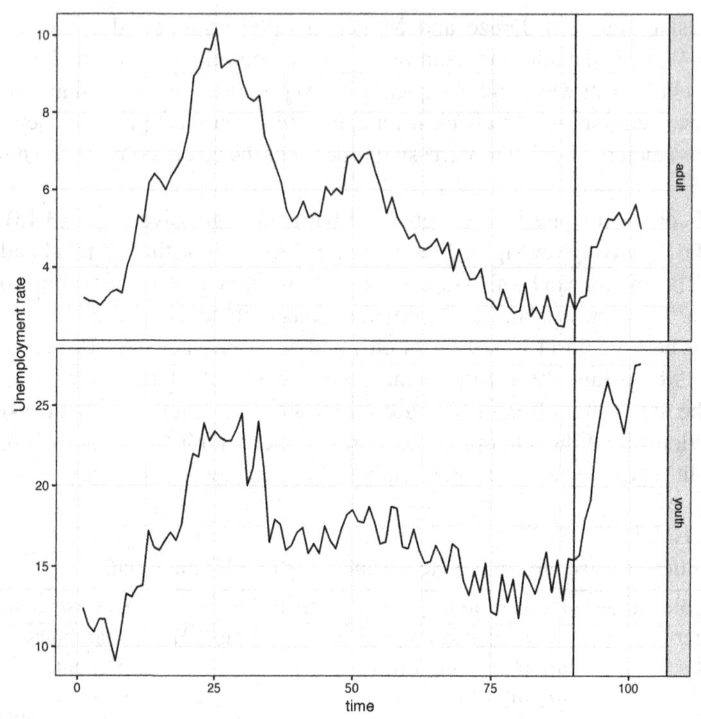

FIGURE 3.8
Time series for the unemployment rate by youth and adult. The vertical lines in both subplots indicate the time of the abolition of the act of the differential youth minimum wage.

```
formula1 <- youth ~ centeredadult*policy
nzunemploy.inla1 <- inla(formula1, data= nzunemploy)
round(nzunemploy.inla1$summary.fixed, 4)
```

```
                             mean     sd 0.025quant 0.5quant 0.975quant    mode kld
(Intercept)               16.2823 0.1536    15.9800  16.2823    16.5843 16.2823   0
centeredadult              1.5333 0.0751     1.3856   1.5333     1.6810  1.5333   0
policyEqual                9.4417 0.5266     8.4056   9.4417    10.4766  9.4418   0
centeredadult:policyEqual  2.8533 0.4622     1.9438   2.8533     3.7617  2.8533   0
```

Regression coefficients are statistically significant (that is, in the Bayesian sense, have high posterior probabilities of being positive). We want to check the Bayesian residuals for the model:

```
nzunemploy.res1 <- bri.lmresid.plot(nzunemploy.inla1, type="o")
```

The graph of the Bayesian residuals from the linear regression suggests that they may be autocorrelated in certain degree (Figure 3.9). Based on Dreze and Mouchart (1990)'s suggestion, we can quickly check the autocorrelation and partial autocorrelation functions. The `acf` function in the R `stats` package computes and plots

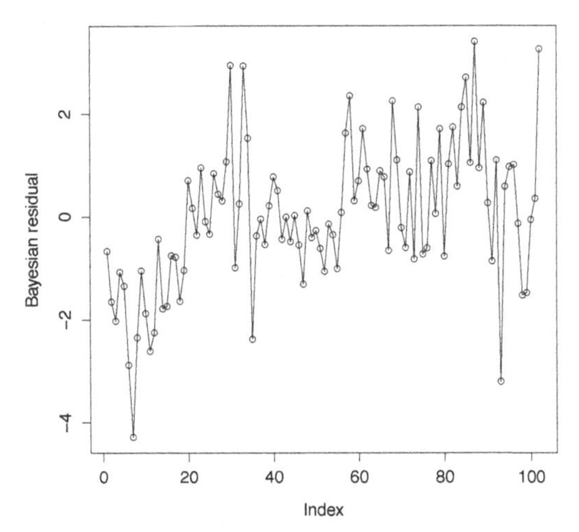

FIGURE 3.9
Bayesian residuals from the standard regression of youth's unemployment rate on the centered adult's unemployment rate, policy group, and their interaction.

(frequentist) estimates of the autocorrelation and partial autocorrelation functions of a time series, here for the Bayesian residuals (Figure 3.10):

```
acf(nzunemploy.res1$resid, main = "")
acf(nzunemploy.res1$resid, type = "partial", main="")
```

The dashed horizontal lines on the plots correspond to 95% confidence bands. The pattern of the autocorrelation function shows an exponential decay, while that of partial autocorrelation function has a high spike at lag 1. These suggest that an AR(1) process would be appropriate for the error term in the regression model.

The following code fits a linear regression with AR(1) error using INLA:

```
formula2 <- youth ~ centeredadult*policy + f(time, model = "ar1")
nzunemploy.inla2 <- inla(formula2, data = nzunemploy, control.family =
    ↪ list(hyper = list(prec = list(initial = 15, fixed = TRUE))))
```

Note that the precision `prec` of the regression model is fixed at $\tau = \exp(15)$ by specifying `control.family = list(hyper = list(prec = list(initial = 15, fixed = TRUE)))` in the call of the `inla` function. This is necessary and important because the uncertainty of regression error has already modeled in `f(time, model = "ar1")` when we define the new formula. We print the result:

```
round(nzunemploy.inla2$summary.fixed, 4)
```

```
                              mean     sd 0.025quant 0.5quant 0.975quant    mode kld
(Intercept)                16.3466 0.3097    15.7645  16.3361    16.9932 16.3229   0
centeredadult               1.5196 0.1368     1.2430   1.5211     1.7869  1.5235   0
policyEqual                 8.9611 0.9979     6.7815   9.0252    10.7431  9.1124   0
centeredadult:policyEqual   2.5083 0.6050     1.2829   2.5196     3.6685  2.5396   0
```

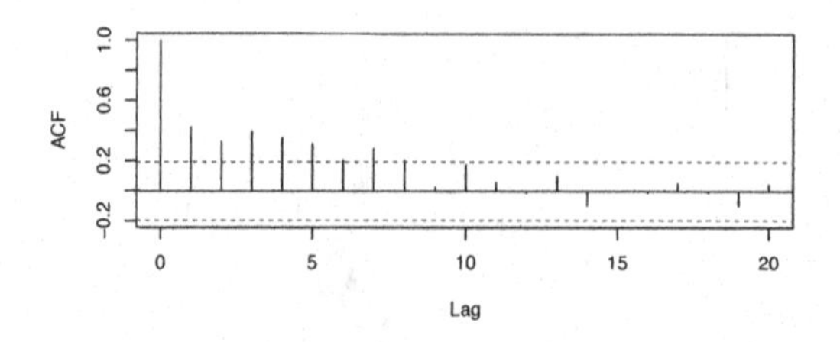

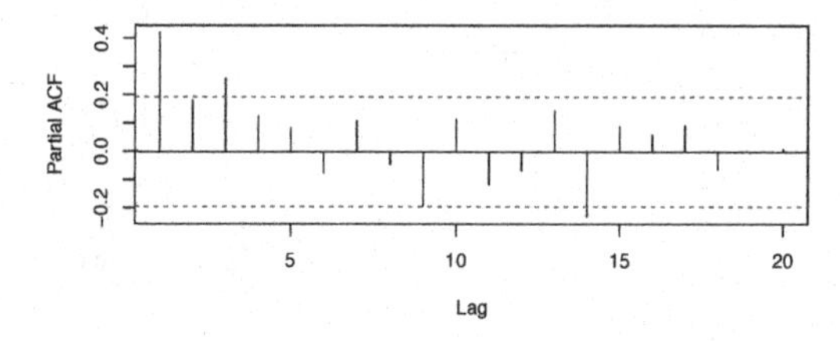

FIGURE 3.10
Autocorrelation and partial autocorrelation functions for the Bayesian residuals from the standard regression of youth's unemployment rate on several predictors.

```
round(nzunemploy.inla2$summary.hyperpar, 4)
```

```
                      mean     sd 0.025quant 0.5quant 0.975quant   mode
Precision for time 0.4542 0.0817     0.3046   0.4516     0.6230 0.4494
Rho for time       0.4953 0.0997     0.3094   0.4916     0.6932 0.4712
```

Comparing with the previous result assuming independent error, the estimates of the regression coefficients are close. However, the standard deviation of the parameters substantially increase. This confirms that the standard errors of the regression coefficients often are seriously underestimated when ignoring the presence of autocorrelation in the dataset. The estimated autocorrelation parameter ρ is 0.4953 with 95% credible intervals $(0.3094, 0.6932)$, indicating the time series have a moderate autocorrelation.

Finally, we want to compare the result using the frequentist approach. The `gls` function in the R `nlme` package fits regression models with a variety of correlated-error variance structures. The following code fits the regression with AR(1) error using generalized least squares:

```
library(nlme)
nzunemploy.gls = gls(youth ~ centeredadult*policy, correlation =
    ↪ corAR1(form=~1), data = nzunemploy)
summary(nzunemploy.gls)
```

```
Generalized least squares fit by REML
  Model: youth ~ centeredadult * policy
  Data: nzunemploy
       AIC      BIC    logLik
  353.0064 368.5162 -170.5032
```

```
Correlation Structure: AR(1)
 Formula: ~1
 Parameter estimate(s):
      Phi
0.5012431

Coefficients:
                              Value Std.Error  t-value p-value
(Intercept)               16.328637 0.2733468 59.73598       0
centeredadult              1.522192 0.1274453 11.94389       0
policyEqual                9.082626 0.8613543 10.54459       0
centeredadult:policyEqual  2.545011 0.5771780  4.40940       0

 Correlation:
                          (Intr) cntrdd plcyEq
centeredadult             -0.020
policyEqual               -0.318  0.007
centeredadult:policyEqual -0.067 -0.155  0.583

Standardized residuals:
        Min          Q1         Med          Q3         Max
-2.89233359 -0.55460580 -0.02419759  0.55449166  2.29571080

Residual standard error: 1.5052
Degrees of freedom: 102 total; 98 residual
```

The results are fairly close to those using the INLA method.

4

Generalized Linear Models

Generalized linear models (GLMs), originally formulated by Nelder and Baker (2004), provide a unifying family of linear models that is widely used in practical regression analysis. The GLMs generalize ordinary linear regression by allowing the models to be related to the response variable via a link function and by allowing the magnitude of the variance of each measurement to be a function of its predicted value. Thus, these models allow for describing response variables that have an error distribution other than normal. They avoid having to select certain transformations of the data to achieve the possibly conflicting objects of normality, linearity and/or homogeneity of variance. Commonly used GLMs include logistic regression for binary data and Poisson regression or negative-binomial regression for count data.

4.1 GLMs

Let us start from a review of the exponential family of distributions. In statistics, the distribution of a random variable Y belongs to an exponential family if its probability density function (or probability mass function for the case of a discrete distribution) can be written in the form

$$f(y|\theta,\phi) = \exp\left\{\frac{y\theta - b(\theta)}{a(\phi)} + c(y,\phi)\right\}, \tag{4.1}$$

where $\theta = g(\mu)$, called the *canonical parameter*, is a function of the expectation $\mu \equiv E(Y)$ of Y, and the canonical link function $g(\cdot)$ does not depend on ϕ. The parameter $\phi > 0$, called the *dispersion parameter*, represents the scale of the distribution. The functions $a(\cdot)$, $b(\cdot)$, and $c(\cdot)$ are known functions that vary from one distribution to another.

The exponential families include many of the common distributions, including the normal, inverse Gaussian, exponential, gamma, Bernoulli, binomial, multinomial, Poisson, chi-squared, Wishart, Inverse Wishart and many others. Here we look at a few typical distributions in detail.

Let Y be normally distributed with mean μ and variance σ^2. Putting the normal distribution into the form of equation (4.1) requires some algebraic manipulation,

$$f(y|\theta,\phi) = \exp\left\{\frac{y\theta - \theta^2/2}{\phi} - \frac{1}{2}\left[\frac{y^2}{\phi} + \log(2\pi\phi)\right]\right\},$$

where $\theta = g(\mu) = \mu$, $\phi = \sigma^2$, $a(\phi) = \phi$, $b(\theta) = \theta^2/2$, and $c(y,\phi) = -(y^2/\phi + \log(2\pi\phi))/2$.

Now let us consider the binomial distribution, where Y is the number of "successes" in n independent binary trials, and μ is the probability of success in an individual trial. The probability mass function of Y is $f(y|\mu) = \binom{n}{y}\mu^y(1-\mu)^{n-y}$. Written as an exponential family, we have

$$f(y|\theta,\phi) = \exp\left\{y\theta - n\log(1+\exp\theta) + \log\binom{n}{y}\right\},$$

where $\theta = g(\mu) = \log\left(\frac{\mu}{1-\mu}\right)$, $\phi = 1$, $a(\phi) = 1$, $b(\theta) = n\log(1+\exp\theta)$, and $c(y,\phi) = \log\binom{n}{y}$.

The third example is Poisson distribution, which is used to model count data. It is appropriate for applications that involve counting the number of times a random event occurs in a given amount of time, distance, area, etc. Its probability mass function is $f(y|\mu) = \exp(-\mu)\mu^y/y!$, which can be rewritten as

$$f(y|\theta,\phi) = \exp(y\theta - \exp(\theta) - \log y!).$$

Here $\theta = \log(\mu)$, $\phi = 1$, $a(\phi) = 1$, $b(\theta) = \exp(\theta)$ and $c(y,\phi) = -\log y!$.

A key property of the exponential families is that the distributions have mean

$$E(Y) \equiv \mu = b'(\theta)$$

and variance

$$Var(Y) = a(\phi)b''(\theta).$$

Note that $b'(\cdot)$ is the inverse of the canonical link function. The mean of the distribution is a function of θ only, while the variance of the distribution is a product of functions of the location parameter θ and the scale parameter ϕ. In GLM, the $b''(\theta)$ is called the *variance function* to describe how the variance relates to the mean. Table 4.1 shows the link functions and their inverses, as well as the variance function for some common distributions.

TABLE 4.1
Link functions, their inverses, and variance functions for some common distributions.

Family	$\theta = g(\mu)$	$\mu = g^{-1}(\theta)$	Variance Function
Normal	μ	θ	1
Poisson	$\log\mu$	$\exp(\theta)$	μ
Binomial	$\log(\mu/(1-\mu))$	$\exp(\theta)/(1+\exp(\theta))$	$\mu(1-\mu)$
Gamma	μ^{-1}	θ^{-1}	μ^2
Inverse Gaussian	μ^{-2}	$\theta^{-1/2}$	μ^3

A GLM provides a unified modeling framework for many commonly used statistical models. Here we define the model in terms of a set of the observations $y_1,...,y_n$ which are regarded as realizations of random variables $Y_1,...,Y_n$. There are the following three components in a GLM:

Random Component: The dependent variables, Y_i's, are assumed to be generated from a particular distribution in the exponential family (4.1).

Linear Predictor: That is a linear combination of the predictors

$$\theta_i = \beta_0 + \beta_1 x_{i1} + \ldots + \beta_p x_{ip} = \mathbf{x}_i^T \boldsymbol{\beta},$$

where $\boldsymbol{\beta} = (\beta_0, \beta_1, \ldots, \beta_p)^T$, $\mathbf{x}_i = (1, x_{1i}, \ldots, x_{ip})^T$, and x_{ij}, $j = 1, \ldots, p$ is the value of the jth covariate for the ith observation, as we have seen in a linear regression in Chapter 3.

Link Function: The expectation of the response variable, $\mu_i \equiv E(Y_i)$ and the linear predictor are related through a link function $g(\cdot)$:

$$g(\mu_i) = \theta_i.$$

Note that in most applications, the so-called *natural link function* is used, i.e., $g(\cdot) = b'(\cdot)$.

The GLM covers a large class of regression models, such as normal linear regression, logistic and probit regression, Poisson regression, negative-binomial regression and gamma regression. The classical estimation method for GLMs is maximum likelihood. There are several excellent textbooks discussing theory and applications for GLMs from a frequentist point of view; see for example, McCullagh and Nelder (1989); Lindsey (1997); Dobson and Barnett (2008). These books provide a rich collection of maximum likelihood estimation methods, hypothesis testing, real case studies.

For Bayesian analysis, MCMC is the common choice, which requires generating samples from posterior distributions. INLA treats a wide range of GLMs in a unified manner, thus allowing for greater automation of the inference process. In the rest of the chapter, we will discuss a few popular GLMs and demonstrate real case studies by applying the INLA method.

4.2 Binary Responses

In many applications, the response variable takes one of only two possible values representing success and failure, or more generally the presence or absence of an attribute of interest. Logistic regression, as a special case of GLMs, has a long tradition with widely varying applications to model such data. It is used to estimate the probability of a binary response based on one or more predictor variables.

Let Y be Bernoulli distributed with success probability $P(Y = 1) = \pi$. Its density is given by

$$f(y) = \exp\left\{ y \log\left(\frac{\pi}{1-\pi}\right) + \log(1-\pi) \right\}.$$

The distribution belongs to the exponential family, with canonical parameter θ equal to the logit of π, i.e., $\log(\pi/(1-\pi))$, dispersion parameter $\phi = 1$. Its mean is π and variance function is $\pi(1-\pi)$. The canonical link function, the logit link, leads to the classical logistic regression model:

$$\begin{cases} Y_i \sim \text{Bernoulli}(\pi_i), \\ \text{logit}(\pi_i) = \beta_0 + \beta_1 x_{i1} + \ldots + \beta_p x_{ip}. \end{cases}$$

Sometimes, the logit link function can be replaced by the probit link, which is the inverse of the standard normal distribution function, $\Phi^{-1}(\cdot)$. It has been shown that the logit and probit link functions behave similarly except the case for extreme probabilities (Agresti, 2012).

Here we analyze low birth weight data to illustrate the use of logistic regression. The dataset has been presented in Hosmer and Lemeshow (2004). The dataset contains information on 189 births to women seen in the obstetric clinic, where data were collected as part of a larger study at Baystate Medical Center in Springfield, Massachusetts. The response variable `LOW` is a binary outcome indicating birth weight less than 2500 grams, which has been of concern to physicians for years. A woman's behavior during pregnancy, such as smoking habits, receiving prenatal care can greatly change the chances of carrying the baby to term, and thus, of delivering a baby of normal birth weight. The variables that are potentially associated with low birth weight are recorded in the study, given in Table 4.2. The goal of this study was to determine whether some or all of these variables were risk factors in the clinic population being treated by the medical center.

TABLE 4.2
Code sheet for the variables in the low birth weight data.

Variable Name	Description	Codes/Values
LOW	indicator of low birth weight	$0 = \geq 2500g$
		$1 = < 2500g$
AGE	age of mother	years
LWT	weight of mother at last menstrual period	pounds
RACE	race of mother	1 = white
		2 = black
		3 = other
SMOKE	smoking status during pregnancy	0=no
		1 = yes
HT	history of hypertension	0 = no
		1 = yes
UI	presence of uterine irritability	0 =no
		1 = yes
FTV	number of physician visits during the first trimester	counts

Data were collected on 189 women, 59 of whom had low birth weight babies and 130 of whom had normal birth weight babies. Seven variables were considered to be of importance: AGE, LWT, RACE, SMOKE, HT, UI, and FTV. For comparison purposes, we begin to fit a logistic regression with conventional maximum-likelihood estimation:

```
data(lowbwt, package = "brinla")
lowbwt.glm1 <- glm(LOW ~ AGE + LWT + RACE + SMOKE + HT + UI + FTV,
    ↪ data=lowbwt, family=binomial())
round(coef(summary(lowbwt.glm1)), 4)
```

```
            Estimate Std. Error z value Pr(>|z|)
(Intercept)   0.4548     1.1854  0.3837   0.7012
AGE          -0.0205     0.0360 -0.5703   0.5684
LWT          -0.0165     0.0069 -2.4089   0.0160
RACE2         1.2898     0.5276  2.4445   0.0145
RACE3         0.9191     0.4363  2.1065   0.0352
SMOKE1        1.0416     0.3955  2.6337   0.0084
HT1           1.8851     0.6948  2.7130   0.0067
UI1           0.9041     0.4486  2.0155   0.0439
FTV           0.0591     0.1720  0.3437   0.7311
```

In the output, the categorical variable RACE has been recoded as the two design variables, RACE2 and RACE3. In general, if a nominal scaled variable has k possible values, then $k-1$ design variables will be needed. Here RACE2 denotes the effect of a black mother relative to a white mother, and RACE3 denotes the effect of a mother in other races relative to a white mother. Similar recoding has been done for the variables, SMOKE, HT, UI. We then fit the logistic regression model using the INLA method:

```
lowbwt.inla1 <- inla(LOW ~ AGE + LWT + RACE + SMOKE + HT + UI + FTV,
    ↪ data=lowbwt, family = "binomial", Ntrials = 1, control.compute
    ↪ = list(dic = TRUE, cpo = TRUE))
round(lowbwt.inla1$summary.fixed, 4)
```

```
               mean     sd 0.025quant 0.5quant 0.975quant    mode kld
(Intercept)  0.5672 1.1853    -1.7289   0.5563     2.9235  0.5347   0
AGE         -0.0207 0.0360    -0.0921  -0.0204     0.0491 -0.0199   0
LWT         -0.0176 0.0069    -0.0317  -0.0174    -0.0047 -0.0169   0
RACE2        1.3405 0.5275     0.3151   1.3370     2.3851  1.3298   0
RACE3        0.9456 0.4362     0.1028   0.9409     1.8151  0.9314   0
SMOKE1       1.0749 0.3954     0.3140   1.0696     1.8664  1.0590   0
HT1          1.9727 0.6946     0.6595   1.9542     3.3909  1.9165   0
UI1          0.9331 0.4485     0.0524   0.9330     1.8130  0.9330   0
FTV          0.0559 0.1720    -0.2891   0.0585     0.3868  0.0635   0
```

We obtain estimates similar to those obtained when using the frequentist method. For example, the posterior mean of the parameter for LWT is -0.0176. Its estimated posterior standard deviation is 0.0069. The 2.5% and 97.5% posterior quantiles are both negative, which indicates with 95% probability that the effect for LWT is negative. Among all other predictors, the 95% credible intervals for AGE and FTV contain zero, while RACE2, RACE3, SMOKE1, HT1, and UI1 are positively associated with the outcome.

The odds ratios of the predictors can be calculated by exponentiating their estimated coefficients. For instance, the odds ratio for LWT is $\exp(-0.0176) = 0.9826$. It is interpreted as we expect to see 1.74% ($= 1 - 0.9826$) decrease in the odds of having a low birth weight baby for a one-unit increase in mother's weight, assuming all other predictors are fixed.

We want to further obtain the reduced model while minimizing the number of parameters. We may perform a backward elimination procedure using DIC (i.e., manually eliminating variables based on DIC; see the definition of DIC in Chapter 1). The reduced model we obtain is the following:

```
lowbwt.inla2 <- inla(LOW ~ LWT + RACE + SMOKE + HT + UI, data=lowbwt,
    ↪ family = "binomial", Ntrials = 1, control.compute = list(dic =
    ↪ TRUE, cpo = TRUE))
round(lowbwt.inla2$summary.fixed, 4)
```

```
               mean     sd 0.025quant 0.5quant 0.975quant    mode kld
(Intercept)  0.1087 0.9378    -1.6834   0.0915     2.0003  0.0567   0
LWT         -0.0175 0.0068    -0.0315  -0.0172    -0.0048 -0.0168   0
RACE2        1.3620 0.5214     0.3476   1.3587     2.3934  1.3522   0
RACE3        0.9486 0.4303     0.1198   0.9431     1.8091  0.9320   0
SMOKE1       1.0619 0.3925     0.3075   1.0563     1.8485  1.0451   0
HT1          1.9342 0.6907     0.6271   1.9163     3.3429  1.8799   0
UI1          0.9250 0.4475     0.0467   0.9249     1.8032  0.9246   0
```

In this reduced model, `LWT` has a negative effect with the estimated coefficient -0.0175; all other predictors have a positive effect on the regression coefficients. We can compare the DICs for the full model and the reduced model:

```
c(lowbwt.inla1$dic$dic, lowbwt.inla2$dic$dic)
```

```
[1] 221.2093 217.7459
```

DIC for the reduced model is less than that for the full model, which indicates that the reduced model has the better trade off between fit and model complexity. So, we prefer to use the reduced model. Its estimated logit is given by the following expression:

$$\begin{aligned}\widehat{\text{logit}(\pi)} =& 0.109 - 0.018 \times \text{LWT} + 1.362 \times \text{RACE\_2} + 0.949 \times \text{RACE\_3} \\ &+ 1.062 \times \text{SMOKE} + 1.934 \times \text{HT} + 0.925 \times \text{UI},\end{aligned}$$

The equation can be used to obtain the fitted values, or make predictions for new observations.

4.3 Count Responses

In many application studies, the response variable of interest is the counted number of occurrences of an event. In this type of data, the observations take only the non-negative integer values $\{0, 1, 2, 3, ...\}$, which arise from counting rather than ranking, or grouping. The distribution of counts is discrete, and typically skewed. Applying an ordinary linear regression model to these data could present at least two problems. First, it is quite likely that the regression model will produce negative predicted values, which are theoretically impossible. Second, many distributions of count data are positively skewed with many observations in the dataset having a value of 0. When one considers a transformation of the response variable (such as the log transformation, $\log(y+c)$, where c is a positive constant), the high number of 0's in the dataset prevents the transformation of a skewed distribution into normal.

4.3.1 Poisson Regression

The basic GLM for count data is the Poisson regression. Let Y be Poisson distributed with mean μ, where its probability mass function is $f(y|\mu) = \exp(-\mu)\mu^y/y!$. The distribution has the canonical parameter $\theta = \log\mu$, the dispersion parameter $\phi = 1$, and its variance function equals μ. The canonical link function, the logarithm link, leads to the Poisson regression model,

$$\begin{cases} Y_i \sim \text{Poisson}(\mu_i), \\ \log(\mu_i) = \beta_0 + \beta_1 x_{i1} + \ldots + \beta_p x_{ip}. \end{cases}$$

Let us consider a classical example for the simple Poisson regression. Whyte et al. (1987) reported the number of deaths due to AIDS in Australia per 3-month period from January 1983 to June 1986. The dataset only contains one predictor and one response with 14 observations, summarized in Table 4.3.

TABLE 4.3
Description for the variables in the `AIDS` data.

Variable Name	Description	Values
TIME	time measured in multiples of 3 months after January 1983	continuous
DEATHS	number of deaths in Australia due to AIDS	counts

Figure 4.1 displays the scatterplot of the data (the left panel) and the histogram for the response variable `DEATHS` (the right panel). We note that there is a nonlinear relationship between `TIME` and `DEATHS`, and `DEATHS` show a right-skewed distribution. A Poisson regression seems to be a reasonable choice to model the data. To fit the Poisson model using INLA, we use the following command:

```
AIDS.inla1 <- inla(DEATHS ~ TIME, data = AIDS, family = "poisson",
    ↪ control.compute = list(dic = TRUE, cpo = TRUE))
round(AIDS.inla1$summary.fixed, 4)
```

```
              mean     sd 0.025quant 0.5quant 0.975quant   mode kld
(Intercept) 0.3408 0.2512    -0.1690   0.3467     0.8183 0.3586   0
TIME        0.2565 0.0220     0.2142   0.2562     0.3008 0.2555   0
```

The coefficient table shows the posterior summary statistics for the unknown parameters in the model. We could write down the estimated equation for the mean response:

$$\hat{\mu} = \exp(0.3408 + 0.2565 \times \text{TIME}).$$

The `TIME` effect can be interpreted as follows: in the period between January 1983 to June 1986, the number of deaths due to AIDS in a year was on average $\exp(0.2565 \times$

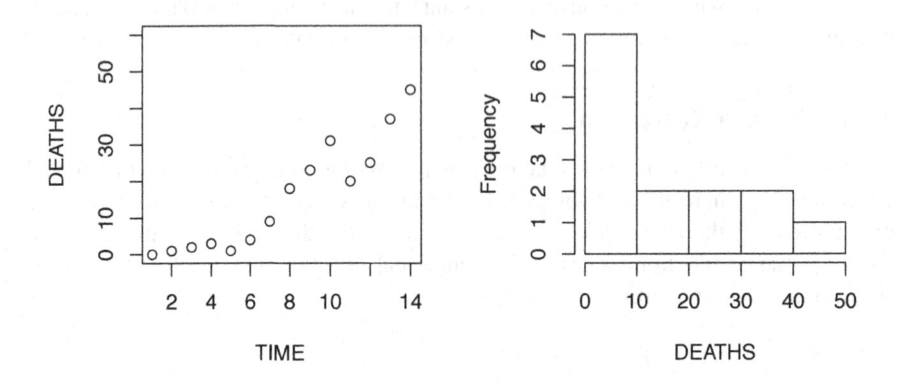

FIGURE 4.1
The scatterplot between TIME and DEATHS; and the histogram of DEATHS, in the AIDS data.

4) = 2.7899 times higher than in the year before. Next we generate the plot of the estimated mean function and its 95% credible interval:

```
plot(DEATHS ~ TIME, data=AIDS, ylim=c(0,60))
lines(AIDS$TIME, AIDS.inla$summary.fitted.values$mean, lwd=2)
lines(AIDS$TIME, AIDS.inla$summary.fitted.values$"0.025quant", lwd=1,
    ↪ lty=2)
lines(AIDS$TIME, AIDS.inla$summary.fitted.values$"0.975quant", lwd=1,
    ↪ lty=2)
```

From Figure 4.2, we note that in the beginning and the end of the time period the observed responses are less than the fitted values, while in the center period they are greater than the corresponding fitted values. This points to the fact that the model seems not entirely appropriate. Now let us consider log(TIME) instead of TIME as the explanatory variable. We fit the following model:

```
AIDS.inla2 <- inla(DEATHS ~ log(TIME), data=AIDS, family = "poisson",
    ↪ control.compute = list(dic = TRUE, cpo = TRUE))
round(AIDS.inla2$summary.fixed, 4)
```

```
                mean     sd 0.025quant 0.5quant 0.975quant    mode kld
(Intercept) -1.9424 0.5116    -2.9899  -1.9272    -0.9792 -1.8963   0
log(TIME)    2.1747 0.2150     1.7675   2.1692     2.6131  2.1581   0
```

The estimated equation is

$$\hat{\mu} = \exp(-1.9424 + 2.1747 \times \log(\text{TIME})).$$

The interpretation of this model is less intuitive: for a 1 unit increase in log(TIME), the estimated count increases by a factor of exp(2.1747) = 8.7995. Let us look the estimated mean function and its credible intervals for this model:

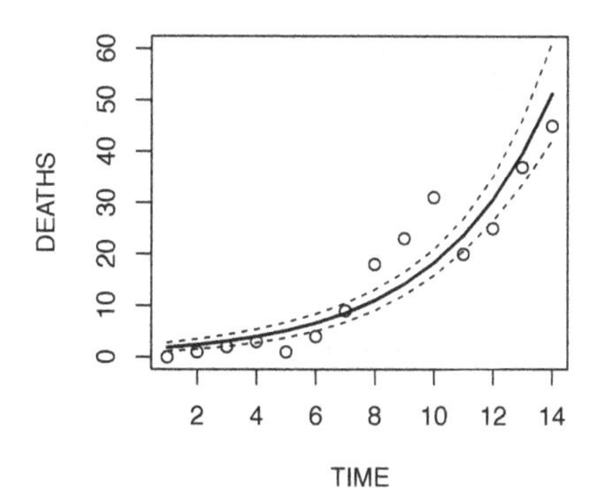

FIGURE 4.2
The estimated posterior mean function and its 95% credible interval for `DEATHS` when considering `TIME` as the explanatory variable.

```
plot(DEATHS ~ log(TIME), data = AIDS, ylim=c(0,60))
lines(log(AIDS$TIME), AIDS.inla2$summary.fitted.values$mean, lwd=2)
lines(log(AIDS$TIME), AIDS.inla2$summary.fitted.values$"0.025quant",
    ↪ lwd=1, lty=2)
lines(log(AIDS$TIME), AIDS.inla2$summary.fitted.values$"0.975quant",
    ↪ lwd=1, lty=2)
```

It appears that we obtain a better fit compared with the previous model. We could further check DICs for the two models:

```
c(AIDS.inla1$dic$dic, AIDS.inla2$dic$dic)
```

```
[1] 86.70308 74.10760
```

The second model has a smaller DIC, which confirms our findings from Figure 4.2 and Figure 4.3.

We may further compare the INLA results with the results using the conventional maximum likelihood estimation:

```
AIDS.glm <- glm(DEATHS ~ log(TIME), family=poisson(), data=AIDS)
round(coef(summary(AIDS.glm)), 4)
```

```
            Estimate Std. Error z value Pr(>|z|)
(Intercept)  -1.9442     0.5116 -3.8003    1e-04
log(TIME)     2.1748     0.2150 10.1130    0e+00
```

Very similar estimates are obtained from both methods.

4.3.2 Negative Binomial Regression

In applied work Poisson regression is restrictive in the analysis of count data. It is recognized that counts often display substantial extra-Poisson variation, or *overdis-*

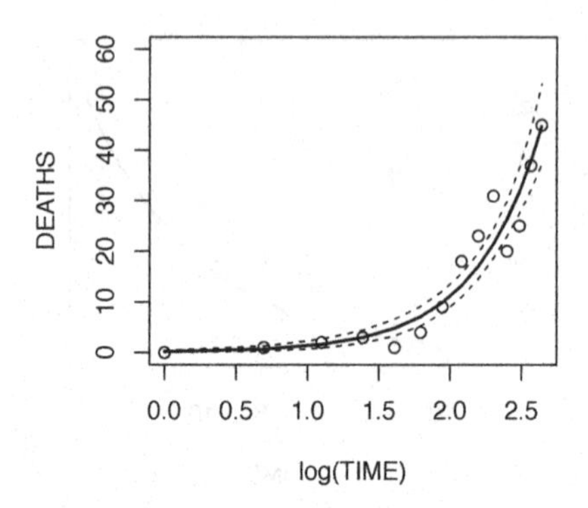

FIGURE 4.3
The estimated posterior mean function and its 95% credible interval for (DEATHS) when considering log(TIME) as the explanatory variable.

persion. Overdispersion refers to the situation when the variance of an observed dependent variable exceeds the nominal variance, given the respective assumed distribution. The assumption in Poisson model that the conditional mean and variance of Y given X are equal may be too strong and thus fail to account for the overdispersion. Inappropriate imposition of this restriction may result in unreasonably small estimated standard errors of the parameter estimates. Negative binomial regression is perhaps the most convenient way to relax the Poisson restriction and deal with the overdispersion.

More specifically, assume that $v_i, i = 1,...,n$ are unobserved random variables that follow a gamma distribution with shape parameter α and rate parameter α, Gamma(α, α); that is, $f(v) \propto x^{\alpha-1}\exp(-\alpha v)I\{v > 0\}$. Conditional on v_i, Y_i has a Poisson distribution with mean $v_i\mu_i$, i.e., $Y_i|v_i \sim \text{Poisson}(v_i\mu_i)$. Then it follows that marginally Y_i has the negative binomial distribution given by

$$P(Y_i = y; \alpha, \mu_i) = \frac{\Gamma(y+\alpha)}{\Gamma(\alpha)y!}\left(\frac{\alpha}{\mu_i+\alpha}\right)^{\alpha}\left(\frac{\mu_i}{\mu_i+\alpha}\right)^{y}, \tag{4.2}$$

where $y \in \{0,1,2,...\}$. The negative binomial distribution (4.2) will be denoted by $Y_i \sim NB(\alpha, \mu_i)$. It can be shown that the marginal mean and variance of Y_i are μ_i and $\mu_i + \mu_i^2/\alpha$, respectively. Thus, the parameter α quantifies the amount of overdispersion. Oftentimes, we define $\varphi = 1/\alpha$ as the dispersion parameter in the negative binomial model. The parameter $\varphi \to 0$ corresponds to no overdispersion. In such a case, the negative binomial model reduced to the Poisson model.

The log link is commonly used in negative binomial regression. Suppose we have a vector of p explanatory variables, $(x_{i1},...,x_{ip})$, that is related to the response Y_i. The model is written as

$$\begin{cases} Y_i \sim \text{NB}(\alpha, \mu_i), \\ \log(\mu_i) = \beta_0 + \beta_1 x_{i1} + \ldots + \beta_p x_{ip}. \end{cases}$$

Let us use an example of nesting horseshoe crabs (Brockmann, 1996) to illustrate negative-binomial regression modeling. Agresti (2012) analyzed the data using the conventional frequentist GLM approach from Section 4.3 of his book. In this study, each female horseshoe crab in the study had a male crab attached to her in her nest. The study investigated factors that affect whether the female crab had any other males, called satellites, residing near her. Explanatory variables that are thought to affect this included the female crab's color (COLOR), spine condition (SPINE), weight (WEIGHT), and carapace width (WIDTH). The response variable for each female crab is her number of satellites (SATELLITES). The code sheet for the variables is displayed in Table 4.4. There are 173 females in this study.

TABLE 4.4
Code sheet for the variables in the crab data.

Variable Name	Description	Codes/Values
SATELLITES	the number of satellites for a female crab	counts
COLOR	crab's color	1 = light medium 2 = medium 3 = dark medium 4 = dark
SPINE	crab's spine condition	1 = both good 2 = one worn or broken 3 = both worn or broken
WEIGHT	crab's weight	kilogram (kg)
WIDTH	crab's carapace width	centimeter (cm)

It is always a good idea to start with descriptive statistics and plots. We first check the unconditional mean and variance of the outcome variable:

```
round(c(mean(crab$SATELLITES), var(crab$SATELLITES)), 4)
```

```
[1] 2.9191 9.9120
```

We note that the sample mean of SATELLITES is much lower than its variance. We further check the means and variances of SATELLITES by the crab's color type.

```
with(crab, tapply(SATELLITES, COLOR, function(x){round(mean(x), 4)}))
```

```
     1      2      3      4
4.0833 3.2947 2.2273 2.0455
```

```
with(crab, tapply(SATELLITES, COLOR, function(x){round(var(x), 4)}))
```

```
      1       2      3       4
 9.7197 10.2739  6.7378 13.0931
```

It seems that the variable COLOR is a good candidate for predicting SATELLITES, since the mean value of the response appears to vary by COLOR. Also, we note that the variances within each level of COLOR are much higher than the means within each level. Let us plot the histogram and conditional histograms by COLOR for the response variable SATELLITES:

```
(p1 <- ggplot(crab, aes(x=SATELLITES)) + geom_histogram(binwidth=1,
    ↪ color="black"))
(p2 <- p1 + facet_wrap(~ COLOR, ncol=2))
```

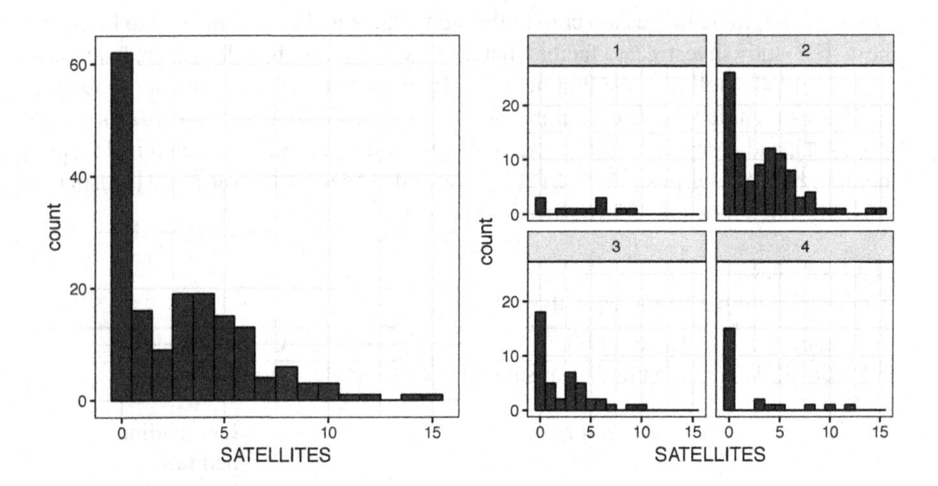

FIGURE 4.4
The histogram and conditional histograms (by COLOR) for the response variable, SATELLITES, in the *crab* data.

Figure 4.4 shows the histogram of SATELLITES (the left panel) as well as the conditional histograms by crab's color type (the right panel). The histograms confirm our findings from summary statistics. These exploratory analysis results suggest that overdispersion is present and that a negative binomial model would be appropriate for the data.

By examining the correlation among the predictors, we notice that WEIGHT and WIDTH are highly correlated:

```
round(cor(crab$WEIGHT, crab$WIDTH),4)
```

```
[1] 0.8869
```

To avoid the multicollinearity problem, we only include the predictors COLOR, SPINE, and WIDTH in the model. We first fit a negative binomial regression with conventional maximum likelihood estimation:

```
library(MASS)
crab.glm <- glm.nb(SATELLITES ~ COLOR + SPINE + WIDTH, data=crab)
round(coef(summary(crab.glm)), 4)
```

```
            Estimate Std. Error z value Pr(>|z|)
(Intercept)  -0.3213     0.5637 -0.5700   0.5687
```

```
COLOR2        -0.3206     0.3725 -0.8607    0.3894
COLOR3        -0.5954     0.4159 -1.4317    0.1522
COLOR4        -0.5788     0.4643 -1.2467    0.2125
SPINE2        -0.2411     0.3934 -0.6130    0.5399
SPINE3         0.0425     0.2479  0.1713    0.8640
WIDTH          0.6925     0.1656  4.1826    0.0000
```

The following command fits the negative binomial regression with INLA:

```
crab.inla1 <- inla(SATELLITES ~ COLOR + SPINE + WIDTH, data = crab,
    ↪ family = "nbinomial", control.compute = list(dic = TRUE, cpo =
    ↪ TRUE))
round(crab.inla1$summary.fixed, 4)
```

```
               mean     sd 0.025quant 0.5quant 0.975quant    mode kld
(Intercept) -0.3158 0.5963    -1.4858  -0.3169     0.8592 -0.3187   0
COLOR2      -0.3216 0.3922    -1.1199  -0.3122     0.4244 -0.2941   0
COLOR3      -0.5988 0.4292    -1.4643  -0.5915     0.2257 -0.5775   0
COLOR4      -0.5814 0.4900    -1.5577  -0.5772     0.3708 -0.5691   0
SPINE2      -0.2467 0.3912    -1.0038  -0.2508     0.5346 -0.2587   0
SPINE3       0.0392 0.2527    -0.4661   0.0419     0.5287  0.0471   0
WIDTH        0.7001 0.1839     0.3457   0.6976     1.0691  0.6927   0
```

The results from the MLE approach and INLA are very close. We see that the posterior mean of the parameter for `WIDTH` is 0.7001 and its posterior standard deviation is 0.1839. Its 0.025 and 0.975 quantiles are both positive, which indicates with high probability that the effect for `WIDTH` is positive. The 95% credible intervals for all other predictors contain zero, so one cannot determine whether those effects are positive or negative based on the data.

In INLA, the size parameter α is represented as $\alpha = \exp(\theta)$ and a diffuse gamma distribution is defined on θ. By default, the summary of the posterior estimate of α is output:

```
round(crab.inla1$summary.hyperpar, 4)
```

```
                                                          mean        sd 0.025quant
size for the nbinomial observations (1/overdispersion) 0.9289    0.1572     0.6612
                                                        0.5quant 0.975quant      mode
size for the nbinomial observations (1/overdispersion) 0.915      1.2703      0.883
```

If we are interested in the overdispersion parameter, φ, the reciprocal of α, we could employ the function `inla.tmarginal`, which applies a transformation on the entire posterior distribution:

```
overdisp_post <- inla.tmarginal(fun=function(x) 1/x, marg=crab.inla1$
    ↪ marginals.hyperpar[[1]])
```

The posterior mean of φ can be obtained by the function `inla.emarginal`, which computes the expected value of a function `fun` applied to the marginal distribution `marg`:

```
round(inla.emarginal(fun=function(x) x, marg=overdisp_post), 4)
```

```
[1] 1.108
```

To obtain the posterior credible interval of φ, we apply the function `inla.qmarginal`:

```
round(inla.qmarginal(c(0.025, 0.975), overdisp_post), 4)
```

```
[1] 0.7610 1.5468
```

The above posterior summary of φ indicates a moderate overdispersion in the data set. We may want to compare the results using Poisson regression and negative binomial model:

```
crab.inla2 <- inla(SATELLITES ~ COLOR + SPINE + WIDTH, data = crab,
    ↪ family = "poisson", control.compute = list(dic = TRUE, cpo =
    ↪ TRUE))
round(crab.inla2$summary.fixed, 4)
```

```
               mean     sd 0.025quant 0.5quant 0.975quant    mode kld
(Intercept) -0.0491 0.2535    -0.5499  -0.0481     0.4457 -0.0461   0
COLOR2      -0.2695 0.1678    -0.5916  -0.2720     0.0673 -0.2770   0
COLOR3      -0.5232 0.1941    -0.9003  -0.5246    -0.1384 -0.5275   0
COLOR4      -0.5434 0.2253    -0.9866  -0.5430    -0.1023 -0.5423   0
SPINE2      -0.1615 0.2115    -0.5892  -0.1571     0.2420 -0.1483   0
SPINE3       0.0924 0.1195    -0.1393   0.0913     0.3297  0.0893   0
WIDTH        0.5475 0.0732     0.4028   0.5478     0.6901  0.5485   0
```

The posterior mean estimates of the predictors do not change too much, though the estimate of the intercept is very different. However, the posterior standard deviations from the Poisson regression are much less than those from the negative binomial regression. For over-dispersed data, Poisson regression underestimates the standard errors of the coefficients, leading to confidence intervals that are too narrow and, potentially leading to incorrect inferences (Wang, 2012). We further compare the DICs for the negative binomial model and the Poisson model:

```
c(crab.inla1$dic$dic, crab.inla2$dic$dic)
```

```
[1] 761.2103 918.9907
```

The DIC of the negative binomial model is much smaller than that of the Poisson model, indicating that the negative binomial model is preferred in fitting the `crab` data.

4.4 Modeling Rates

In many applications, the count of an event is observed over a period or amount of exposure, for example, traffic accidents per year, or count of deaths per age group. We often call the type of data, *rates*. A rate is a count of events divided by some measure of that unit's *exposure* (a particular unit of observation). Unlike a proportion, which ranges from 0 to 1, a rate could have any nonnegative value. Poisson or negative-binomial regression are often appropriate for modeling rate data. In Poisson or negative-binomial model, this is handled as an *offset*, where the exposure variable enters on the right-hand side of the equation, but with a parameter estimate (for log(exposure)) constrained to 1.

The following is a log-linked model for a rate as a function of the predictor variables, $(x_1,...,x_p)$:

$$\log(\mu_i/e_i) = \beta_0 + \beta_1 x_{i1} + ... + \beta_p x_{ip},$$

where μ_i is the mean event count and e_i is the exposure for the ith observation. Note that the above equation can be rewritten as

$$\log(\mu_i) = \log(e_i) + \beta_0 + \beta_1 x_{i1} + ... + \beta_p x_{ip}.$$

The model becomes a Poisson or negative binomial model in which the additional term on the right-hand side, $\log(e_i)$, is the offset, the log of the exposure.

Let us take a look at an example of car insurance claims (Aitkin et al., 2005). The data consist of the numbers of policyholders of an insurance company who were exposed to risk, and the numbers of car insurance claims made by those policyholders in the third quarter of 1973. The data include three four-level categorical predictors. The code sheet for the variables is displayed in Table 4.5.

TABLE 4.5
Code sheet for the variables in the insurance claim data.

Variable Name	Description	Codes/Values
District	the district of residence of policyholder	1 = rural; 2 = small towns; 3 = large towns; 4 = major cities.
Group	group of cars based on the engine capacity	<1 liter; 1–1.5 liter; 1.5–2 liter; >2 liter.
Age	age group of the policyholders	<25; 25–29; 30–35; > 35.
Holders	numbers of policyholders	counts
Claims	numbers of claims	counts

We want to model the relation between the rate of claims and the three explanatory variables, `District`, `Group`, and `Age`. We fit the rate data using Poisson regression with offset. Let us start the analysis with conventional maximum likelihood approach:

```
library(MASS)
data(Insurance, package = "MASS")
insur.glm <- glm(Claims ~ District + Group + Age + offset(log(Holders)
    ↪ ), data = Insurance, family = poisson)
round(summary(insur.glm)$coefficients, 4)
```

```
            Estimate Std. Error  z value Pr(>|z|)
(Intercept)  -1.8105     0.0330 -54.9102   0.0000
District2     0.0259     0.0430   0.6014   0.5476
District3     0.0385     0.0505   0.7627   0.4457
District4     0.2342     0.0617   3.7975   0.0001
```

```
Group.L       0.4297   0.0495   8.6881   0.0000
Group.Q       0.0046   0.0420   0.1103   0.9121
Group.C      -0.0293   0.0331  -0.8859   0.3757
Age.L        -0.3944   0.0494  -7.9838   0.0000
Age.Q        -0.0004   0.0489  -0.0073   0.9942
Age.C        -0.0167   0.0485  -0.3452   0.7299
```

Note that, to specify the model correctly in `glm` function, we must include the term log(`Holders`) as an explanatory variable with a coefficient of 1. That is, log(`Holders`) is taken as an offset for the model by specifying "`offset(log(Holders))`" in the model formula.

When we fit a Bayesian model using `inla` function, the offset term needs to be specified by the argument `E = Holders`:

```
insur.inla1 <- inla(Claims ~ District + Group + Age, data = Insurance,
    ↪ family = "poisson", E = Holders)
round(insur.inla1$summary.fixed, 4)
```

```
               mean     sd 0.025quant 0.5quant 0.975quant    mode kld
(Intercept) -1.8122 0.0330    -1.8774  -1.8120    -1.7479 -1.8117   0
District2    0.0259 0.0430    -0.0588   0.0259     0.1101  0.0261   0
District3    0.0385 0.0505    -0.0612   0.0387     0.1372  0.0391   0
District4    0.2342 0.0617     0.1118   0.2346     0.3541  0.2355   0
Group.L      0.4296 0.0495     0.3320   0.4298     0.5262  0.4301   0
Group.Q      0.0043 0.0420    -0.0787   0.0044     0.0862  0.0047   0
Group.C     -0.0294 0.0331    -0.0943  -0.0294     0.0356 -0.0295   0
Age.L       -0.3943 0.0494    -0.4900  -0.3947    -0.2961 -0.3956   0
Age.Q       -0.0002 0.0489    -0.0964  -0.0001     0.0956  0.0000   0
Age.C       -0.0164 0.0485    -0.1116  -0.0164     0.0787 -0.0163   0
```

From the output above, we note that the effect of major cities, `District4`, is 0.2342, with the 95% credible interval $(0.1118, 0.3541)$. The results can be interpreted as follows: the estimated rate of claims for major cities is 26.36% = exp(0.234) – exp(0), with credible levels (11.83%, 42.49%) = (exp(0.1118) – exp(0), exp(0.3541) – exp(0)), higher than that of claims for rural areas, assuming the group and age effects are fixed. Similar statements can be made for the two other significant effects, `Group.L` and `Age.L`.

To check the validation of a Poisson model, Pearson residuals are often used for model diagnostics in frequentist analysis,

$$\hat{\varepsilon}_i = (y_i - \hat{\mu}_i)/\sqrt{\hat{\mu}_i}, i = 1, \ldots, n,$$

where $\hat{\mu}_i$ is the maximum likelihood estimate. The residuals approximately follow standard normal distribution if the model is correctly specified. By analogy with classical model checking, we define Bayesian Pearson residuals as

$$r_i = (y_i - \mu_i)/\sqrt{\mu_i} = \frac{y_i - g^{-1}(\mathbf{x}_i, \beta)}{\sqrt{g^{-1}(\mathbf{x}_i, \beta)}}, i = 1, \ldots, n.$$

Each r_i is just a function of unknown parameters, and its posterior distribution is thus straightforward to calculate (see more discussions regarding Bayesian residuals in Chapter 3). Plotting the posterior mean or median of the r_i's versus the index of the observations or the fitted values might reveal failure in model assumptions.

We have written a convenience function, `bri.Pois.resid`, to calculate Bayesian Pearson residuals (posterior means) for Poisson regression in our `brinla` library. When the argument `plot = TRUE` in the function is specified, a residual plot by case is output:

```
insur.bresid <- bri.Pois.resid(insur.inla1, plot = TRUE)
abline(1.96, 0, lty=2); abline(-1.96, 0, lty=2)
qqnorm(insur.bresid$resid); qqline(insur.bresid$resid)
```

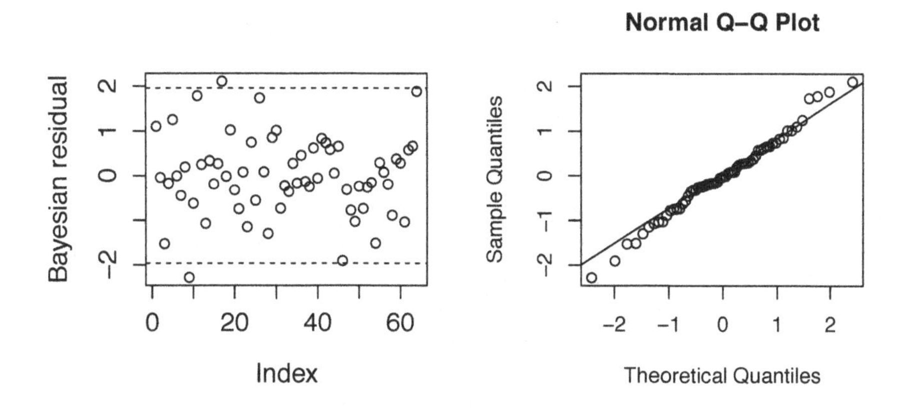

FIGURE 4.5
Bayesian Pearson residual plots of Poisson model for the insurance claim data. The left panel is the plot of residuals by case, and the right panel is the normal QQ plot for the residuals.

The left panel in Figure 4.5 shows the plot of residuals by case (index plot). We observe a horizontal band with points that vary at random. The right panel in Figure 4.5 shows a normal QQ plot for the residuals, indicating the residuals fit to a standard normal distribution very well. Thus, the assumption of a Poisson model for the insurance claim data is well supported.

4.5 Gamma Regression for Skewed Data

Poisson and negative binomial models are very popular in practice, but there are a number of other GLMs which are useful for particular types of data. The gamma GLM can be used for continuous but skewed responses. The most common way to analyze such data is to log transform the responses. However, modeling the skewed data with gamma distribution in GLM framework may give better interpretability, since gamma regression parameters are interpretable in terms of the mean of the response. The density of the gamma distribution is usually given by:

$$f(y) = \frac{b^a}{\Gamma(a)} y^{a-1} \exp(-by), \qquad a > 0, b > 0, y > 0,$$

where a is the shape parameter and b is the scale parameter of the distribution. So, $\mathrm{E}(y) = a/b$, and $\mathrm{Var}(y) = a/b^2$. For the purpose of a GLM, INLA uses the following reparameterization:

$$\mu = a/b, \quad \phi = \mu \frac{b^2}{a},$$

where μ is the mean parameter, and ϕ is the precision parameter (or $\varphi = 1/\phi$ is the dispersion parameter). The corresponding density is

$$f(y) = \frac{1}{\Gamma(\phi)} \left(\frac{\phi}{\mu}\right)^{\phi} y^{\phi-1} \exp\left(-y\phi/\mu\right).$$

The linear predictor η is linked to the mean μ using a default log-link, $\mu = \exp(\eta)$. The hyperparameter is the precision parameter ϕ, which is represented as $\phi = \exp(\theta)$ and the diffuse gamma prior is defined on θ.

Myers and Montgomery (1997) presented data from a step in the manufacturing process for semiconductors. Four factors are believed to influence the resistivity of the wafer and so a full factorial experiment with two levels of each factor was run. Faraway (2016b) analyzed the data with different frequentist models. Table 4.6 presents the variable code for the wafer data.

TABLE 4.6
Code sheet for the variables in the wafer data.

Variable Name	Description	Codes/Values
x1	a factor with levels '-' '+'	'-' = level 1; '+' = level 2
x2	a factor with levels '-' '+'	'-' = level 1; '+' = level 2
x3	a factor with levels '-' '+'	'-' = level 1; '+' = level 2
x4	a factor with levels '-' '+'	'-' = level 1; '+' = level 2
resist	resistivity of the wafer	number

We start with a look at the distribution of the response variable, `resist`:

```
library(faraway)
data(wafer)
hist(wafer$resist, prob=T, col="grey", xlab= "resist", main = "")
lines(density(wafer$resist), lwd=2)
```

Figure 4.6 shows that the `resist` variable has clearly a skewed distribution. Now let us fit a gamma GLM with conventional maximum-likelihood estimation:

```
formula <- resist ~ x1 + x2 + x3 + x4
wafer.glm <- glm(formula, family = Gamma(link = log), data = wafer)
round(coef(summary(wafer.glm)), 4)
```

```
            Estimate Std. Error t value Pr(>|t|)
(Intercept)   5.4455     0.0586 92.9831   0.0000
```

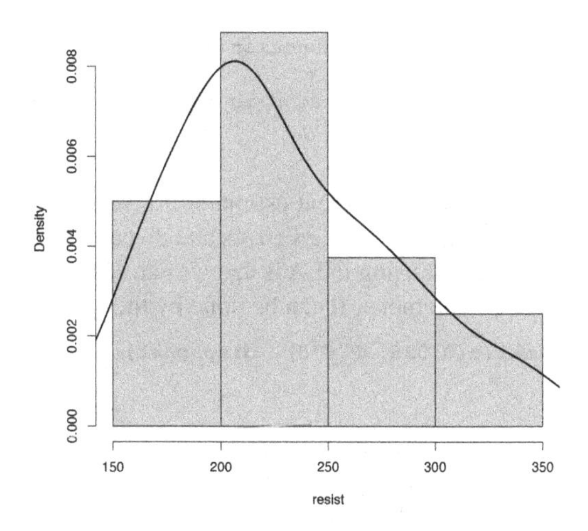

FIGURE 4.6
Histogram and nonparametric density estimate for the response variable, `resist`, in the wafer dataset.

```
x1+           0.1212    0.0524  2.3129   0.0411
x2+          -0.3005    0.0524 -5.7364   0.0001
x3+           0.1798    0.0524  3.4323   0.0056
x4+          -0.0576    0.0524 -1.0990   0.2952
```

The estimate of the dispersion parameter is:

```
round(summary(wafer.glm)$dispersion, 4)
```

```
[1] 0.011
```

Now let us fit the gamma GLM using INLA:

```
wafer.inla1 <- inla(formula, family = "gamma", data = wafer)
round(wafer.inla1$summary.fixed, 4)
```

```
               mean     sd 0.025quant 0.5quant 0.975quant    mode kld
(Intercept)  5.4465 0.0656     5.3170   5.4461     5.5785  5.4454   0
x1+          0.1212 0.0571     0.0073   0.1212     0.2349  0.1212   0
x2+         -0.3005 0.0572    -0.4144  -0.3005    -0.1867 -0.3005   0
x3+          0.1798 0.0573     0.0657   0.1798     0.2938  0.1798   0
x4+         -0.0576 0.0572    -0.1715  -0.0576     0.0562 -0.0576   0
```

We output the results for the precision parameter:

```
round(wafer.inla1$summary.hyperpar, 4)
```

```
                                             mean      sd 0.025quant 0.5quant
Precision parameter for the Gamma observations  90.5665 34.0182    37.3194  86.0918
                                          0.975quant    mode
Precision parameter for the Gamma observations 168.9319 76.625
```

We can extract the posterior mean for the dispersion parameter φ by the following command:

```
disp_post <- inla.tmarginal(fun=function(x) 1/x, wafer.inla1$marginals
    ↪ .hyperpar[[1]])
round(inla.emarginal(function(x) x, marg = disp_post),4)
```

```
0.0129
```

In this example, the regression coefficient estimates are very similar using both frequentist approach and INLA. The standard errors and dispersion parameter estimates are also close. An advantage of using INLA is that we can easily compute the credible interval for the dispersion parameter. It can be done by the following code:

```
round(inla.qmarginal(c(0.025, 0.975), disp_post), 4)
```

```
[1] 0.0058 0.0275
```

The usual way to model skewed continuous data is the linear regression with log-transformation. Let us also compare the linear model:

```
wafer.inla2 <- inla(log(resist) ~ x1 + x2 + x3 + x4, family = "
    ↪ gaussian", data = wafer)
round(wafer.inla2$summary.fixed, 4)
```

	mean	sd	0.025quant	0.5quant	0.975quant	mode	kld
(Intercept)	5.4405	0.0598	5.3213	5.4405	5.5595	5.4405	0
x1+	0.1228	0.0535	0.0162	0.1228	0.2293	0.1228	0
x2+	-0.2999	0.0535	-0.4064	-0.2999	-0.1934	-0.2999	0
x3+	0.1784	0.0535	0.0719	0.1784	0.2849	0.1784	0
x4+	-0.0561	0.0535	-0.1627	-0.0561	0.0503	-0.0561	0

We see that the posterior estimates of the coefficients are remarkably similar to the previous two models. However, the interpretation of the linear regression with log-transformation and that of the gamma GLM are very different. In the linear regression with log-transformation,

$$E[\log(Y)] = \beta_0 + \beta_1 X_1 + \beta_2 X_2 + \beta_3 X_3 + \beta_4 X_4.$$

It is interpretable in terms of the arithmetic mean change on the log scale. For example, having $x_1 = +$ changes the expected value of log outcome by 0.1228. In the gamma GLM model

$$\log(E[Y]) = \beta_0 + \beta_1 X_1 + \beta_2 X_2 + \beta_3 X_3 + \beta_4 X_4.$$

That means that $E[Y] = \exp(\beta_0 + \beta_1 X_1 + \beta_2 X_2 + \beta_3 X_3 + \beta_4 X_4)$, which assumes multiplicative effects on the original outcome by the predictors. Having $x_1 = +$ increases the log arithmetic mean outcome by 0.1212. The exponentiated coefficient $\exp(0.1212) = 1.1289$ indicates that the mean outcome on the original scale, when $x_1 = +$, is 1.1289 times as high as the mean on the original scale when $x_1 = -$, assuming other predictors are fixed.

4.6 Proportional Responses

Many studies in different fields involve the regression analysis of proportions observed in the open interval $(0,1)$. Such data are often continuous but heteroscedastic on the unit interval: they display more variation around the mean and less variation as we approach the lower and upper limits of the interval. The typical approach to model the data is to transform the response variable and then apply a standard linear regression analysis. A commonly used transformation is the logit transformation, $\tilde{y} = \log(y/(1-y))$. However, the logit transfer of rates or proportions are often asymmetric, and thus Gaussian-based inference for interval estimation and hypothesis testing can be inaccurate in small samples. Also, the regression parameters are interpretable in terms of the mean of $\tilde{y}$, and not in terms of the mean of y, like in the case of gamma GLM.

Beta regression model is in the framework of GLMs, which is a natural way to model continuous responses that assume values in the open interval $(0,1)$. The regression parameters in beta regression are interpretable in terms of the mean of y and the model is naturally heteroscedastic and easily accommodates asymmetries (Ferrari and Cribari-Neto, 2004; Wang, 2012).

The density function of the beta distribution is given by $f(y;p,q) = y^{p-1}(1-y)^{q-1}/B(p,q)$, where $p,q > 0$, and $B(p,q) = \Gamma(p)\Gamma(q)/\Gamma(p,q)$ is the beta function. If we let $\mu = p/(p+q)$ and $\tau = p+q$. The density can be reparameterized as

$$f(y;\mu,\tau) = \frac{1}{B(\mu\tau,(1-\mu)\tau)} y^{\mu\tau-1}(1-y)^{(1-\mu)\tau-1}, \tag{4.3}$$

where $\mu \in (0,1)$, $\tau > 0$. We denote a random variable Y that follows a beta distribution with the density form (4.3) by $Y \sim Beta(\mu,\tau)$. It can be shown that $\mathrm{E}(Y) = \mu$, and $\mathrm{Var}(Y) = \mu(1-\mu)/(1+\tau)$. The parameter τ quantifies the amount of overdispersion, since the dispersion of the distribution increases as τ decreases.

Figure 4.7 displays a few different beta density functions along with the corresponding values of mean and dispersion parameters. It is noted that the beta distribution family covers quite different shapes depending on the two parameters. The beta density can be "U-shaped" or "J-shaped," and same cases are not displayed in Figure 4.7. Thus, beta regression provides a flexible way to model the continuous data bounded in $(0,1)$.

There are a few possible choices for the link function in beta regression. The most common one is the logit link, $g(\mu) = \log(\mu/(1-\mu))$. One could also use the probit function, $g(\mu) = \Phi^{-1}(\mu)$, where $\Phi(\cdot)$ is the standard normal cumulative distribution function, or the log-log link, $g(\mu) = -\log(-\log(\mu))$. The particularly useful link function is the logit link, since the coefficients in beta regression could have the odds ratio interpretation. In the INLA package, the logit link function is used in beta regression.

Specifically, suppose we have a vector of p explanatory variables, $(x_{i1},\ldots,x_{ip})$,

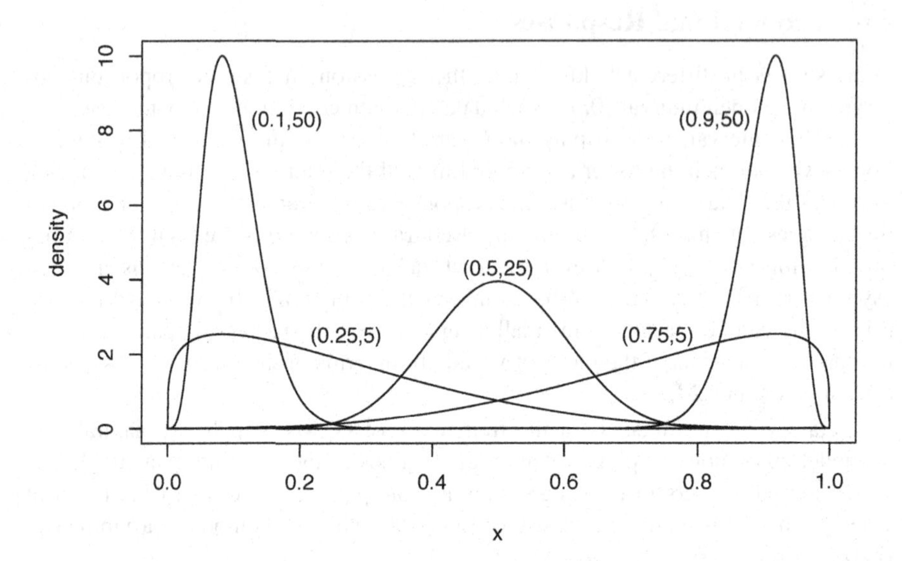

FIGURE 4.7
Beta density functions for different combinations of mean and dispersion parameters.

that is related to the response $Y_i \in (0,1)$. The beta model is given by

$$\begin{cases} Y_i \sim \text{Beta}(\mu_i, \tau), \\ \text{logit}(\mu_i) = \beta_0 + \beta_1 x_{i1} + \ldots + \beta_p x_{ip}. \end{cases}$$

Let us employ the gasoline yield data set taken from Prater (1956). It has been well-investigated by Ferrari and Cribari-Neto (2004). The dependent variable is the proportion of crude oil converted to gasoline after distillation and fractionation (`yield`). We study here four explanatory variables: crude oil gravity (`gravity`), vapor pressure of crude oil (`pressure`), temperature at which 10 percent of crude oil has vaporized (`temp10`), temperature at which all gasoline has vaporized (`temp`). The code sheet for the variables in the gasoline yield data is displayed in Table 4.7.

We first look at the distribution of the response variable:

```
library(betareg)
data(GasolineYield)
hist(GasolineYield$yield, prob=T, col="grey", xlab= "yield", main = ""
    ↪ , xlim =c(0,1))
lines(density(GasolineYield$yield), lwd=2)
```

Figure 4.8 shows the histogram and nonparametric density estimate for `yield`. We see that the variable is right-skewed distributed and its range is from 0 to about 0.5. Since the response is a proportion, a beta regression model is rather natural. Let us fit a beta regression with conventional maximum-likelihood estimation:

TABLE 4.7
Code sheet for the variables in the gasoline yield data.

Variable Name	Description	Codes/Values
yield	proportion of crude oil converted to gasoline after distillation and fractionation	proportion
gravity	crude oil gravity	API scale
pressure	vapor pressure of crude oil	lbf/in2
temp10	temperature at which 10 percent of crude oil has vaporized	degrees F
temp	temperature at which all gasoline has vaporized	degrees F
batch	factor indicating unique batch of conditions gravity, pressure, and temp10	levels 1 to 10

```
gas.glm <- betareg(yield ~ gravity + pressure + temp10 + temp, data =
    ↪ GasolineYield)
coef(summary(gas.glm))
```

```
$mean
                Estimate   Std. Error    z value     Pr(>|z|)
(Intercept) -2.694942189 0.7625693428 -3.5340290 4.092761e-04
gravity      0.004541209 0.0071418995  0.6358545 5.248712e-01
pressure     0.030413465 0.0281006512  1.0823046 2.791172e-01
temp10      -0.011044935 0.0022639670 -4.8785761 1.068544e-06
temp         0.010565035 0.0005153974 20.4988154 2.205984e-93

$precision
      Estimate Std. Error  z value     Pr(>|z|)
(phi) 248.2419    62.0162 4.002856 6.258231e-05
```

Now, we fit the beta regression with INLA by specifying `family = "beta"` in `inla` call.

```
gas.inla <- inla(yield ~ gravity + pressure + temp10 + temp, data =
    ↪ GasolineYield, family = "beta")
round(gas.inla$summary.fixed, 4)
```

```
               mean     sd 0.025quant 0.5quant 0.975quant    mode kld
(Intercept) -2.6607 1.3041    -5.2760  -2.6476    -0.1210 -2.6235   0
gravity      0.0045 0.0122    -0.0195   0.0045     0.0286  0.0045   0
pressure     0.0297 0.0480    -0.0634   0.0290     0.1265  0.0278   0
temp10      -0.0109 0.0039    -0.0184  -0.0110    -0.0031 -0.0110   0
temp         0.0104 0.0009     0.0087   0.0104     0.0121  0.0104   0
```

```
round(gas.inla$summary.hyperpar, 4)
```

```
                                             mean      sd 0.025quant 0.5quant
precision parameter for the beta observations  88.9334 22.1599     50.869   87.077
                                        0.975quant    mode
precision parameter for the beta observations 137.2354 83.0305
```

The estimates of regression coefficients using INLA are almost the same as those using the maximum likelihood approach. Noticeably, the estimate of the dispersion parameter using INLA is much smaller than that using the maximum likelihood. The

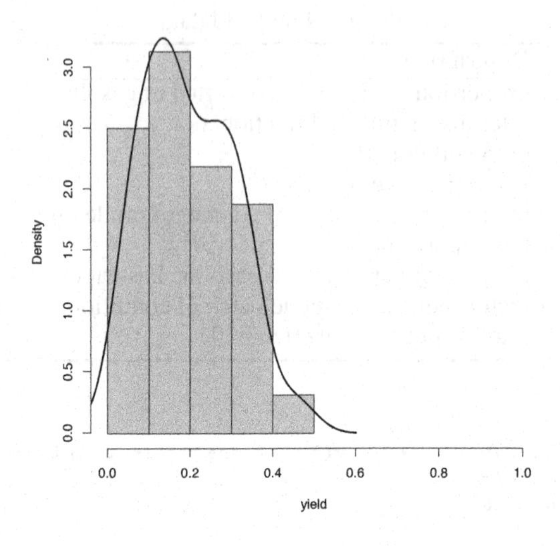

FIGURE 4.8
Histogram and nonparametric density estimate for the response variable, `yield`, in the gasoline yield dataset.

reason for the differences needs to be further investigated. We leave it as an open problem for the readers.

After fitting the model, it is important to perform diagnostic analyses to check the goodness-of-fit of the model. We here introduce a Bayesian residual graphical tool for detecting departures from the postulated model and influential observations. Ferrari and Cribari-Neto (2004) defined the standardized residuals for beta regression:

$$\varepsilon_i = \frac{y_i - \hat{\mu}_i}{\sqrt{\widehat{\mathrm{Var}}(y_i)}}, i = 1, \ldots, n,$$

where $\hat{\mu}_i = g^{-1}(\mathbf{x}_i^T \hat{\beta})$ and $\widehat{\mathrm{Var}}(y_i) = \hat{\mu}_i(1 - \hat{\mu}_i)/(1 + \hat{\tau})$. Here $(\hat{\beta}, \hat{\tau})$ are the maximum likelihood estimates. Similar to Poisson regression, we can define Bayesian standardized residuals for beta regression:

$$r_i = \frac{y_i - \mu_i}{\sqrt{\mathrm{Var}(y_i)}} = \frac{y_i - g^{-1}(\mathbf{x}_i^T \beta)}{\sqrt{g^{-1}(\mathbf{x}_i^T \beta)(1 - g^{-1}(\mathbf{x}_i^T \beta))/(1 + \tau)}}, i = 1, \ldots, n.$$

The posterior samples of the residuals are obtained by substituting the samples from the posterior distributions of β and τ. A plot of the posterior means or medians of the r_i's against the index of the observations can then be examined. We have written a convenience function, `bri.beta.resid`, for calculating Bayesian Pearson residuals for beta regression in our `brinla` library. When the argument `plot = TRUE` in this function is specified, a residual plot by case is output:

```
gas.inla1.resid <- bri.beta.resid(gas.inla1, plot = TRUE, ylim = c
    ↪ (-2,2))
abline(1.96, 0, lty=2); abline(-1.96, 0, lty=2)
```

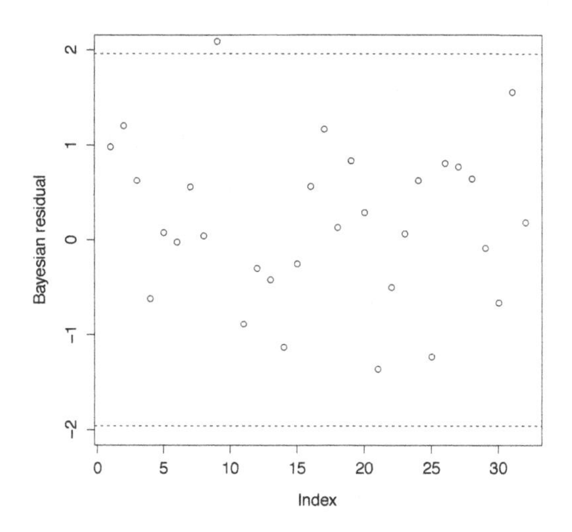

FIGURE 4.9
Bayesian residual plot of the beta model for the gasoline yield data.

Figure 4.9 shows Bayesian residuals versus the index of the observations for the beta regression. We observe a horizontal band with points that vary at random. There is only one data point out of the band.

In this fitted model, we note that the 95% credible intervals for `temp10` and `temp` do not contain zero, while those for `gravity` and `pressure` do contain zero. Let us build a reduced model with the significant variables (in the Bayesian sense), `temp10` and `temp`:

```
gas.inla2 <- inla(yield ~ temp10 + temp, data = GasolineYield, family
    ↪ = "beta", control.compute = list(dic = TRUE, cpo = TRUE))
round(gas.inla2$summary.fixed, 4)
```

```
                mean     sd 0.025quant 0.5quant 0.975quant    mode kld
(Intercept) -1.7802 0.3695    -2.5082  -1.7809    -1.0498 -1.7821   0
temp10      -0.0134 0.0015    -0.0164  -0.0134    -0.0105 -0.0134   0
temp         0.0105 0.0008     0.0089   0.0105     0.0121  0.0105   0
```

```
round(gas.inla2$summary.hyperpar, 4)
```

```
                              mean      sd 0.025quant 0.5quant 0.975quant    mode
precision parameter
for the beta observations 92.9781 22.4935    54.1396  91.1616    141.828 87.1999
```

There is a negative relationship between the mean response (proportion of crude oil converted to gasoline) and the temperature at which 10 percent of crude oil has vaporized, but there is a positive relationship between the mean response and the temperature at which all gasoline has vaporized. Let us compare the DICs of the previous model and the current model:

```
c(gas.inla1$dic$dic, gas.inla2$dic$dic)
```

```
[1] -126.8995 -131.2635
```

Thus, we prefer the reduced model because of the smaller DIC. We further examine the Bayesian residual plot for the reduced model:

```
gas.inla2.resid <- bri.beta.resid(gas.inla2, plot = TRUE, ylim = c
    ↪ (-2,2))
abline(1.96, 0, lty=2); abline(-1.96, 0, lty=2)
```

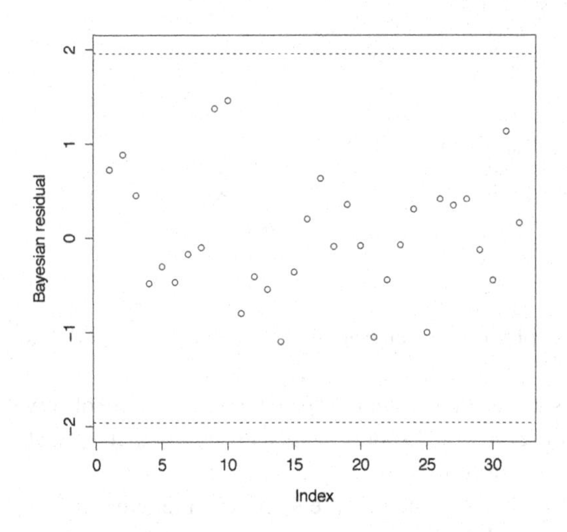

FIGURE 4.10
Bayesian residual plot of the reduced beta model for the gasoline yield data.

Figure 4.10 also shows a random pattern. All residuals are within the horizontal band. Thus, the assumption of the beta model for the gasoline yield data is valid.

4.7 Modeling Zero-Inflated Data

In statistics, a zero-inflated regression is often used for the problem of modeling data with excess zeros. The model is based on a zero-inflated probability distribution, i.e., a distribution that allows for frequent zero-valued observations. Zero-inflated situation occurs often in modeling count data. For example, the number of insurance claims within a population for some risk would be zero-inflated by those people who have not taken out insurance against the risk and thus are unable to claim. INLA allows users to fit zero-inflated models with Poisson, binomial, negative-binomial and beta-binomial distributions.

Zero-inflated models provide a mixture-modeling approach to model the excess zeros in addition to allowing for overdispersion. In particular, there are two possible data generation processes for each observation, in which the result of a Bernoulli trial is used to determine the process. For i^{th} observation, the first process generates only zeros with probability ϕ, whereas the second process generates the outcomes from a specific parametric distribution, such as Poisson or negative binomial, with probability $1-\phi$ (Lambert, 1992). Specifically, a zero-inflated model can be expressed as

$$Y_i \sim \begin{cases} 0 & \text{with probability } \phi, \\ g(Y_i|\mathbf{x}_i) & \text{with probability } 1-\phi, \end{cases} \tag{4.4}$$

where $g(\cdot)$ is a certain probability mass or density function. The corresponding probability of $p(Y_i = y_i|\mathbf{x}_i)$ is

$$p(Y_i = y_i|\mathbf{x}_i) = \phi \cdot I_{\{y_i=0\}} + (1-\phi) \cdot g(y_i|\mathbf{x}_i).$$

In INLA, the above zero-inflated models (4.4) are implemented as "Type 1" zero-inflated models. INLA also provides the options for "Type 0" zero-inflated models, in which the likelihood is defined as

$$p(Y_i = y_i|\mathbf{x}_i) = \phi \cdot I_{\{y_i=0\}} + (1-\phi) \cdot g(y_i|y_i > 0, \mathbf{x}_i).$$

Here we only focus on an example of modeling count data with excess zeros using "Type 1" zero-inflated Poisson and negative-binomial models, as other cases are similar. The `articles` data are taken from Long (1997). This study examines how factors such as gender (`fem`), marital status (`mar`), number of young children (`kid5`), prestige of the graduate program (`phd`), and number of articles published by a scientist's mentor (`ment`) affect the number of articles (`art`) published by the scientist. Table 4.8 displays the code sheet for the variables in the articles data.

TABLE 4.8
Code sheet for the variables in the articles data.

Variable Name	Description	Codes/Values
fem	gender	0 = male; 1 = female
ment	number of articles published by a scientist's mentor	integers
phd	prestige of the graduate program	numbers
mar	marital status	0 = No; 1 = Yes
kid5	number of young children	integers
art	number of articles	integers

Let us first take a look at the frequency and proportion of scientists who publish each observed number of articles. We use the following R commands:

```
data(articles, package = "brinla")
table(articles$art)
```

```
  0   1   2   3   4   5   6   7   8   9  10  11  12  16  19
275 246 178  84  67  27  17  12   1   2   1   1   2   1   1
```

```
round(prop.table(table(articles$art)),3)
```

```
    0     1     2     3     4     5     6     7     8     9    10    11    12
0.301 0.269 0.195 0.092 0.073 0.030 0.019 0.013 0.001 0.002 0.001 0.001 0.002
   16    19
0.001 0.001
```

The observed proportion of scientists who publish no articles is 0.301, indicating there is a large number of zeros in the response variable. Thus, we fit zero-inflated Poisson and zero-inflated negative binomial models. In the following statements, `family = "zeroinflatedpoisson1"` requests the zero-inflated Poisson model:

```
articles.inla1 <- inla(art ~ fem + mar + kid5 + phd + ment, data =
    ↪ articles, family = "zeroinflatedpoisson1", control.compute =
    ↪ list(dic = TRUE, cpo = TRUE))
round(articles.inla1$summary.fixed, 4)
```

```
               mean     sd 0.025quant 0.5quant 0.975quant    mode kld
(Intercept)  0.5504 0.1139     0.3262   0.5506     0.7734  0.5509   0
fem1        -0.2315 0.0586    -0.3468  -0.2315    -0.1167 -0.2313   0
mar1         0.1322 0.0661     0.0026   0.1321     0.2619  0.1320   0
kid5        -0.1706 0.0433    -0.2559  -0.1705    -0.0860 -0.1703   0
phd          0.0028 0.0285    -0.0531   0.0028     0.0587  0.0027   0
ment         0.0216 0.0022     0.0173   0.0216     0.0258  0.0216   0
```

```
round(articles.inla1$summary.hyperpar, 4)
```

```
                                      mean     sd 0.025quant 0.5quant
zero-probability parameter
for zero-inflated poisson_1         0.1556 0.0204     0.1156   0.1555
                                0.975quant   mode
zero-probability parameter
for zero-inflated poisson_1          0.196 0.1554
```

From the results, the variable, phd, is the only variable that is not significant in the Bayesian sense. The estimated zero-probability parameter is 0.1556 with the 95% credible interval (0.1156, 0.196).

The zero-inflated negative binomial model can be fit similarly by specifying `family = "zeroinflatednbinomial1"`:

```
articles.inla2 <- inla(art ~ fem + mar + kid5 + phd + ment, data =
    ↪ articles, family = "zeroinflatednbinomial1", control.compute =
    ↪ list(dic = TRUE, cpo = TRUE))
round(articles.inla2$summary.fixed, 4)
```

```
               mean     sd 0.025quant 0.5quant 0.975quant    mode kld
(Intercept)  0.2745 0.1391     0.0008   0.2747     0.5470  0.2750   0
fem1        -0.2177 0.0724    -0.3598  -0.2177    -0.0757 -0.2177   0
mar1         0.1493 0.0818    -0.0112   0.1493     0.3099  0.1493   0
kid5        -0.1761 0.0529    -0.2802  -0.1760    -0.0727 -0.1758   0
phd          0.0147 0.0359    -0.0558   0.0146     0.0851  0.0146   0
ment         0.0288 0.0034     0.0221   0.0287     0.0356  0.0287   0
```

```
round(articles.inla2$summary.hyperpar, 4)
```

```
                                                mean     sd 0.025quant 0.5quant
size
for nbinomial zero-inflated observations      2.3923 0.2973     1.8543   2.3773
zero-probability parameter
for zero-inflated nbinomial_1                 0.0110 0.0093     0.0010   0.0085
                                          0.975quant   mode
size
for nbinomial zero-inflated observations      3.0209 2.3510
zero-probability parameter
for zero-inflated nbinomial_1                 0.0353 0.0028
```

Comparing the results from two models, the posterior mean estimates of the predictors are close, but their posterior SDs from the zero-inflated negative binomial model are generally larger than those from the zero-inflated Poisson model. In particular, the 95% credible interval for the variable, `mar`, is (0.0025, 0.2619) in the zero-inflated Poisson model, while it is (-0.0127, 0.3117) in the zero-inflated negative binomial model. The `mar` becomes a non-significant variable (in the Bayesian sense) in the zero-inflated negative binomial model. The estimated zero-probability parameter becomes 0.0110, which is much smaller than that from the zero-inflated Poisson model.

As we have discussed in the previous sections, for over-dispersed count data, the estimated standard errors of the parameters are often too low in Poisson models, because Poisson distribution assumes that its variance equals its mean. Negative binomial distribution is often a better choice in practice. Let's further check the DICs for the two models:

```
c(articles.inla1$dic$dic, articles.inla2$dic$dic)
```

```
3255.584 3137.209
```

The zero-inflated negative binomial model is favorable based on the DICs. We further extract the posterior mean and the 95% credible interval for the overdispersion parameter of negative binomial distribution:

```
overdisp_post <- inla.tmarginal(fun = function(x) 1/x, marg = articles
    ↪ .inla2$marginals.hyperpar[[1]])
round(inla.emarginal(fun=function(x) x, marg=overdisp_post), 4)
```

```
[1] 0.4457
```

```
round(inla.qmarginal(c(0.025, 0.975), overdisp_post), 4)
```

```
[1] 0.3507 0.5590
```

The posterior mean is 0.4457, indicating a mild to moderate overdispersion in this dataset.

In conclusion from the zero-inflated negative binomial modeling results, we find that female scientists published fewer articles than male scientists, the number of young children is negatively associated with the number of articles, and the number of articles published by a scientist's mentor positively affects the outcomes. However, the marital status and the prestige of the graduate program are not associated with the number of articles by the scientist at the 95% credible level.

5

Linear Mixed and Generalized Linear Mixed Models

In linear and generalized linear models, we assume that the responses, conditional on the linear predictor, are independent. When data have a grouping or hierarchical structure or we have multiple observations on individuals, we will have dependent responses. This chapter is about how to extend the LMs and GLMs to hande this type of response. We start by extending LMs with a simple grouping structure that can be generalized to more complex models. We look at an example where individuals are measured longitudinally. The GLM also has its extension to generalized linear mixed models (GLMM).

5.1 Linear Mixed Models

Linear mixed models (LMM) are characterized by a Gaussian response, y, together with a mix of fixed and random components. They can be written in the form:

$$y = \mathbf{X}\beta + \mathbf{Z}u + \varepsilon$$

where $\mathbf{X}$ is a matrix whose columns are predictors, usually including the intercept. The parameters β are called fixed effects. Without the $\mathbf{Z}u$ term, this would be just a linear model. The $\mathbf{Z}$ is also a matrix of predictors, some of which may be in common with $\mathbf{X}$. The u are the random effects. In the frequentist form of the model, the β are fixed parameters while the u would be random with a multivariate normal distribution with zero mean and covariance that we would wish to estimate. This mixture of fixed and random components suggests the mixed model name. From the Bayesian perspective, all the parameters have distributions so they are all random and mixed model terminology is not as appropriate. Nevertheless, we will treat the two sets of parameters differently so we will keep the fixed and random names.

Usually, it is not convenient to use the $\mathbf{Z}u$ form for the random component when constructing or interpreting the model. Instead, we will use the latent Gaussian model (LGM) formulation, where $EY_i = \mu_i$, of:

$$\mu_i = \alpha + \sum_j \beta_j x_{ij} + \sum_k f^{(k)}(u_{ij}). \tag{5.1}$$

The errors, ε_i are i.i.d. normal. We can construct the $f^{(k)}(u_{ij})$ terms in various ways that introduce different patterns of correlation in the response as appropriate for the particular application. Common structures are:

- *Grouping or clustering*: The cases are divided into groups within which the responses are correlated. Sometimes these represent the same individual being measured repeatedly or different individuals that belong to some common group.
- *Hierarchy*: The cases belong to groups. The groups belong to broader groups. For example, students belong to a class and a school consists of different classes. There can be several layers of hierarchy as the schools may belong to a district and so on.
- *Longitudinal*: An individual is followed over time and measured on different occasions. This is an example of grouping already mentioned but longitudinal data requires a particular specification that we will explore in this chapter.

It is difficult to precisely define scope of the predictor side of a LMM but we can be definite about the response. This should be Gaussian. If you want a binomial, Poisson or other distribution for the response, you need to use a generalized LMM or GLMM, as described later in this chapter. We develop the methodology of INLA for LMMs with a series of examples, starting with the simplest kind of mixed model.

5.2 Single Random Effect

Hawkins (2005) reports data on the nitrogen content of reeds at three different locations in Cambridgeshire, England. The reeds are part of an ecosystem supporting a variety of moth and there is some interest in how the nitrogen content might vary from location to location. We start by loading the data and summarizing the data:

```
data(reeds, package="brinla")
summary(reeds)
```

```
 site     nitrogen
 A:5   Min.    :2.35
 B:5   1st Qu.:2.62
 C:5   Median :3.06
       Mean    :3.03
       3rd Qu.:3.27
       Max.    :3.93
```

There are five observations per site. The sites are labeled, A, B and C, which is an indication that we do not take much interest in what the nitrogen level is at a particular site. We are more interested in the variation across the whole region and in what nitrogen content we might find if we sample from a new site. For this reason, it makes more sense to regard the site as a random effect than a fixed effect. This means there is no β term in the LGM from (5.1). We use a model of the form:

$$y_{ij} = \alpha + u_i + \varepsilon_{ij} \qquad i = 1, \ldots, a \qquad j = 1, \ldots, n,$$

where the u_i and ε_{ij}s are normal with mean zero, but variances σ_u^2 and σ_ε^2, respectively. The σ_u^2 and σ_ε^2 are the hyperparameters and there is a single "fixed" parameter, α.

For comparison purposes, we make the standard likelihood-based analysis using the `lme4` R package:

```
library(lme4)
mmod <- lmer(nitrogen ~ 1+(1|site), reeds)
summary(mmod)
```

```
Random effects:
 Groups   Name        Variance Std.Dev.
 site     (Intercept) 0.1872   0.433
 Residual             0.0855   0.292
Number of obs: 15, groups:  site, 3

Fixed effects:
            Estimate Std. Error t value
(Intercept)    3.029      0.261    11.6
```

We can see that the estimated variation between sites ($\hat{\sigma}_u = 0.433$) is somewhat larger than that seen within sites ($\hat{\sigma}_\varepsilon = 0.292$).

We will need the `INLA` package throughout this chapter and will also use our `brinla` package. From now on, we will assume these have already been loaded. If you forget, you will get a "function not found" error message.

```
library(INLA); library(brinla)
```

There are three parameters, α, σ_u^2 and σ_ε^2 for which we must specify prior distributions. We will discuss the default priors used by INLA but these are essentially uninformative. We also need to specify that the u_i are independent and identically distributed which we achieve in the model formula using `model="iid"`. Let's see what we get with these priors:

```
formula <- nitrogen ~ 1 + f(site, model="iid")
imod <- inla(formula,family="gaussian", data = reeds)
summary(imod)
```

```
Fixed effects:
              mean     sd 0.025quant 0.5quant 0.975quant   mode kld
(Intercept) 3.0293 0.1829     2.6627   3.0293     3.3958 3.0293   0

Random effects:
Name          Model
 site   IID model

Model hyperparameters:
                                 mean     sd 0.025quant 0.5quant 0.975quant   mode
Precision for the Gaussian obs 13.25  5.205      5.486    12.49      25.55 10.972
Precision for site             20.35 25.119      1.936    12.81      84.52  5.086

Expected number of effective parameters(std dev): 1.912(0.7592)
Number of equivalent replicates : 7.845

Marginal log-Likelihood:  -20.88
```

The posterior mean of α is 3.03 and a 95% credibility interval is $[2.66, 3.40]$. INLA works with precision, which is the inverse of the variance. This is convenient for the theory and computation, but not familiar for interpretation. We would like to obtain summary statistics on the posteriors of σ_u and σ_ε by transforming to the SD scale:

```
invsqrt <- function(x) 1/sqrt(x)
sdt <- invsqrt(imod$summary.hyperpar[,-2])
row.names(sdt) <- c("SD of epsilson","SD of site")
sdt
```

```
                  mean 0.025quant 0.5quant 0.975quant    mode
SD of epsilson 0.27468    0.42695  0.28294    0.19785 0.30190
SD of site     0.22169    0.71862  0.27942    0.10877 0.44343
```

Converting the SD of the precision to the SD of the standard deviation requires a different transformation so we have omitted this from the computation. Unfortunately, the mean and the mode are not invariant to the transform in scale. Quantiles, such as the median, are invariant to monotone transformation. This means that the mean and mode above are incorrect and we must first derive the transformed posterior distributions and then recompute these summary statistics. We can do this for the means:

```
prec.site <- imod$marginals.hyperpar$"Precision for site"
prec.epsilon <- imod$marginals.hyperpar$"Precision for the Gaussian
    ↪ observations"
c(epsilon=inla.emarginal(invsqrt,prec.epsilon),
  site=inla.emarginal(invsqrt,prec.site))
```

```
epsilon    site
0.29083 0.31355
```

We can also compute the posterior modes:

```
sigma.site <- inla.tmarginal(invsqrt, prec.site)
sigma.epsilon <- inla.tmarginal(invsqrt, prec.epsilon)
c(epsilon=inla.mmarginal(sigma.epsilon),
  site=inla.mmarginal(sigma.site))
```

```
epsilon    site
0.26655 0.22174
```

The posterior distributions of these two components are quite asymmetrical (in contrast to α) so the mean, median and mode show some differences. If we compare this to the `lme4` output above, where maximum likelihood estimates are used, the mode is natural comparison. We see some differences in the outcomes. If we use flat priors, we expect posterior modes and maximum likelihood estimates to be similar under most circumstances. The site SD is smaller from the Bayes calculation indicating the default prior is giving some preference to smaller values.

The INLA package contains a function to make the computation of summary statistics for SDs easier:

```
inla.contrib.sd(imod)$hyper
```

```
                                   mean       sd     2.5%   97.5%
sd for the Gaussian observations 0.29196 0.059868 0.202465 0.44965
sd for site                      0.31652 0.162710 0.098207 0.71926
```

The function works by sampling from the posterior distribution for the precision, transforming the samples, and then computing the summary statistics. This is less efficient than directly constructing the posteriors for the SDs, but the technique can be useful for other quantities. For example, the intraclass correlation coefficient (ICC) is defined as

$$\rho = \frac{\sigma_u^2}{\sigma_u^2 + \sigma_\varepsilon^2}.$$

This is used as a measure of how much the response varies within groups compared to between groups. We sample from the joint posterior using `inla.hyperpar.sample()`, invert to get variances and then compute the ICC for each sample. We construct summary statistics of the 1000 sampled ICCs.

```
sampvars <- 1/inla.hyperpar.sample(1000,imod)
sampicc <- sampvars[,2]/(rowSums(sampvars))
quantile(sampicc, c(0.025,0.5,0.975))
```

```
    2.5%      50%    97.5%
0.096973 0.493498 0.891028
```

We see the median of the posterior distribution of the ICC is 0.49 but the 95% credible interval is wide. Note that although an MCMC-based modeling approach would also naturally use samples, this sampling based on INLA is a quite different matter as it can be done quickly and easily. The samples are independent and quickly generated.

Since we often want the hyperparameters summarized as SDs rather than precisions and the results can be computed exactly as shown above, we have written a convenience function for producing this summary in our `brinla` package:

```
bri.hyperpar.summary(imod)
```

```
                                 mean       sd  q0.025    q0.5  q0.975    mode
SD for the Gaussian observations 0.2907 0.057994 0.19832 0.28274 0.42528 0.26654
SD for site                      0.3128 0.155914 0.10906 0.27884 0.71091 0.22178
```

Although summary statistics are useful, there is no good substitute for looking at plots of the posterior distributions. The fixed component of the model has a single component, α:

```
alpha <- data.frame(imod$marginals.fixed[[1]])
library(ggplot2)
ggplot(alpha, aes(x,y)) + geom_line() + geom_vline(xintercept = c
  ↪ (2.66, 3.40)) +
     xlim(2,4)+xlab("nitrogen")+ylab("density")
```

The plot, including the 95% credibility interval, is shown in the first panel of Figure 5.1. We see a symmetric, normally-shaped, posterior distribution. The two random components of the model can be extracted and transformed to the SD scale:

```
x <- seq(0,1,len=100)
d1 <- inla.dmarginal(x,sigma.site)
d2 <- inla.dmarginal(x,sigma.epsilon)
rdf <- data.frame(nitrogen=c(x,x),sigma=gl(2,100,
     labels=c("site","epsilon")),density=c(d1,d2))
ggplot(rdf, aes(x=nitrogen, y=density, linetype=sigma))+geom_line()
```

The resulting plot is seen in the right panel of Figure 5.1. We see that the posterior for σ_ε is more concentrated. This is not surprising since we have 15 observations that provide information about this parameter in contrast to only three sites used for σ_u.

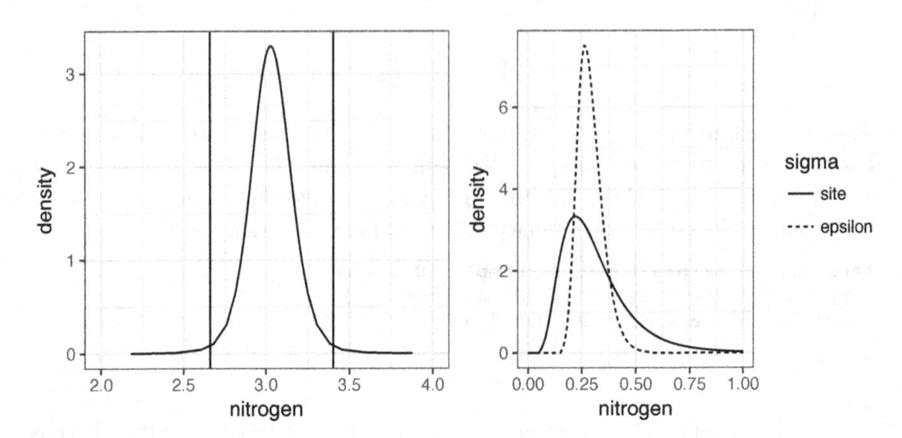

FIGURE 5.1
Posterior densities for single random effect model of nitrogen content in reeds. The intercept, α, is shown on the left with a 95% credibility interval and the two SDs for the site and error (σ_ε and σ_u) are shown on the right.

We might ask whether there is any difference between the sites. In the frequentist paradigm, this question would be formulated as the null hypothesis that $\sigma_u = 0$. As it turns out, we can see from Figure 5.1 that the posterior distribution has a support separated from $\sigma_u = 0$. We are very confident that there is a difference between the sites. Given the closeness of the two SDs, we might say that the difference between two sites is about the same as the difference between two samples from the same site.

We have written a convenience function for producing this plot more directly:

```
bri.hyperpar.plot(imod)
```

You may need to customize the plot so it is still worth understanding the details of its construction as shown above.

5.2.1 Choice of Priors

Thus far we have relied on the default priors, but these deserve a closer look. These are:

- The intercept α has the improper prior $N(0,\infty)$. It's called improper because $N(0,\infty)$ has infinite variance and is not a real distribution. This corresponds to the prior view that the intercept is equally likely to be anything.
- The fixed-effect parameters β have prior distribution $N(0,1000)$.
- The priors for σ_u^2 and σ_ε^2 are defined internally in terms of logged precision. They follow log gamma distribution. The corresponding gamma distribution $\Gamma(a,b)$ has

mean a/b and variance a/b^2. The values used for the default prior are $a = 1$ and $b = 10^{-5}$.

In this example, we could be more definite in specifying the prior on the fixed effect. The response is a percentage and so lies in $[0, 100]$. Some basic biological knowledge will suggest a reasonable range for the nitrogen content of the plant. We could develop a prior that would encompass this information. Experience suggests, however, that unless we specify a sharp prior that it is in contradiction with the data, this will not make much difference to the posterior distribution. One is usually satisfied with the default flat prior for the intercept in such models.

In this particular model, we have no fixed effects other than the intercept. If we did, it is worth noticing that for data where the variables do not have magnitudes greatly different from one, the prior of $N(0, 1000)$ for the coefficients β will be effectively flat. If the data have widely different magnitudes, large values of β might not be unreasonable and the default prior would be informative, perhaps in an undesirable way. In such examples, one would either scale the data or modify the prior appropriately.

The prior specification for σ_u^2 is important. Each observation contributes to the information about σ_ε^2 so the choice of prior for this hyperparameter is less crucial. In contrast, there may be relatively few groups in the data so there is less information about σ_u^2. This means that the prior for this parameter will have a greater influence on the outcome and requires our greatest attention among all the priors. The choice of $a = 1$ for the shape parameter of the gamma reduces the prior to a simple exponential distribution. The small value of b means that the prior for the precision has both a large mean and variance.

Penalized complexity prior: In Simpson et al. (2017), a class of penalized complexity (PC) priors is introduced. These are based on general principles concerning the construction of priors. For a Gaussian random effect, these take the form of an exponential distribution on the standard deviation (in contrast to the exponential on the precision seen in the default). The principle of Occam's razor, as used in statistics, urges us to prefer the simpler model. In this context, a model without a random effect, i.e., $\sigma_u = 0$, should be preferred. An exponential prior on the SD puts the greatest weight on this choice. In contrast, the default prior puts no weight on infinite precision (which is $\sigma_u = 0$). Also, PC priors require that we specify some scaling. If we took the default approach of making the prior very flat, we would lose the principle of a preference for simplicity embodied in the exponential prior on the SD. We need to use the information about the scale of the response to provide a sensible scaling for the prior. We wish to be *weakly informative* — we do not want to strongly influence the posterior but we should use available information to get an effective prior.

To calibrate the scaling of the random effects prior, we set U and p so that $P(\sigma_u > U) = p$. We set $p = 0.01$ and fix $U = 3SD(y)$. Ideally, we would choose U based on contextual knowledge but, failing this, we take the residual SD of the model without fixed effects. In this case, this is the standard deviation of the response. Thus, we calibrate the prior so that probability that the SD of the random effect is three times the SD of the response is quite small. Hence, we allow for the possibility that σ_u

might be larger than the data suggest but not substantially larger. We can reasonably call this weakly informative. We implement this:

```
sdres <- sd(reeds$nitrogen)
pcprior <- list(prec = list(prior="pc.prec", param = c(3*sdres,0.01)))
formula <- nitrogen ~ f(site, model="iid", hyper = pcprior)
pmod <- inla(formula, family="gaussian", data=reeds)
```

The fixed effects summary is:

```
pmod$summary.fixed
```

```
              mean     sd 0.025quant 0.5quant 0.975quant   mode        kld
(Intercept) 3.0293 0.3029     2.4097   3.0293     3.6496 3.0293 2.6993e-08
```

This is very similar to the output for the default prior model. Now consider the hyperparameters:

```
bri.hyperpar.summary(pmod)
```

```
                                     mean       sd  q0.025    q0.5  q0.975    mode
SD for the Gaussian observations 0.28643 0.055213 0.19808 0.27897 0.41419 0.26369
SD for site                      0.46552 0.217727 0.18267 0.41674 1.02417 0.33783
```

The error component SD is only slightly different from the default prior-based model. But we see a substantial difference for the site SD. We plot the posteriors as seen in the first panel of Figure 5.2. The posterior distribution of σ_u is stochastically larger than for the default prior case.

```
bri.hyperpar.plot(pmod)
```

User defined prior: We can also define our own prior using INLA. A half Cauchy distribution for the standard deviation is a popular choice but is not among those coded in INLA. The half Cauchy density, with scale parameter λ, is

$$p(\sigma|\lambda) = \frac{2}{\pi\lambda(1+(\sigma/\lambda)^2)}, \qquad \sigma \geq 0.$$

We need to make a suitable choice of λ. If we use the same reasoning as for the PC prior, this leads to:

```
(lambda <- 3*sdres/tan(pi*0.99/2))
```

```
[1] 0.022066
```

INLA works with the precision and the calculation requires the log density of the precision, τ, which is

$$\log p(\tau|\lambda) = -\frac{3}{2}\log\tau - \log(\pi\lambda) - \log(1+1/(\tau\lambda^2)).$$

The half Cauchy prior is then defined using:

```
halfcauchy <- "expression:
              lambda = 0.022;
              precision = exp(log_precision);
              logdens = -1.5*log_precision-log(pi*lambda)-
                        log(1+1/(precision*lambda^2));
              log_jacobian = log_precision;
              return(logdens+log_jacobian);"
```

The code within the quotes looks like R but is written in muparser. This library parses mathematical expressions for implementation in C++ within INLA. The set of expressions it can understand is much more limited than R but is enough for our purposes. If you want to define another prior, you will need to refer to the muparser website to discover which functions are available. We also need to hardcode the scale λ into the expression. The internal representation uses $\log \tau$ so we must account for the change of variables by including a Jacobian.

We now use this prior for the random effect SD:

```
hcprior <- list(prec = list(prior = halfcauchy))
formula <- nitrogen ~ f(site, model="iid", hyper = hcprior)
hmod <- inla(formula, family="gaussian", data=reeds)
bri.hyperpar.summary(hmod)
```

```
                                   mean       sd  q0.025    q0.5  q0.975    mode
SD for the Gaussian observations 0.28763 0.056198 0.19809 0.28006 0.41769 0.26436
SD for site                      0.42336 0.231601 0.14712 0.36566 1.03014 0.28080
```

The results are comparable to the penalized complexity prior.

We cannot claim that this prior, the PC prior, the default prior, or any number of other priors we might reasonably propose are definitively best. In some cases, we can argue that the result is insensitive to the choice of the prior. For the intercept and the error SD, this is a reasonable claim. For the random effect SD, we see that the prior does make a difference. This is not surprising since we have only three groups and estimating an SD with so little information is sure to be subject to considerable uncertainty. The selection of the prior and the likelihood model requires subjective judgment. This is unavoidable. Since we must choose, we choose the PC prior on the grounds that it uses information and principles that are defensible. The half Cauchy is similar to the PC prior and is a good alternative. In contrast, the default prior is harder to defend.

5.2.2 Random Effects

Although we did not express a particular interest in the three sites, we can extract the posterior distributions of u_1, u_2 and u_3. These are sometimes called the random effects and can be extracted from the INLA model object. We compute the densities on a common grid so that they can be plotted together as seen in Figure 5.2. We can see that the nitogren content at the sites is in the order $B < A < C$. We are reasonably confident of this ordering but there is some overlap in the distributions. We use the PC prior version of the model.

```
reff <- pmod$marginals.random
x <- seq(-1.5,1.5,len=100)
d1 <- inla.dmarginal(x, reff$site[[1]])
d2 <- inla.dmarginal(x, reff$site[[2]])
d3 <- inla.dmarginal(x, reff$site[[3]])
rdf <- data.frame(nitrogen=x,density=c(d1,d2,d3),site=gl(3,100,
                  labels=LETTERS[1:4]))
ggplot(rdf, aes(x=nitrogen, y=density, linetype=site))+geom_line()
```

Almost the same plot may be obtained as

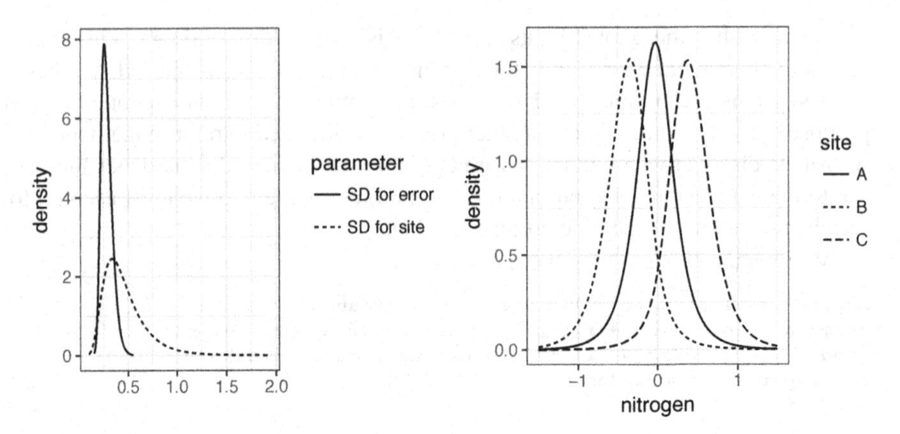

FIGURE 5.2
Posterior densities of the SDs for the random components are shown on the left. Posterior densities of the random effects for the nitrogen content at the three sites are shown on the right.

```
bri.random.plot(pmod)
```

We might be interested in whether site C has more nitrogen content than site A. We might hope to calculate the probability of this using the marginal distributions of the random effects as seen in Figure 5.2. But there is some correlation between the random effect posteriors and so this calculation would be incorrect. We can answer questions of this nature by drawing samples from the joint posterior. Earlier we used `inla.hyperpar.sample()` to obtain samples of the hyperparameters. But now we need samples from the latent variables as well. We take 1000 samples. It is necessary to recompute the model to obtain the information necessary for resampling:

```
sdres <- sd(reeds$nitrogen)
pcprior <- list(prec = list(prior="pc.prec", param = c(3*sdres,0.01)))
formula <- nitrogen ~ f(site, model="iid", hyper = pcprior)
pmod <- inla(formula, family="gaussian", data=reeds,
        control.compute=list(config = TRUE))
psamp <- inla.posterior.sample(n=1000, pmod)
psamp[[1]]
```

```
$hyperpar
Precision for the Gaussian observations                    Precision for site
                                13.5261                                8.5909


$latent
                sample1
Predictor:01  3.07707
Predictor:02  3.08278
...excised...
Predictor:15  3.26561
site:A        0.19414
site:B       -0.31424
```

```
site:C          0.37918
(Intercept)     2.88533
```

The latent variables contain all 15 predictors but we are interested in the sites for our problem:

```
lvsamp <- t(sapply(psamp, function(x) x$latent))
colnames(lvsamp) <- row.names(psamp[[1]]$latent)
mean(lvsamp[,'site:C'] > lvsamp[,'site:A'])
```

```
[1] 0.987
```

We see that there is very strong probability that site C contains more nitrogen than site A. The probability is larger than one might expect from the plot as there is a positive correlation between the two random effects. This technique can be used to answer various questions about the hyperparameters and the latent variables.

5.3 Longitudinal Data

Data where observations are collected on individuals over time are called longitudinal. Observations on a particular individual will not be independent but may show some common pattern of variation over time. The response for an individual may depend on observed characteristics but may also show unexplained variation specific to that individual. In Singer and Willett (2003), reading scores on a Peabody Individual Achievement test (PIAT) for 89 children, measured at 6.5, 8.5 and 10.5 years of age, are reported. We load this data and plot:

```
data(reading, package="brinla")
ggplot(reading, aes(agegrp, piat, group=id)) + geom_line()
```

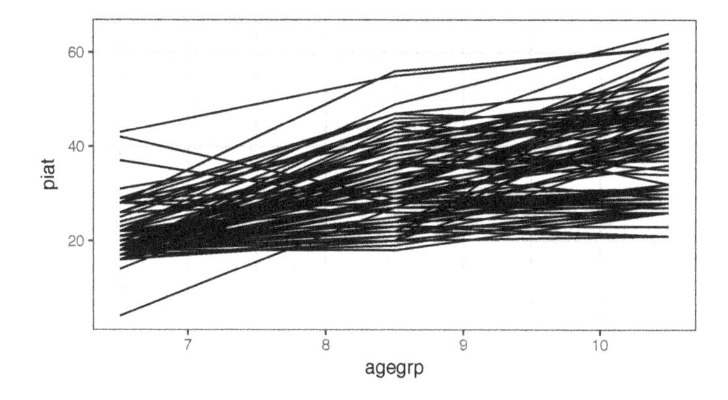

FIGURE 5.3
Reading scores of 89 students as they vary across three ages of measurement.

As can be seen in Figure 5.3, reading scores increase over time. Although there is some variation, we see that children who score well initially, continue to score well and vice versa. This indicates that the observations within an individual are not independent. We need a model that reflects this structure. The simplest way to do this is a model with a random intercept.

5.3.1 Random Intercept

The model is:

$$y_{ij} = \beta_0 + \beta_1 t_j + \alpha_i + \varepsilon_{ij} \qquad i = 1, \ldots, 89. \quad j = 1, 2, 3.$$

Time (or age in the example) is measured at three times, t_1, t_2, t_3. The parameters β_0 and β_1 are the so-called fixed effects and are the common intercept and slope for all students. A Gaussian prior is used for these. The random components are $\alpha_i \sim N(0, \sigma_\alpha^2)$ and $\varepsilon_{ij} \sim N(0, \sigma_\varepsilon^2)$. The hyperparameters σ_α^2 and σ_ε^2 require the specification of priors.

Before using INLA to fit the model, it is worth considering the corresponding likelihood-based model:

```
library(lme4)
lmod <- lmer(piat ~ agegrp + (1|id), reading)
summary(lmod)
```

```
Random effects:
 Groups   Name        Variance Std.Dev.
 id       (Intercept) 29.8     5.46
 Residual             44.9     6.70
Number of obs: 267, groups:  id, 89

Fixed effects:
            Estimate Std. Error t value
(Intercept)  -11.538      2.249   -5.13
agegrp         5.031      0.251   20.04
```

Now, we fit the model using INLA. We use the default priors (and subsequent analysis suggests these are adequate):

```
formula <- piat ~ agegrp + f(id, model="iid")
imod <- inla(formula, family="gaussian", data=reading)
```

The random intercept term is represented using `f(id, model="iid")` where `id` distinguishes the students. We use `model="iid"` because we believe that the students are independent and have a common variance. We can look at the fixed effect summary as:

```
imod$summary.fixed
```

```
                mean      sd 0.025quant 0.5quant 0.975quant     mode        kld
(Intercept) -11.5350 2.25815   -15.9727 -11.5351    -7.1013 -11.5350 5.8942e-13
agegrp        5.0306 0.25258     4.5341   5.0306     5.5265   5.0306 7.3789e-13
```

Because the fixed effect posterior distributions are approximately Gaussian, the mean, median and mode are identical. We see that the mean and SD are virtually the same as the likelihood model as would be expected. We see that reading scores increase about five points a year and we are quite sure about this given the small SD. The hyperparameters are more interesting. We use the `bri.hyperpar.summary()` function from our `brinla` package which converts the precisions used internally by INLA onto the SD scale which is interpretable:

```
bri.hyperpar.summary(imod)
```

```
                                   mean      sd q0.025   q0.5 q0.975   mode
SD for the Gaussian observations 6.7131 0.35476 6.0481 6.7001 7.4415 6.6752
SD for id                        5.2960 0.62761 4.1430 5.2681 6.6050 5.2263
```

These posterior distributions are not quite symmetrical so there are some differences with mean, median and mode. Maximum likelihood is a mode so we use this for comparison to the `lme4` output. We see the INLA values are slightly smaller. We have three observations per individual and hence about three times as much information for the estimation of σ^2_ε than σ^2_α. We see the distance between the 2.5th and 97.5th percentiles is smaller for the former. The posterior mean of σ_α is about five — similar to that for β_1. So we see that variation between students at a given point in time is comparable to the amount they might be expected to improve over one year. Finally, we see that σ^2_α is clearly bigger than zero so we cannot claim the 89 students are all the same with some measurement error thrown in. We can look at the posterior densities (plot not shown) with:

```
bri.hyperpar.plot(imod)
```

We might be interested in individual students. Information about individuals can be found in the posterior distributions of the α_i. These are sometimes called the random effects and can be found in `imod$summary.random$id`. We look at the summary of the posterior means of these:

```
summary(imod$summary.random$id$mean)
```

```
  Min. 1st Qu.  Median    Mean 3rd Qu.    Max.
-7.890  -2.940  -0.145   0.000   2.650  14.000
```

The mean is zero as would be expected given the specification of the model. From the quartiles, we see that about half the students have mean scores within about $[-3,3]$ of the overall mean. One student is 14 points above the mean. We can examine the whole posterior distributions for the random effects:

```
bri.random.plot(imod)
```

The plot (not shown) has too many individuals to separately identify but it will be difficult to claim differences between most students.

5.3.2 Random Slope and Intercept

We can see from Figure 5.3 that the slopes also vary and we would want to incorporate this into our model. We can do this as:

$$y_{ij} = \beta_0 + \beta_1 t_j + \alpha_{0i} + \alpha_{1i} t_j + \varepsilon_{ij}, \qquad i = 1,\ldots,89, \quad j = 1,2,3,$$

where we have added the slope variation as $\alpha_{1i} \sim N(0, \sigma^2_{\alpha_1})$. For comparison purposes, the `lme4`-based model is:

```
reading$cagegrp <- reading$agegrp - 8.5
lmod <- lmer(piat ~ cagegrp + (cagegrp|id), reading)
summary(lmod)
```

```
Random effects:
 Groups   Name        Variance Std.Dev. Corr
 id       (Intercept) 35.72    5.98
          cagegrp      4.49    2.12     0.83
 Residual             27.04    5.20
Number of obs: 267, groups:  id, 89

Fixed effects:
            Estimate Std. Error t value
(Intercept)   31.225      0.709    44.0
cagegrp        5.031      0.297    16.9
```

We have centered the age by its mean value of 8.5. This makes the interpretation of the intercept more useful since it represents the response at age 8.5 and not age zero as before. The model contains a correlation between the slope and intercept random effects, estimated at 0.83. Indeed, we can see from the plot of the data that individuals with higher intercepts tend also to have greater slopes so this is a necessary component of the model. We must ensure that our INLA model has this same feature and this requires some preparation. We must create another set of student labels, one for each but with a different label from the original set.

```
nid <- length(levels(reading$id))
reading$numid <- as.numeric(reading$id)
reading$slopeid <- reading$numid + nid
```

The students are labelled 1 to 89 which we will use for the intercept component. We create another set, labelled from 90 to 178, corresponding to the same students in the same order that we will use for the slope component. Now we are ready to fit the INLA model:

```
formula <- piat ~ cagegrp + f(numid, model="iid2d", n = 2*nid) +
              f(slopeid, cagegrp, copy="numid")
imod <- inla(formula, family="gaussian", data=reading)
```

We want to specify $(\alpha_{0i}, \alpha_{1i})$ as bivariate normal which is achieved using `model="iid2d"`. The total number of random effect terms is twice the number of students as stated by the `n=2*nid` part. The `f(slopeid, cagegrp, copy="numid")` term achieves the linear dependence on `cagegrp`. The `copy` part ensures that a correlation is modeled between the two random components. If we omit this part, no correlation will be included. Note that the term `copy` is somewhat of a misnomer since the two random components will be different, just correlated.

First we examine the fixed effects terms:

```
imod$summary.fixed
```

```
               mean      sd 0.025quant 0.5quant 0.975quant    mode        kld
(Intercept) 31.2241 0.69742    29.8516  31.2241    32.5950 31.2241 1.1102e-12
cagegrp      5.0305 0.29246     4.4551   5.0305     5.6052  5.0305 1.1838e-12
```

We can compare this to the result for the random intercept only model. The fixed effect intercept is different because we have centered on age. The posterior mean is about the same as is commonly the case when we change only the random structure. Notice that the posterior distribution is a little more dispersed due to the use of more parameters in this model. We can also obtain the summary for the random component:

```
bri.hyperpar.summary(imod)
```

```
                                 mean       sd  q0.025   q0.5  q0.975    mode
SD for the Gaussian observations 5.5126 0.304548 4.94350 5.5008 6.13965 5.47820
SD for numid (component 1)       5.6930 0.556589 4.67245 5.6657 6.85768 5.61643
SD for numid (component 2)       1.9150 0.258807 1.44888 1.9000 2.46409 1.87329
Rho1:2 for numid                 0.9439 0.047314 0.81548 0.9572 0.98661 0.97448
```

The posterior mean of the SD for the slope variation among individuals is 1.92. We can compare this to the overall slope of 5.03. This gives us a sense of how much variation there is between individual students in the rate at which their reading scores improve. We notice that the credible interval for this SD bounds the distribution well away from zero so we are sure there is variation between individuals in this respect.

The correlation between the slope and intercept SDs is 0.94. This means that students who are above average will tend to increase their reading score over time at a greater rate than students who are less able readers. We see that the credible interval for this term means we are sure this correlation is strong.

We can find information about the posterior distributions of $(\alpha_{0i}, \alpha_{1i})$ in `imod$summary.random$numid`. It is interesting to combine this with information found in the posterior distribution of the fixed effects found in `imod$summary.fixed` to obtained predicted means for individual students. We extract, combine and plot these as follows:

```
postmean <- matrix(imod$summary.random$numid[,2],nid,2)
postmean <- sweep(postmean,2,imod$summary.fixed$mean,"+")
p <- ggplot(reading, aes(cagegrp, piat, group=id)) +
     geom_line(col=gray(0.95)) + xlab("centered age")
p+geom_abline(data=postmean,intercept=postmean[,1],slope=postmean[,2])
```

The resulting plot is shown in Figure 5.4. We see how students who score well initially tend to improve at a faster rate than those who do not. We could also extract the full posterior for particular students if we so desired. We can also plot the distributions of the hyperparameters:

```
library(gridExtra)
sd.epsilon <- bri.hyper.sd(imod$internal.marginals.hyperpar[[1]],
              internal=TRUE)
sd.intercept <- bri.hyper.sd(imod$internal.marginals.hyperpar[[2]],
                internal=TRUE)
sd.slope <- bri.hyper.sd(imod$internal.marginals.hyperpar[[3]],
            internal=TRUE)
p1 <- ggplot(data.frame(sd.epsilon),aes(x,y))+geom_line()+
      ggtitle("Epsilon")+xlab("piat")+ylab("density")
p2 <- ggplot(data.frame(sd.intercept),aes(x,y))+geom_line()+
      ggtitle("Intercept")+xlab("piat")+ylab("density")
p3 <- ggplot(data.frame(sd.slope),aes(x,y))+geom_line()+
      ggtitle("Slope")+xlab("piat")+ylab("density")
p4 <- ggplot(data.frame(imod$marginals.hyperpar[[4]]),aes(x,y))+
      geom_line()+ggtitle("Rho")+ylab("density")
```

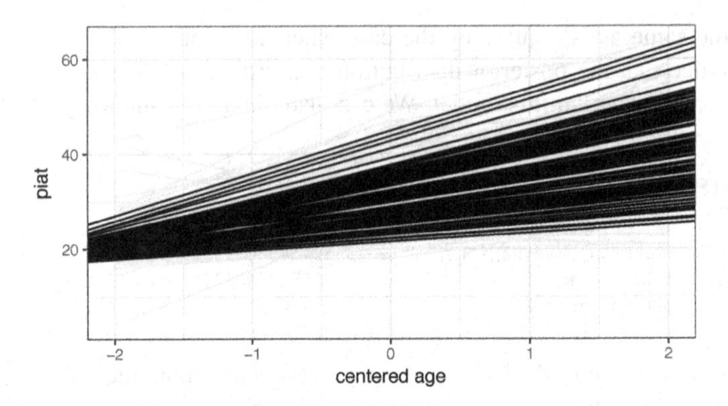

FIGURE 5.4
Posterior predicted mean profiles for individuals are shown as solid lines. The data are shown as gray lines.

```
grid.arrange(p1,p2,p3,p4,ncol=2)
```

Almost the same plot can be obtained more quickly with:

```
bri.hyperpar.plot(imod, together=FALSE)
```

The resulting plots are seen in Figure 5.5. The SDs are on different scales so we have not plotted them together. In this dataset, we have 89 individuals so we are not surprised to see posteriors which are relatively compact and normally shaped. We see clearly that the correlation is strong. We have seen that strong students get even stronger as they age when considering the reading scores in an absolute sense. We can ask the same question in a relative sense by considering a logged response. We refit the model and look at the posterior distribution of the correlation:

```
formula <- log(piat) ~ cagegrp + f(numid, model="iid2d", n = 2*nid) +
          f(slopeid, cagegrp, copy="numid")
imod <- inla(formula, family="gaussian", data=reading)
bri.density.summary(imod$marginals.hyperpar[[4]])
```

```
    mean        sd    q0.025      q0.5    q0.975      mode
0.059009  0.112198 -0.162033  0.058883  0.275997  0.060206
```

We see that the 95% credible interval constructed from the two extreme quantiles includes zero. Although there is some weak suggestion of a positive correlation, we cannot be sure of this. This is in contrast to the previous unlogged response model where the correlation was very strong. Based on this, we can reasonably claim that stronger students maintain a constant relative advantage over weaker students. There is no sense that the rich get richer.

5.3.3 Prediction

We can generate predictions for new cases by appending the known predictor values, together with missing values for the response, to the data frame for the original data.

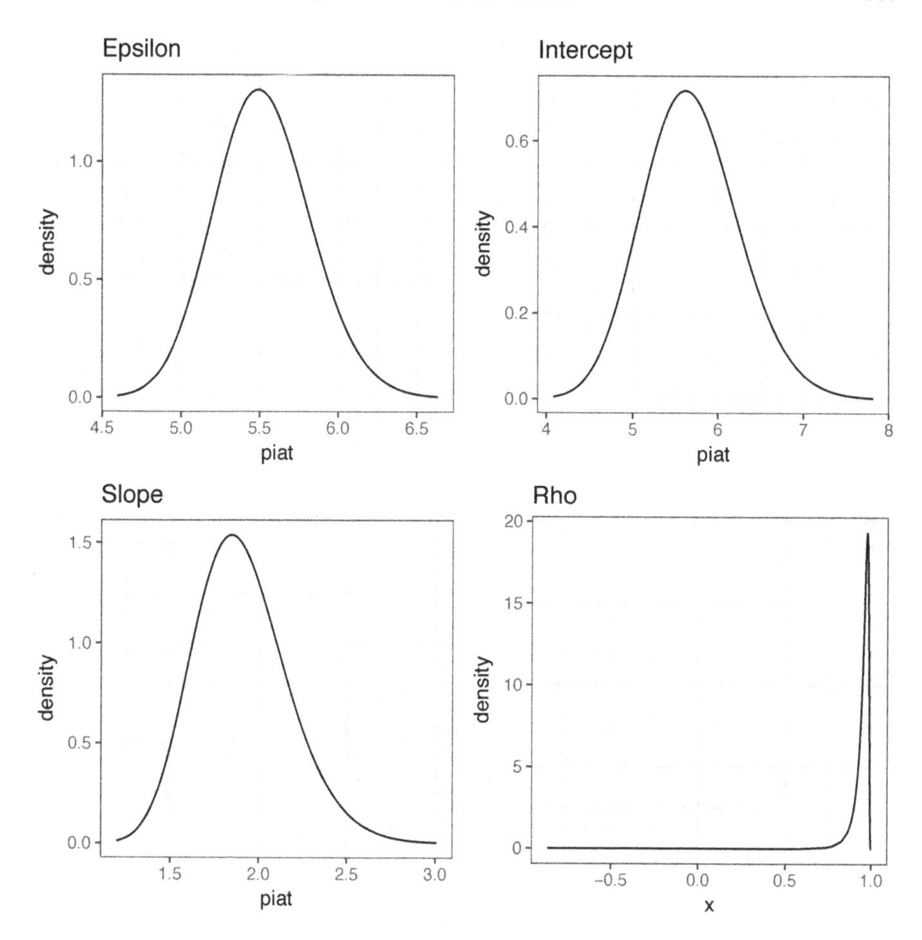

FIGURE 5.5
Posterior distributions of the hyperparameters.

Suppose we have a new student who has scored 18 and 25 at the first two points of measurement and we are interested in what will happen at the final point of measurement. We create a small data frame with this information and append it to the original dataset (which we reload to clear out the additional variables we created above):

```
data(reading, package="brinla")
reading$id <- as.numeric(reading$id)
newsub <- data.frame(id=90, agegrp = c(6.5,8.5,10.5),
          piat=c(18, 25, NA))
nreading <- rbind(reading, newsub)
```

Since we are adding a new individual with the label 90, it is easier if we specify the `id` for the students as numeric rather than factor. Now we refit the model as before:

```
formula <- piat ~ agegrp + f(id, model="iid")
```

```
imod <- inla(formula, family="gaussian", data=nreading,
        control.predictor = list(compute=TRUE))
pm90 <- imod$marginals.fitted.values[[270]]
p1 <- ggplot(data.frame(pm90),aes(x,y))+geom_line()+xlim(c(20,60))
```

There are $90 \times 3 = 270$ cases in the data. Only the last of these has an unknown response. The posterior distribution for this fitted value (the linear predictor) is saved and plotted (not shown yet as we plan to add another prediction). Notice that the two new cases are used in the construction of the model. If we believe that the student is exchangeable with the original set, then this is sensible. But in some situations, we might wish to base the prediction solely on the original data (without the partial information from the new individual). We can achieve this by the appropriate use of the `weights` argument to the `inla()` function. In this example, because we have 89 students in the original data and are considering just one more, it will not make much difference which option we take.

Another kind of prediction would involve a new student for whom we have no information. We can achieve this by setting all three responses for this student as missing values:

```
newsub=data.frame(id=90, agegrp = c(6.5,8.5,10.5), piat=c(NA, NA, NA))
nreading <- rbind(reading, newsub)
```

We fit the model again and construct the prediction. We put this together with the previous prediction and show this in Figure 5.6:

```
formula <- piat ~ agegrp + f(id, model="iid")
imodq <- inla(formula, family="gaussian", data=nreading, control.
    ↪ predictor = list(compute=TRUE))
qm90 <- imodq$marginals.fitted.values[[270]]
p1+geom_line(data=data.frame(qm90),aes(x,y),linetype=2)+
   xlab("PIAT")+ylab("density")
```

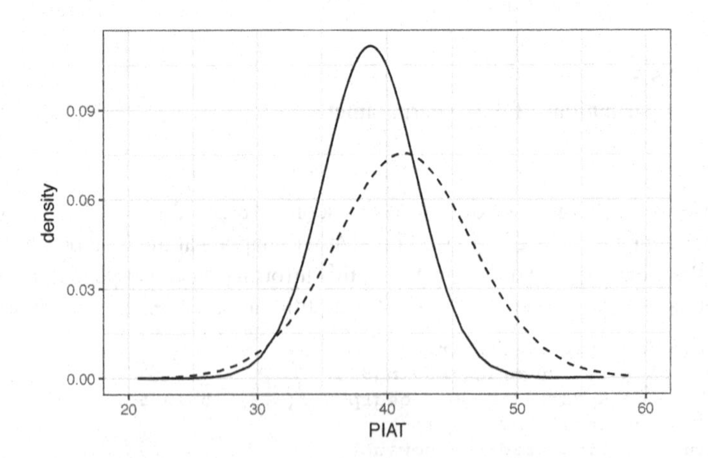

FIGURE 5.6
Predictive distributions for individual with partial response information (solid) and no response information (dashed).

We see that for the first individual, who has achieved relatively weak reading scores at the first two timepoints, the predictive distribution is shifted below the overall mean of about 41 expected at age 10.5years. In contrast, we see that the predictive distribution for the individual without known response information is centered on this expected mean. For the former distribution, we have some information about the student so the prediction is better (has a more concentrated distribution). We have no information in the latter case, so the predictive distribution has greater spread.

The predictive distributions above refer only to the linear predictor so we are expressing only the uncertainty in this component. A future observation will add an ε to the linear predictor which will require knowledge of the uncertainty in ε. There are two ways in which we might accomplish this. Conveniently, the future ε should be independent of the linear predictor and we know its distribution via the posterior on the corresponding hyperparameter for its precision. In principle, we can combine the two densities to get the full predictive distribution. In practice, this requires some effort to implement and so we might resort to sampling. We can sample from the posteriors for the linear predictor and for ε, add the samples and estimate the density from these. This is inelegant but effective.

We draw 10,000 samples. This is more than necessary but the computation is fast so we can be extravagant. We draw samples from the posterior for the precision of the error: $1/\sigma_\varepsilon^2$. We convert this to an SD and then sample from normal densities with these randomly drawn SDs. We combine these randomly generated new εs with samples from posterior for the mean response that we computed earlier (where we have partial information about the student). We compute the density and add it to the plot of the density for the mean response that we computed earlier as seen in Figure 5.7.

```
nsamp <- 10000
randprec <- inla.hyperpar.sample(nsamp, imod)[,1]
neweps <- rnorm(nsamp, mean=0, sd=1/sqrt(randprec))
newobs <- inla.rmarginal(nsamp, pm90) + neweps
dens <- density(newobs)
p1 + geom_line(data=data.frame(x=dens$x,y=dens$y),aes(x,y),linetype=2)
```

We see that there is greater variation in the predictive density for the new response. In this case, we could approximate the predictive density with a normal distribution but in general, one needs to be careful. The distribution of the new ε is not normal since the SD is random.

5.4 Classical Z-Matrix Model

In Section 5.1, we introduced the representation of the model as

$$y = \mathbf{X}\beta + \mathbf{Z}u + \varepsilon. \tag{5.2}$$

In the examples that followed, we did not directly use the Z matrix but used a more explicit representation of the random effects. Let's see how we can specify the Z

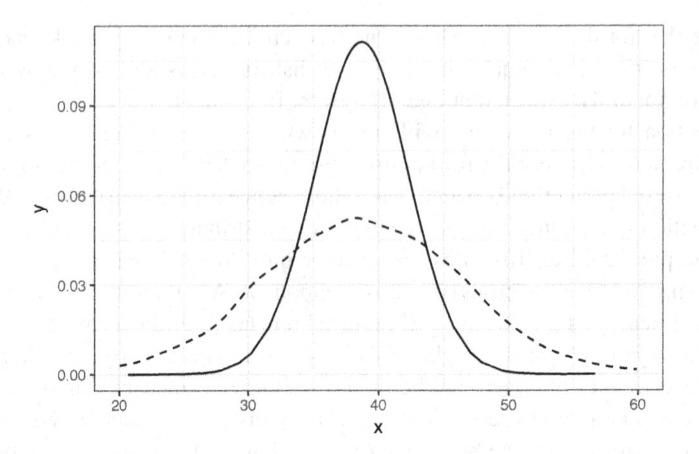

FIGURE 5.7
Posterior predictive densities for an individual with partial response information. The density for the mean response is shown as a solid line and the density for a new observation is shown with a dashed line.

matrix directly to fit some mixed effect models. We have called it the classical approach because it dates back to Henderson (1982) who introduced this formulation. The `lme4` package provides a useful way to construct the X and Z matrices which we demonstrate using the first example of the chapter:

```
library(lme4)
mmod <- lmer(nitrogen ~ 1+(1|site), reeds)
Z <- getME(mmod, "Z")
X <- getME(mmod, "X")
```

We need to create an indicator variable for the cases which we call `id.z`. The Z model is selected using the `model="z"` option to `f()`.

```
n <- nrow(reeds)
formula <- y ~ -1 + X +  f(id.z, model="z",  Z=Z)
imodZ <- inla(formula, data = list(y=reeds$nitrogen, id.z = 1:n, X=X))
```

We have used the default setting although more structure can be supplied for the random effects. The u is a vector of length m which is distributed $N(0, \tau C)$ where τ is the precision and C is a fixed $m \times m$ matrix. In this example, C is the identity matrix. As usual, we find it easier to look at the SDs rather than the precisions:

```
bri.hyperpar.summary(imodZ)
```

```
                                   mean      sd  q0.025    q0.5  q0.975    mode
SD for the Gaussian observations 0.29079 0.05833 0.19817 0.28283 0.42608 0.26647
SD for id.z                      0.31313 0.15740 0.10893 0.27909 0.71369 0.22175
```

which is the same result as seen before in Section 5.1. We can extract summary information regarding the random effects as:

```
imodZ$summary.random
```

```
$id.z
   ID       mean        sd 0.025quant    0.5quant 0.975quant        mode       kld
1   1 -0.0041086 0.085306  -0.209637  0.00016443   0.156661 -0.00015857 0.0052616
... ditto ...
6   6 -0.0504086 0.147214  -0.530901 -0.00406940   0.025792 -0.00097854 0.0010167
... ditto ...
11 11  0.0533301 0.154185  -0.027216  0.00339512   0.557732  0.00069156 0.0015510
... ditto ...
16 16 -0.0041052 0.085314  -0.209677  0.00018201   0.156701 -0.00016665 0.0052413
17 17 -0.0503191 0.147077  -0.530555 -0.00406628   0.025763 -0.00099760 0.0010821
18 18  0.0531454 0.153943  -0.027167  0.00337379   0.557112  0.00069253 0.0016927
```

We have deleted all but the first of each set of five random effects because they are the same within each group. Internally, INLA uses an "augmented" model where the first n values are the random effects $v \sim N_n(Zu, \kappa I)$ where κ is a fixed high precision. The remaining m values are the u. Thus in this example, the last three cases essentially reiterate the random effects seen previously. In other examples, this may not be so.

We have not achieved anything new over the analysis of this data seen in Section 5.1 but we now have a different tool for fitting mixed effects models that allows some new functionality. We can also implement the random intercept model described in Section 5.3.1 but not the the random slope and intercept model of Section 5.3.2 which requires hyperparameters for the intercept and slope along with a correlation between them. The Z matrix formulation allows only a single hyperparameter so we are stymied. Nonetheless, we have other uses for this model as we see in the next section.

5.4.1 Ridge Regression Revisited

In the linear regression model, $y = Zu + \varepsilon$, least squares estimates of the parameters u can be unstable when Z has a high degree of correlation. To reduce this instability, we can penalize the size of u by minimizing:

$$(y - Zu)^T (y - Zu) + \lambda \sum_{j=1}^{p} u_j^2.$$

The second part is the penalty term which prevents the u from being too large. This method is called *ridge regression*. The parameter λ controls the degree of penalization and can be chosen using cross-validation in the frequentist framework. We have already seen one approach to ridge regression in Section 3.7.

We can achieve a similar effect by putting informative priors on the u. In the mixed effects model seen in (5.2), the Zu is the usual linear model predictor while $X\beta$ becomes simply an intercept term. We set priors $u \sim N(0, \sigma_u^2 I)$ and $\varepsilon \sim N(0, \sigma^2 I)$ which corresponds to $\lambda = \sigma^2/\sigma_u^2$. Let's see how this works with an example.

A total of 215 samples of finely chopped meat were measured. For each sample, we have the fat content as the response with 100 predictors which are absorbances across a range of frequencies. Since determining the fat content via analytical chemistry is time consuming, we would like to build a model to predict the fat content of new samples using the 100 absorbances which can be measured more easily. See

Thodberg (1993) for more about the origin of the data and Faraway (2014) for other analyses of the same data.

The true performance of any model is hard to determine based on just the fit to the available data. We need to see how well the model does on new data not used in the construction of the model. For this reason, we will partition the data into two parts — a *training sample* consisting of the first 172 observations that we will use to build and estimate the models and a *testing sample* of the remaining 43 observations.

```
data(meatspec, package="brinla")
trainmeat <- meatspec[1:172,]
testmeat <- meatspec[173:215,]
wavelengths <- seq(850, 1050, length=100)
```

By way of comparison, we fit a standard linear model to the training data and evaluate the RMSE of prediction on the test data:

```
modlm <- lm(fat ~ ., trainmeat)
rmse <- function(x,y) sqrt(mean((x-y)^2))
rmse(predict(modlm,testmeat), testmeat$fat)
```

```
[1] 3.814
```

We plot the estimated coefficients of this linear model as seen in the first panel of Figure 5.8.

```
plot(wavelengths,coef(modlm)[-1], type="l",ylab="LM Coefficients")
```

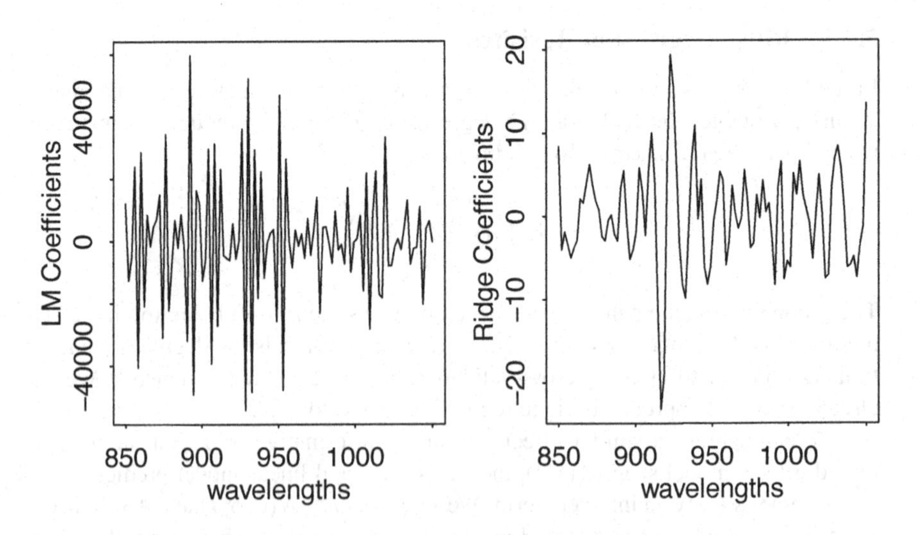

FIGURE 5.8
Coefficients for models fitting the meat spectroscopy data. The linear model is shown on the left and the ridge regression on the right.

We would expect some continuity in the effect as the wavelength varies but the least squares estimates vary greatly, indicating the instability caused by the strong collinearity.

We can implement ridge regression using the Bayesian model by creating the respective X and Z matrices. In this case, X is just a column of ones representing an intercept. The formula contains a `-1` because this intercept is already included. Z is the matrix of predictors. We assume no knowledge of the response in the test set so y has missing values for these cases. We need to include the test set in the computation so that predictive distributions will be generated for these cases using the option `control.predictor = list(compute=TRUE)`. The test set is not used in actually fitting the model. The method performs better if the response is scaled into roughly a $[-1,1]$ range.

```
n <- nrow(meatspec)
X <- matrix(1,nrow = n, ncol= 1)
Z <- as.matrix(meatspec[,-101])
y <- meatspec$fat
y[173:215] <- NA
scaley <- 100
formula <- y ~ -1 + X +  f(idx.Z, model="z", Z=Z)
zmod <- inla(formula, data = list(y=y/scaley, idx.Z = 1:n, X=X),
    ↪ control.predictor = list(compute=TRUE))
```

We now extract the posterior means and compute the RMSE of prediction (taking into account the scaling):

```
predb <- zmod$summary.fitted.values[173:215,1]*scaley
rmse(predb, testmeat$fat)
```

```
[1] 1.9028
```

The RMSE is half the size it was for the ordinary linear model. We can also plot the posterior means for the coefficients as seen in the second panel of Figure 5.8.

```
rcoef <- zmod$summary.random$idx.Z[216:315,2]
plot(wavelengths, rcoef, type="l", ylab="Ridge Coefficients")
```

Compared to the linear model, we see the variation has been much reduced (even allowing for scaling) and there is some continuity in the coefficient as wavelength varies.

Linear combination of a linear predictor. We can implement ridge regression using a different approach within INLA. The approach is worth knowing as it can be used in other situations. Usually we have a response depending on a linear predictor. But suppose that linear predictor depended on another linear combination. Concretely,

$$EY = \eta' \text{ and } \eta' = A\eta,$$

where Y and η' are vectors of length n, η is a vector of length m and A is an $n \times m$ matrix. For the ridge regression example, we can set:

$$A = [X : Z] \text{ and } \eta = \begin{pmatrix} \beta_0 \\ u \end{pmatrix}.$$

We need priors for β_0, u and ε which we set via the precisions:

```
int.fixed <- list(prec = list(initial = log(1.0e-9), fixed=TRUE))
u.prec <- list(prec = list(param = c(1.0e-3, 1.0e-3)))
epsilon.prec <- list(prec = list(param = c(1.0e-3, 1.0e-3)))
```

For the intercept, the prior is set on the log precision. We set this precision to be very small indicating very little information about the intercept. By fixing this, we generate no additional hyperparameter and accept this precision as fixed. For u and ε we specify weakly informative log-gamma priors.

We now create index sets for β_0 and u. η is length 101 with the first element corresponding to β_0 and the remaining 100 corresponding to u.

```
idx.X <- c(1, rep(NA,100))
idx.Z <- c(NA, 1:100)
```

We fit the model with:

```
scaley <- 100
formula <- y ~ -1 + f(idx.X,  model="iid", hyper = int.fixed) + f(idx.
    ↪ Z,  model="iid", hyper = u.prec)
amod <- inla(formula, data = list(y=y/scaley, idx.X=idx.X, idx.Z=idx.Z
    ↪ ),
        control.predictor = list(A=cbind(X, Z),compute=TRUE),
        control.family = list(hyper = epsilon.prec))
```

and check the prediction performance with:

```
predb <- amod$summary.fitted.values[173:215,1]
rmse(predb, testmeat$fat/scaley)*scaley
```

```
[1] 1.9019
```

The performance is very similar to that seen for the mixed effects model based approach although the priors are not entirely in agreement between the two.

Ridge regression can also be implemented in INLA with a linear regression model where more informative priors are directly imposed on the coefficients. The drawback is that we must specify how informative these priors should be, whereas in the approaches we have taken, the amount of shrinkage in the coefficients is built into the model fitting process. For another use of the mixed effects Z model, see Section 11.1.1.

5.5 Generalized Linear Mixed Models

Generalized linear mixed models (GLMM) combine the idea of GLMs as seen in Chapter 4 with the mixed modeling ideas seen earlier in this chapter. Suppose we have a response Y_i, $i = 1, \ldots, n$, that follows an exponential family distribution with $\mathrm{E}(Y_i) = \mu_i$. This is connected to the linear predictor η using a link function g by $\eta_i = g(\mu_i)$.

Let the random effects, u, have corresponding design matrix $\mathbf{Z}$, then

$$\eta = \mathbf{X}\beta + \mathbf{Z}u.$$

Provided we assign Gaussian priors to β and assume a Gaussian distribution for the random effects u, this falls with the latent Gaussian model framework required to use INLA. GLMM models are not necessarily easy to fit using maximum likelihood and we may find the Bayesian approach finds acceptable solutions where the purely likelihood method may struggle.

5.6 Poisson GLMM

In Davison and Hinkley (1997), the results of a study on Nitrofen, a herbicide, are reported. Due to concern regarding the effect on animal life, 50 female water fleas were divided into five groups of ten each and treated with different concentrations of the herbicide. The number of offspring in three subsequent broods for each flea was recorded. We start by loading the data from the `boot` package:

```
data(nitrofen, package="boot")
head(nitrofen)
```

```
  conc brood1 brood2 brood3 total
1    0      3     14     10    27
2    0      5     12     15    32
3    0      6     11     17    34
4    0      6     12     15    33
5    0      6     15     15    36
6    0      5     14     15    34
```

It is more convenient to construct a data frame with one response value per line. We drop the total variable and add an identifier of the flea:

```
library(dplyr)
library(tidyr)
lnitrofen <- select(nitrofen, -total) %>%
             mutate(id=1:nrow(nitrofen)) %>%
             gather(brood,live,-conc,-id) %>%
             arrange(id)
lnitrofen$brood <- factor(lnitrofen$brood,labels=1:3)
head(lnitrofen)
```

```
  conc id brood live
1    0  1     1    3
2    0  1     2   14
3    0  1     3   10
4    0  2     1    5
5    0  2     2   12
6    0  2     3   15
```

We construct a plot of the data as seen in Figure 5.9. We need to offset the concentrations horizontally a little to distinguish the broods. Some vertical jittering is also needed to avoid overplotting cases with the same response. We see that for the first brood, the number of offspring remains relatively constant while for the second and third broods, numbers appear to decrease with increasing concentration of Nitrofen.

```
lnitrofen$jconc <- lnitrofen$conc + rep(c(-10,0,10),50)
ggplot(lnitrofen, aes(x=jconc,y=live, shape=brood)) +
       geom_point(position = position_jitter(w = 0, h = 0.5)) +
       xlab("Concentration")
```

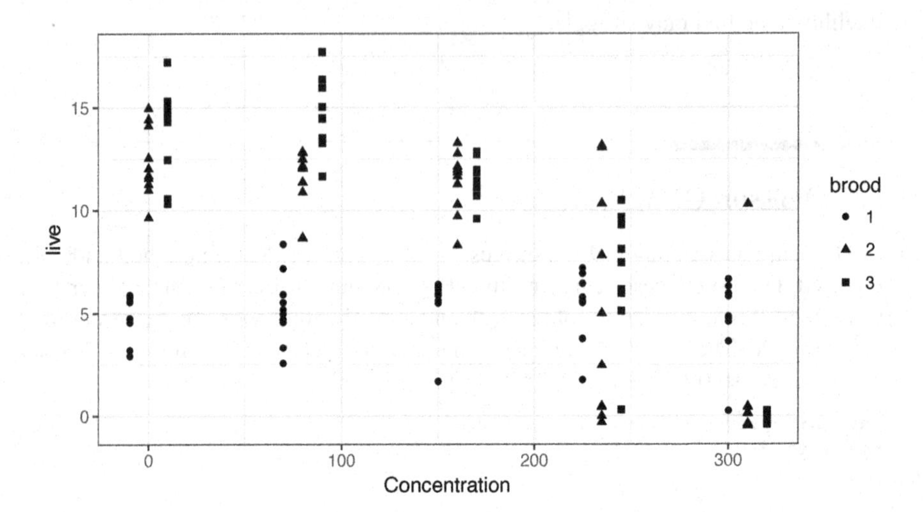

FIGURE 5.9
The number of live offspring varies with the concentration of Nitrofen and the brood number.

Since the response is a small count, a Poisson model is a natural choice. We expect the rate of the response to vary with the brood and concentration level. The plot of the data suggests these two predictors may have an interaction. The three observations for a single flea are likely to be correlated. We might expect a given flea to tend to produce more, or less, offspring over a lifetime. We can model this with an additive random effect. The linear predictor is:

$$\eta_i = x_i^T \beta + u_{j(i)}, \quad i = 1,\ldots,150. \quad j = 1,\ldots 50,$$

where x_i is a vector from the design matrix encoding the information about the ith observation and u_j is the random affect associated with the jth flea. The response has distribution $Y_i \sim Poisson(\exp(\eta_i))$.

For comparison purposes, we fit a model using penalized quasi-likelihood (PQL) using the `lme4` package:

```
library(lme4)
glmod <- glmer(live ~ I(conc/300)*brood + (1|id), nAGQ=25,
          family=poisson, data=lnitrofen)
summary(glmod)
```

```
Random effects:
 Groups Name        Variance Std.Dev.
 id     (Intercept) 0.0911   0.302
Number of obs: 150, groups:  id, 50
```

```
Fixed effects:
                    Estimate Std. Error z value Pr(>|z|)
(Intercept)           1.6386     0.1367   11.99  < 2e-16
I(conc/300)          -0.0437     0.2193   -0.20     0.84
brood2                1.1688     0.1377    8.48  < 2e-16
brood3                1.3512     0.1351   10.00  < 2e-16
I(conc/300):brood2  -1.6730      0.2487   -6.73  1.7e-11
I(conc/300):brood3  -1.8312      0.2451   -7.47  7.9e-14
```

We scaled the concentration by dividing by 300 (the maximum value is 310) to avoid scaling problems encountered with `glmer()`. This is helpful in any case since it puts all the parameter estimates on a similar scale. The first brood is the reference level so the slope for this group is estimated as -0.0437 and is not statistically significant, confirming the impression from the plot. We can see that numbers of offspring in the second and third broods start out significantly higher for zero concentration of the herbicide, with estimates of 1.1688 and 1.3512. But as concentration increases, we see that the numbers decrease significantly, with slopes of -1.6730 and -1.8312 relative to the first brood. The individual SD is estimated at 0.302 which is noticeably smaller than the estimates above, indicating that the brood and concentration effects outweigh the individual variation.

The same model, with default priors, can be fitted with INLA as:

```
formula <- live ~ I(conc/300)*brood + f(id, model="iid")
imod <- inla(formula, family="poisson", data=lnitrofen)
```

The fixed effects summary is:

```
imod$summary.fixed
```

```
                        mean      sd 0.025quant  0.5quant 0.975quant      mode
(Intercept)         1.639493 0.13601    1.36772  1.640988    1.90288  1.644054
I(conc/300)        -0.041413 0.21791   -0.47273 -0.040485    0.38426 -0.038687
brood2              1.164071 0.13757    0.89773  1.162776    1.43766  1.160183
brood3              1.346245 0.13496    1.08541  1.344819    1.61515  1.341962
I(conc/300):brood2 -1.664137 0.24824   -2.15576 -1.662707   -1.18086 -1.659822
I(conc/300):brood3 -1.821494 0.24470   -2.30637 -1.819992   -1.34536 -1.816964
```

The posterior means are very similar to the PQL estimates. We can also see the summary for the random effect SD:

```
bri.hyperpar.summary(imod)
```

```
              mean       sd  q0.025    q0.5  q0.975    mode
SD for id 0.29399 0.056598 0.19169 0.29058 0.41462 0.28463
```

Again the result is very similar to the PQL output although notice that INLA provides some assessment of uncertainty in this value in contrast to the PQL result. We can make some conclusions about the strength of these effects from just looking at the numerical summaries but it is better to check the posterior densities as seen in Figure 5.10.

```
library(reshape2)
mf <- melt(imod$marginals.fixed)
cf <- spread(mf,Var2,value)
names(cf)[2] <- 'parameter'
```

```
ggplot(cf,aes(x=x,y=y)) + geom_line()+facet_wrap(~ parameter,
       scales="free") + geom_vline(xintercept=0) + ylab("density")
```

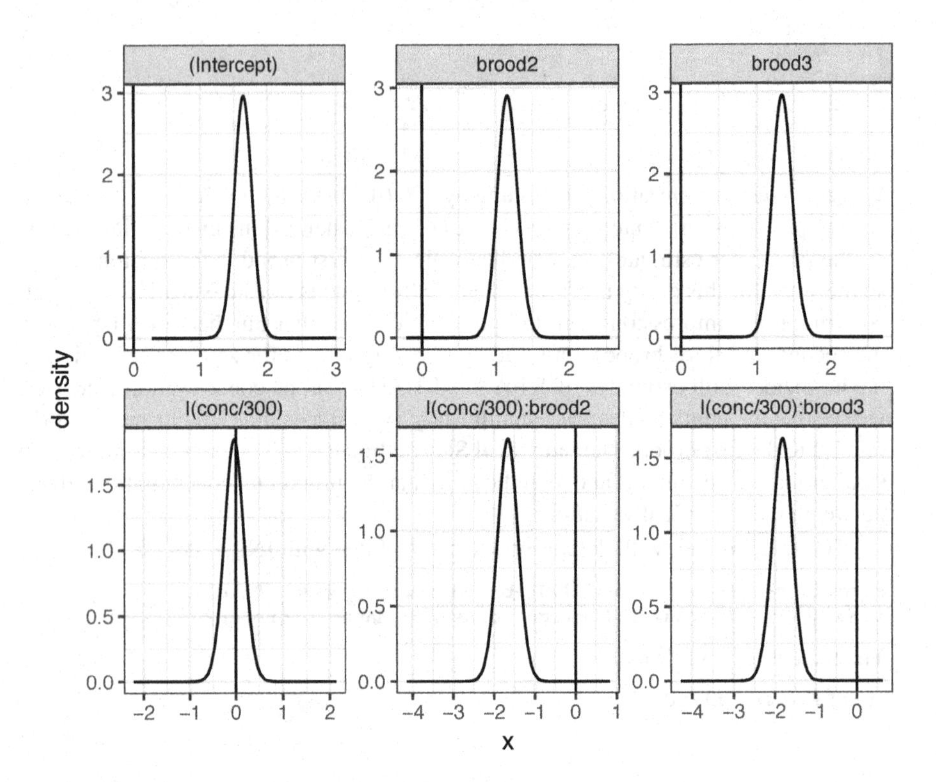

FIGURE 5.10
Posterior densities of the fixed effects model for the Nitrofen data.

Almost the same plot can be produced by: (plots not shown)

```
bri.fixed.plot(imod)
```

or on a single panel as (which works because the scales of the parameters are similar):

```
bri.fixed.plot(imod,together=TRUE)
```

We see that, at concentration zero, broods two and three clearly have more offspring. The densities are well separated from zero so there is no doubt about this. We also see that the slopes for broods two and three and clearly negative, indicating that the offspring decrease substantially relative to brood one as concentration increases. We also see that zero falls right in the middle of the density for the slope for brood one, indicating that offspring in this brood remain about constant as the concentration changes.

Since the Poisson model uses the log link by default, it can be helpful to exponentiate the parameters to interpret the scale of the effects:

```
multeff <- exp(imod$summary.fixed$mean)
```

```
names(multeff) <- imod$names.fixed
multeff[-1]
```

```
  I(conc/300)          brood2          brood3 I(conc/300):brood2 I(conc/300):brood3
      0.95943         3.20294         3.84296            0.18935            0.16178
```

We see that there are three to four times as many offspring in the second and third broods while numbers of offspring for these broods drop more than 80% from no Nitrofen to the highest used concentration.

We should also examine the posterior for the sole hyperparameter in this model that is the precision of the flea random effect. As usual, it is more convenient to express this as an SD as seen in the first panel of Figure 5.11:

```
sden <- data.frame(bri.hyper.sd(imod$marginals.hyperpar[[1]]))
ggplot(sden,aes(x,y)) + geom_line() + ylab("density") +
       xlab("linear predictor")
```

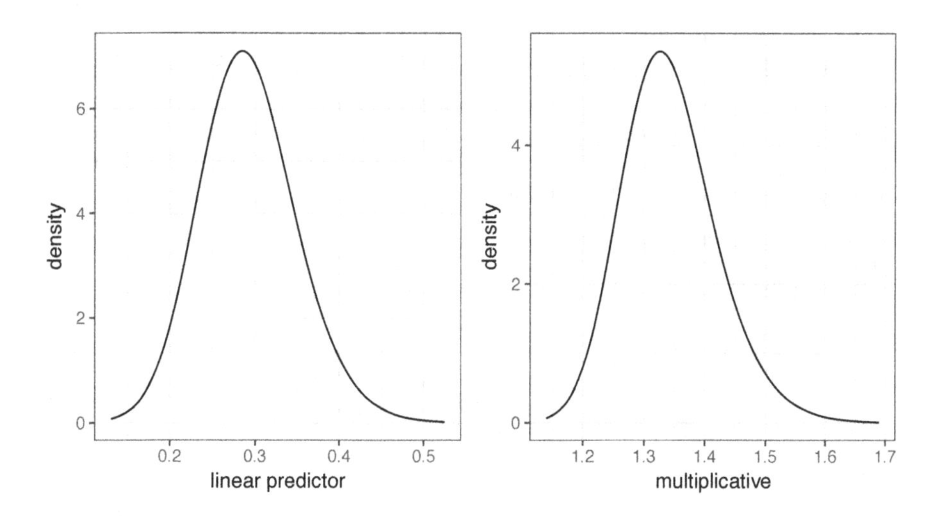

FIGURE 5.11
Posterior densities of the hyperparameter. SD on the linear predictor scale is shown on the left. SD on the rate multiplicative scale is shown on the right.

Or more simply as:

```
bri.hyperpar.plot(imod)
```

As with the fixed effects, it can be easier to interpret this by converting it to a multiplier on the Poisson rate. We can compute and plot this posterior as seen in the second panel of Figure 5.11:

```
mden <- data.frame(inla.tmarginal(function(x) exp(1/sqrt(x)),
        imod$marginals.hyperpar[[1]]))
ggplot(mden,aes(x,y)) + geom_line() + ylab("density") +
       xlab("multiplicative")
```

We see that a typical flea effect would be about 30% more or less offspring.

There seems little justification to attempt a simplification of the model given the salience of the effects we see in the posterior densities. Even so, we might question the linearity of the concentration effect and consider whether a model using a log scale for this predictor might be preferable. We can investigate this possibility by computing the DIC for the competing models:

```
formula <- live ~ I(conc/300)*brood + f(id, model="iid")
imod <- inla(formula, family="poisson", data=lnitrofen,
        control.compute=list(dic=TRUE))
formula <- live ~ log(conc+1)*brood + f(id, model="iid")
imod2 <- inla(formula, family="poisson", data=lnitrofen,
         control.compute=list(dic=TRUE))
c(imod$dic$dic,  imod2$dic$dic)
```

```
[1] 785.90 841.97
```

We see that the original model gives a smaller DIC and so we shall stick with that. Note that it is necessary to specifically ask for the computation of the DIC as this is not computed by default.

Are there any observations or fleas which are unusual? Each u_j has a complete posterior distribution which we can examine but for simplicity we can extract the posterior means. We need to check for values which are unexpectedly large or small. As we expect these posterior means to be approximately normal, a QQ plot is a natural graphical check. This is shown in Figure 5.12:

```
mreff <- imod$summary.random$id$mean
qqnorm(mreff)
qqline(mreff)
```

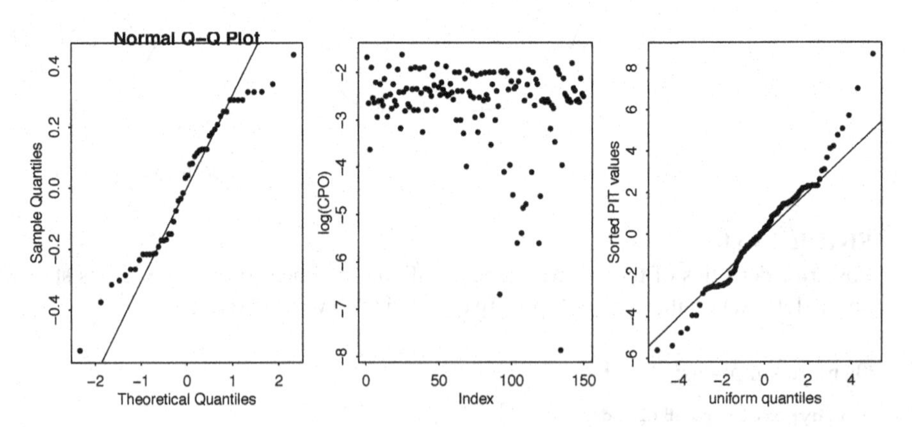

FIGURE 5.12
Diagnostic checks on the Nitrofen model: First panel shows a QQ plot of the posterior mean random effects. The second panel shows an index plot of the CPO statistics on a log scale. The third panel shows a uniform QQ plot for the PIT statistics on a logit scale.

We see that there are no particularly unusual fleas. Some produce more or less offspring but no point is outstanding.

We can check the individual observations in two ways. First consider the conditional predictive ordinate (CPO) statistics. This is $P(y_i|y_{-i})$ and is a "leave-out-one" style predictive measure of fit. Low values of this statistic should draw attention. INLA requires that we specifically ask for the computation of these statistics. We plot these on a log scale so that values particularly close to zero will be more distinguished:

```
formula <- live ~ I(conc/300)*brood + f(id, model="iid")
imod <- inla(formula, family="poisson", data=lnitrofen,
        control.compute=list(cpo=TRUE))
plot(log(imod$cpo$cpo),ylab="log(CPO)")
```

The plot, shown in the second panel of Figure 5.12, reveals several points with low probability, the smallest of which is this point:

```
lnitrofen[which.min(imod$cpo$cpo),]
```

```
    conc id brood live
134  310 45     2   10
```

We consider this case among all the other high concentration, second brood observations:

```
lnitrofen %>% filter(brood == 2, conc==310)
```

```
   conc id brood live
1   310 41     2    0
2   310 42     2    0
3   310 43     2    0
4   310 44     2    0
5   310 45     2   10
6   310 46     2    0
7   310 47     2    0
8   310 48     2    0
9   310 49     2    0
10  310 50     2    0
```

We see that the other nine fleas have zero offspring while this flea has ten. This is certainly unusual.

We may also compute the Probability Integral Transform (PIT) statistics, $P(y_i^{new} \leq y_i|y_{-i})$. We would expect these values to be approximately uniformly distributed under the supposed model. We are particularly interested in values close to zero or one. To magnify the salience of these points of interest, we plot the PIT values against the uniform order statistics on a logit scale:

```
pit <- imod$cpo$pit
n <- length(pit)
uniquant <- (1:n)/(n+1)
logit <- function(p) log(p/(1-p))
plot(logit(uniquant), logit(sort(pit)), xlab="uniform quantiles",
     ylab="Sorted PIT values")
abline(0,1)
```

One advantage of the PIT relative to the CPO is that the deviations have a direction. In this case, we can see that more of the unusual cases occur where there are more offspring than expected. In particular, the most extreme PIT value is:

```
which.max(pit)
```

```
[1] 134
```

This is the same case as for the CPO statistics.

Some care is necessary when using CPO or PIT statistics. When computing leave-out-one statistics, one wants to avoid refitting the model n times to explicitly compute these. But this does mean some approximation is necessary. We can check the quality of these approximations using `imod$cpo$failure` where non-zero values indicate some degree of suspicion. In our example, these statistics are all zero so we have no worries. In case of a problem, use the `inla.cpo()` function to improve the approximation.

We also can investigate the effect of changing the priors. It is usually not worth the trouble of altering the default flat priors on the fixed effect parameters. We might consider something other than the default priors on the hyperparameters. Let's see the effect of using a penalized complexity prior here:

```
sdu <- 0.3
pcprior <- list(prec = list(prior="pc.prec", param = c(3*sdu,0.01)))
formula <- live ~ I(conc/300)*brood + f(id, model="iid",
           hyper = pcprior)
imod2 <- inla(formula, family="poisson", data=lnitrofen)
bri.hyperpar.summary(imod2)
```

```
                mean       sd  q0.025   q0.5  q0.975    mode
SD for id 0.30993 0.056161 0.20885 0.3064 0.42992 0.30017
```

We have calibrated the prior by putting in a guess for the standard deviation of the random effect. The exponential prior is then adjusted so that there is 0.01 probability that the actual SD is more than three times larger. We have chosen our guess of 0.3 based on the likelihood model fit earlier. In practice, it would be better to generate this guess from expert opinion sought prior to data collection.

As it happens, the posterior density is very similar to the previous result derived from the default prior. A little experimentation reveals that the posterior is really quite insensitive to the prior. This is reassuring. The choice of this prior is usually not too important unless we have a small dataset or when the random effect variation is relatively small.

We can generalize from the Poisson model by adding some overdispersion. We modify the linear predictor to:

$$\eta_i = x_i^T \beta + u_{j(i)} + \varepsilon_i, \quad i = 1, \ldots, 150, \quad j = 1, \ldots 50,$$

where $\varepsilon_i \sim N(0, \sigma_\varepsilon^2)$. Hence each linear predictor term has an added independent random component ε. We create an index variable for the observation and incorporate this into the model:

```
lnitrofen$obsid <- 1:nrow(nitrofen)
formula <- live ~ I(conc/300)*brood + f(id, model="iid") +
           f(obsid, model="iid")
imodo <- inla(formula, family="poisson", data=lnitrofen)
bri.hyperpar.summary(imodo)
```

```
                  mean       sd   q0.025      q0.5   q0.975      mode
SD for id     0.293803 0.056124 0.195348 0.2898453 0.415022 0.2835296
SD for obsid 0.010284 0.006119 0.003764 0.0085126 0.026903 0.0062088
```

We see that the SD of the added random component is much smaller than the flea SD. Although we have some assurance that it is not zero, it is small enough that we could reasonably ignore it. Hence no substantial overdispersion appears to be present.

In 1996, the use of Nitrofen was banned in the USA and the EU because of its teratogenic effects.

5.7 Binary GLMM

In Fitzmaurice and Laird (1993), data on 537 children aged 7–10 in six Ohio cities are reported. The response is binary — does the child suffer from wheezing (indication of a pulmonary problem) where one indicates yes and zero no. This status is reported for each of four years at ages 7, 8, 9 and 10. There is also an indicator variable for whether the mother of the child is a smoker. Because we have four binary responses for each child, we expect these to be correlated and our model needs to reflect this.

We sum the number of smoking and non-smoking mothers:

```
data(ohio, package="brinla")
table(ohio$smoke)/4
```

```
  0   1
350 187
```

We use this to produce the proportion of wheezing children classified by age and maternal smoking status:

```
xtabs(resp ~ smoke + age, ohio)/c(350,187)
```

```
     age
smoke    -2    -1     0     1
    0 0.160 0.149 0.143 0.106
    1 0.166 0.209 0.187 0.139
```

Age has been adjusted so that nine years old is zero. We see that wheezing appears to decline with age and that there may be more wheezing in children with mothers who smoke. But the effects are not clear and we need modeling to be sure about these conclusions.

A plausible model uses a logit link with a linear predictor of the form:

$$\eta_{ij} = \beta_0 + \beta_1 age_j + \beta_2 smoke_i + u_i, \quad i = 1, \ldots, 537, \quad j = 1, 2, 3, 4,$$

with

$$P(Y_{ij} = 1) = \frac{\exp(\eta_{ij})}{1 + \exp(\eta_{ij})}.$$

The random effect u_i models the propensity of child i to wheeze. Children are likely to vary in their health condition and this effect enables us to include this unknown variation in the model. Because u_i is added to all four observations for a child, we induce a positive correlation among the four responses as we might naturally expect. The response is Bernoulli or, in other words, binomial with trial size one.

For reference, here is the model fit penalized quasi-likelihood using the `lme4` package:

```
library(lme4)
modagh <- glmer(resp ~ age + smoke + (1|id), nAGQ=25,
          family=binomial, data=ohio)
summary(modagh)
```

```
Random effects:
 Groups Name        Variance Std.Dev.
 id     (Intercept) 4.69     2.16
Number of obs: 2148, groups:  id, 537

Fixed effects:
            Estimate Std. Error z value Pr(>|z|)
(Intercept)  -3.1015     0.2191  -14.16   <2e-16
age          -0.1756     0.0677   -2.60   0.0095
smoke         0.3986     0.2731    1.46   0.1444
```

As with INLA, this method also requires some numerical integration, so the computation is not as fast as likelihood methods that need only optimization.

We can fit this model in INLA as:

```
formula <- resp ~ age + smoke + f(id, model="iid")
imod <- inla(formula, family="binomial", data=ohio)
```

The `id` variable represents the child and we use an `iid` model indicating that the u_i variables should be independent and identically distributed between children. A summary of the posteriors for the fixed effect components can be obtained as:

```
imod$summary.fixed
```

```
                mean       sd 0.025quant 0.5quant 0.975quant     mode        kld
(Intercept) -2.99250 0.201804  -3.409624 -2.98495  -2.618752 -2.96974 1.5266e-13
age         -0.16665 0.062817  -0.290565 -0.16646  -0.043913 -0.16607 1.7805e-14
smoke        0.39133 0.239456  -0.077747  0.39069   0.863404  0.38946 1.2226e-12
```

The posteriors for the fixed effects tend to be approximately normal so there is little difference between mean, median and mode. The posterior means are also quite similar to the PQL output. We see from the 2.5% and 97.5% percentiles that the risk of wheezing clearly decreases with age. The parameters are on a logit scale so interpretation can be aided by some transformation. For example, the intercept represents the response for `age=0, smoke=0`, i.e., a nine year old with a non-smoking mother. We can invert the logit transform as:

```
ilogit <- function(x) exp(x)/(1 + exp(x))
ilogit(imod$summary.fixed[1,c(3,4,5)])
```

```
            0.025quant 0.5quant 0.975quant
(Intercept)   0.031996  0.04811   0.067941
```

We can compare this to the proportion seen in the data as calculated earlier of 0.143. It seems our model does not fit particularly well in this instance. A technical concern here is that a monotone transformation, such as the inverse logit, will preserve the quantiles of the transformed density. This would not work for the mean or the mode. We would need to recompute the density first on the transformed scale, then find the mean or the mode. Typically, there will not be much difference with transforming the original summary statistics but some caution is advisable.

We can better interpret the regression parameters by exponentiating as:

```
exp(imod$summary.fixed[-1, c(3,4,5)])
```

```
      0.025quant 0.5quant 0.975quant
age      0.74784  0.84666    0.95704
smoke    0.92520  1.47800    2.37122
```

We see that an additional year of age multiplies the odds of wheezing by 0.85 or, reduces it by 15%. The two outer quantiles provide a measure of uncertainty in this statement. Nevertheless, we see that the upper quantile is still below one, indicating that we are reasonably sure that wheezing decreases with age. A mother who smokes increases the odds of wheezing by a factor of 1.48 or 48%. However, the lower quantile lies below one, indicating we are not entirely sure whether smoking increases the odds of wheezing. But we also see the upper quantile is quite large, indicating an increase in odds of 137%. So while we cannot dismiss the possibility that a smoking mother is not harmful, we should also be concerned that it may be quite damaging. We need more data. Of course, we should also be cautious about claiming smoking causes the wheezing as it may simply be associated with other risk factors that are the true causal variables.

The posterior distribution of σ_u is also of interest:

```
bri.hyperpar.summary(imod)
```

```
              mean      sd q0.025 q0.5 q0.975   mode
SD for id 1.9256 0.15987 1.6269 1.92 2.2554 1.9127
```

As with the fixed effects, this is easier to interpret on an odds scale (only transforming the quantiles:)

```
exp(bri.hyperpar.summary(imod)[3:5])
```

```
[1] 5.0879 6.8213 9.5395
```

We see that one standard deviation would multiply the odds by about 7 (or equally possible divide them by 7). This means that the individual effects are very strong in this example. We can construct a table of the total number of times recorded as wheezing over the four timepoints for all 537 individuals:

```
table(xtabs(resp ~ id, ohio))
```

```
  0   1   2   3   4
355  97  44  23  18
```

We see that 355 children never wheeze. The model is consistent with this because the large random effect means some children will have a consistently low probability of wheezing while others will have only a modest probability of wheezing.

We can plot all the posteriors on a single display as seen in Figure 5.13.

```
library(gridExtra)
p1 <- ggplot(data.frame(imod$marginals.fixed[[1]]),aes(x,y)) +
      geom_line()+xlab("logit")+ylab("density")+ggtitle("Intercept")
```

```
p2 <- ggplot(data.frame(imod$marginals.fixed[[2]]),aes(x,y)) +
      geom_line()+xlab("logit")+ylab("density")+ggtitle("age")
p3 <- ggplot(data.frame(imod$marginals.fixed[[3]]),aes(x,y)) +
      geom_line()+xlab("logit")+ylab("density")+ggtitle("smoke")
sden <- data.frame(bri.hyper.sd(imod$marginals.hyperpar[[1]]))
p4 <- ggplot(sden,aes(x,y)) + geom_line() + xlab("logit") +
      ylab("density")+ggtitle("SD(u)")
grid.arrange(p1,p2,p3,p4,ncol=2)
```

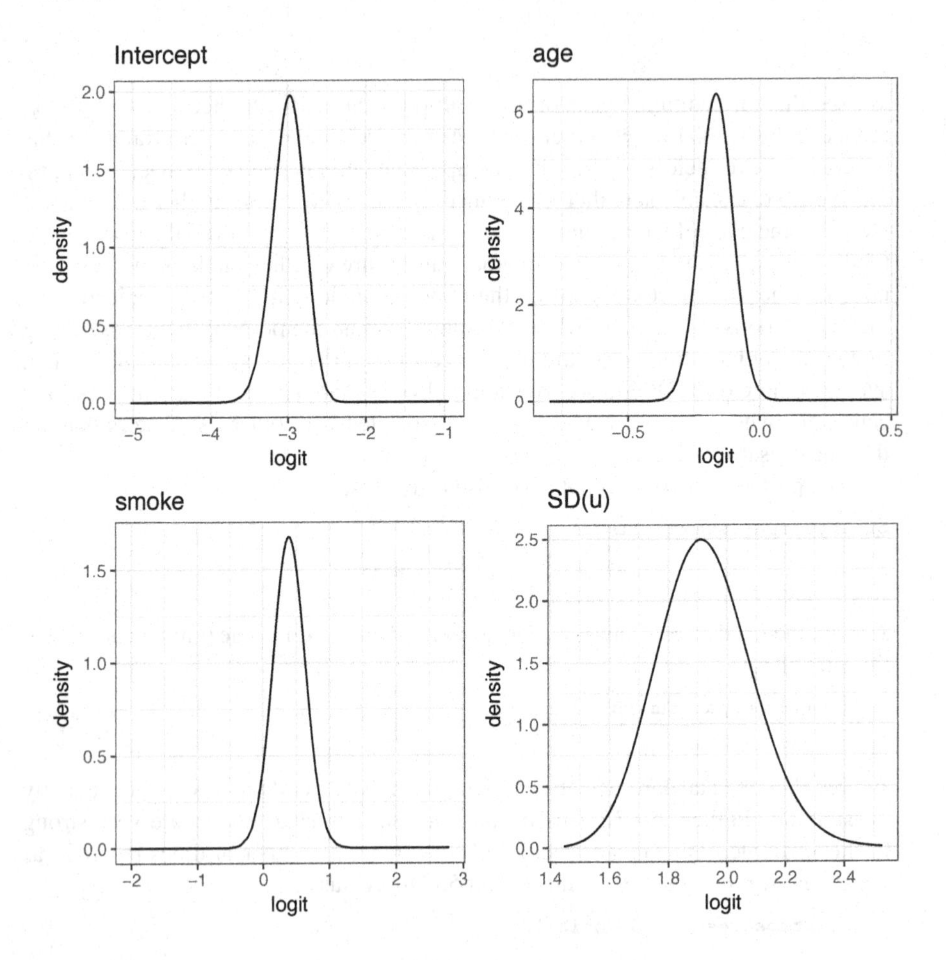

FIGURE 5.13
Posterior densities of the fixed effects and the random effect SD.

We can see that all the posterior densities are approximately normal so that the previous numerical summaries were adequate for interpretation purposes. Nevertheless, it is worth checking these to verify this. We can see that a small area of the density for the smoke coefficient extends below zero. We can calculate this area as:

```
inla.pmarginal(0,imod$marginals.fixed$smoke)
```

```
[1] 0.051348
```

Despite the apparent similarity, this is not a p-value. One objection is that, with the usual two-sided alternate hypothesis, we would need to double this value. But the more important objection concerns the meaning of this probability. It is the probability that the association between maternal smoking and wheezing status is negative. This is a more transparent statement than the meaning of a p-value which concerns the probability of observing this outcome, or one more extreme, under the null hypothesis of no smoking effect.

Given that there is no good reason to emulate a frequentist conclusion under a Bayesian model, one might wonder why we have computed this probability. A measure of how far the posterior density is from zero is a useful summary given the special meaning of zero of "no effect." In this case, the probability is a useful summary of our assessment of the association of smoking with wheezing. Indeed, some have called this a "Bayesian p-value" although definitions differ so it is best to be clear about what has been computed when communicating this information.

We noticed earlier that the predicted probability for one combination of inputs did not agree with the observed proportion. Indeed, simply looking at all the observed proportions for the eight combinations of age and smoke makes us suspect that the effect of age is not linear. We can investigate this by fitting more complex models and comparing these using criteria such as DIC and WAIC. We consider a model that treats age as a factor so that the effect of each age level can vary freely. We also consider a model which allows for an interaction between age and smoke. This exhausts the possibilities for the fixed effects as anything more complex would saturate the model. INLA requires that we specifically ask for the computation of DIC and WAIC:

```
formula <- resp ~ age + smoke + f(id, model="iid")
imod <- inla(formula, family="binomial", data=ohio,
        control.compute=list(dic=TRUE,waic=TRUE))
formula <- resp ~ factor(age) + smoke + f(id, model="iid")
imod1 <- inla(formula, family="binomial", data=ohio,
         control.compute=list(dic=TRUE,waic=TRUE))
formula <- resp ~ factor(age)*smoke + f(id, model="iid")
imod2 <- inla(formula, family="binomial", data=ohio,
         control.compute=list(dic=TRUE,waic=TRUE))
```

The first model is the same as our original but we need to recompute to get the DIC and WAIC. First we consider the DIC for the three models:

```
c(imod$dic$dic, imod1$dic$dic, imod2$dic$dic)
```

```
[1] 1466.2 1463.1 1463.4
```

and now the WAIC:

```
c(imod$waic$waic, imod1$waic$waic, imod2$waic$waic)
```

```
[1] 1418.9 1416.2 1417.6
```

We see that, in both cases, the minimum value is obtained for the model which treats age as a factor but with no smoke interaction. Fortunately, there is no disagreement between the two criteria but if a choice must be made, Gelman et al. (2014) argues that WAIC should be preferred. Let's examine the output for our chosen model. For reasons explained earlier, we transform to the odds scale:

```
exp(imod1$summary.fixed[-1, c(3,4,5)])
```

```
                0.025quant 0.5quant 0.975quant
factor(age)-1     0.74516  1.08376    1.57721
factor(age)0      0.65627  0.95966    1.40220
factor(age)1      0.38380  0.57798    0.86368
smoke             0.92485  1.48199    2.38534
```

Age 7 is the baseline and for ages 8 and 9, the median odds relative change is close enough to one (meaning no change). At age 10, we see a 42% drop in the odds of wheezing. The result and interpretation for smoke is almost the same as the original model. This effect is curious and deserves some explanation.

Given that we do not find the age effect to be linear, it makes little sense to consider generalizing the random effect term from the current constant effect to include a linear effect. Even so, we might wonder whether the subject random effect does vary much from year to year and we would like a model which investigates this effect. We can do this introducing an autoregressive model for the subject random effect:

$$u_{it} = \rho u_{i,t-1} + \varepsilon_t, \quad \varepsilon_t \sim N(0, \sigma^2), \quad t = 2,3,4.$$

See Section 3.8 for another autoregressive model. For the first time period, we have

$$u_{i1} \sim N(0, \sigma^2/(1-\rho^2)).$$

Thus each subject will have its own short, independent, time series but will share the same hyperparameters, σ and ρ. We can fit this model with:

```
ohio$obst <- rep(1:4,537)
ohio$repl <- ohio$id + 1
formula <- resp ~ factor(age) + smoke + f(obst, model="ar1",
           replicate = repl)
imod <- inla(formula, family="binomial", data=ohio)
```

We need to create the t index for each subject which is held in the variable `obst`. The `model="ar1"` creates the time series indexed by `obst`. The argument `replicate = repl` ensures that the short, four value, time series is replicated. Without this term, each subject would generate its own independent (σ, ρ) pair. We do not want this.

We can look at the fixed effects, using the odds as before:

```
exp(imod$summary.fixed[-1, c(3,4,5)])
```

```
                0.025quant 0.5quant 0.975quant
factor(age)-1     0.74423  1.08460    1.58173
factor(age)0      0.65320  0.95871    1.40587
factor(age)1      0.37913  0.57424    0.86225
smoke             0.92542  1.48603    2.39704
```

These are very similar to the previous model. Now the random effects:

```
bri.hyperpar.summary(imod)
```

```
                mean       sd  q0.025    q0.5  q0.975   mode
SD for obst  1.97247 0.169661 1.66322 1.96316 2.32947 1.9445
Rho for obst 0.98471 0.014741 0.94559 0.98874 0.99854 0.9966
```

The SD is about the same as before. The posterior for ρ is close to one. This indicates that the subject effect does not change too much over the four-year time period. We could interpret this as saying that the underlying health of a child remains relatively constant over the time period.

5.7.1 Improving the Approximation

INLA can be inaccurate for binary data as discussed in Fong et al. (2010). Improvements to rectify this problem can be found in Ferkingstad and Rue (2015). The method uses a copula-based correction to the Laplace approximations used in INLA. We can apply it to our binary GLMM example:

```
cmod <- inla(formula, family="binomial", data=ohio,
       control.inla = list(correct = TRUE))
cmod$summary.fixed
```

```
                mean       sd 0.025quant 0.5quant 0.975quant     mode        kld
(Intercept) -3.10775 0.213101  -3.548285 -3.09976  -2.713114 -3.08369 1.1735e-13
age         -0.17115 0.063615  -0.296643 -0.17096  -0.046861 -0.17056 1.6932e-14
smoke        0.40486 0.252137  -0.088977  0.40416   0.902016  0.40280 1.2035e-12
```

We see a small difference in the fixed effects. The effect of the correction is more noticeable on the posterior of σ_u where the previous value was 1.93.

```
bri.hyperpar.summary(cmod)
```

```
             mean      sd q0.025   q0.5 q0.975   mode
SD for id 2.0618 0.16657 1.7498 2.0562 2.4043 2.0486
```

The correction will only make much of a difference where approximation is difficult due to sparsity. Binary GLMMs are particularly challenging. But in other cases, such as the Poisson GLMM example in this chapter, the correction makes very little difference. Since the correction bears very little extra computational cost and will tend to improve the results (though sometimes not by much), we recommend that you apply the correction for all GLMMs (but not for other models). It is expected that the correction will be the default in the future, so you may find this advice redundant by the time you read it.

6

Survival Analysis

Survival analysis is a class of statistical methods for analyzing data where the outcome variable is the time until the occurrence of an event of interest. It is extremely useful for studying many different kinds of events in medicine, economics, engineering, sociology, including death, the onset of disease, equipment failures, job terminations, retirements, marriages, etc. One may be interested in characterizing the distribution of the "time-to-event" data for a given population as well as comparing the outcome among different groups, or modeling the relationship of the outcome to other covariates.

In most applications, the survival data are collected over a finite period of time. For example, in a cancer study, some patients may not have reached the endpoint of interest (death). Consequently, the exact survival times of these patients are not known. The only information is that the survival times are greater than the amount of time the patient has been in the study. The survival times of these patients are said to be *censored*, which creates the difficulty in the analysis of such data.

Since having been introduced by Cox (1972), the proportional hazard regression, also known as Cox regression model, is the default choice when dealing with time-to-event data. In its basic form, it describes the probability of the endpoint (known as *hazard*) as expressed by the baseline hazard function (unspecified) and a set of covariates that have linear effects. While traditional analysis relies on parameter estimation based on partial likelihood, Bayesian approaches for time-to-event data allow us to use the full likelihood to estimate all unknown elements in the model.

6.1 Introduction

Assume that the survival time, T, is a continuous random variable. The distribution of T can be described by the usual cumulative distribution function

$$F(t) = P(T \leq t), \quad t \geq 0,$$

which is the probability that a subject from the population will die (or a specific event of interest for a subject has occurred) before time t. The corresponding density function of T is, $f(t) = dF(t)/dt$.

In survival analysis, it is common to use the *survival function*

$$S(t) = 1 - F(t) = P(T > t),$$

which is the probability that a randomly selected subject will survive to time t or beyond. The survival function $S(t)$ is a non-increasing function over time taking on the value 1 at $t = 0$. For a random variable T, $S(\infty) = 0$, which means that everyone will eventually experience the event. Obviously, if T is a continuous random variable, there is a one-to-one correspondence between $S(t)$ and $f(t)$:

$$f(t) = -\frac{dS(t)}{dt}.$$

It is also of interest, in analyzing survival data, to assess which periods having high or low chances of the event among those still active at the certain time. A suitable method to characterize such risks is the *hazard function*, $h(t)$, defined by the following equation

$$h(t) = \lim_{s \to 0} \frac{P(t \leq T \leq t + s | T \geq t)}{s}.$$

It is the instantaneous rate of failure (experiencing the event) at the time t given that a subject is alive at the time t. The definition of the hazard function implies that

$$h(t) = \frac{f(t)}{S(t)} = -\frac{d}{dt} \log S(t).$$

A related quantity is the *cumulative hazard function*, $H(t)$, defined by

$$H(t) = \int_0^t h(u)du = -\log(S(t)).$$

And thus,

$$S(t) = \exp\{-H(t)\} = \exp\left\{-\int_0^t h(u)du\right\}.$$

When the response variable of interest is a possibly censored survival time, the most widely used regression techniques are Cox's proportional hazards models (Cox, 1972). In a typical survival study, the data, based on a sample size of n, consist of the triple $(t_i, \delta_i, \mathbf{x}_i)$, $i = 1, \ldots, n$, where t_i is the time on the study for the ith subject, δ_i is the event indicator for the ith subject ($\delta_i = 0$ if the time is right-censored, $\delta_i = 1$ if the event has occurred), and $\mathbf{x}_i = (x_{i1}, \ldots, x_{ip})^T$ is the vector of p-dimensional covariate values for the ith subject.

Let $h(t|\mathbf{x})$ be the hazard function at time t for a subject given the covariate vector $\mathbf{x} = (x_1, \ldots, x_p)^T$. The basic model proposed by Cox (1972) is as follows:

$$h(t|\mathbf{x}) = h_0(t) \exp(\beta_1 x_1 + \ldots + \beta_p x_p), \tag{6.1}$$

where $h_0(t)$ is the baseline hazard function and β_i's are the unknown regression parameters to be estimated. It is easy to see that the model (6.1) forces the hazard ratio between two subjects to be constant over time since

$$\frac{h(t|\mathbf{x}_a)}{h(t|\mathbf{x}_b)} = \frac{\exp(\beta_1 x_{a1} + \ldots + \beta_p x_{ap})}{\exp(\beta_1 x_{b1} + \ldots + \beta_p x_{bp})},$$

where $\mathbf{x}_a$ and $\mathbf{x}_b$ are vectors of covariate values for two subjects, a and b. Hence, the model (6.1) is called the *proportional hazards model.*

Depending on the assumptions about the baseline hazard function $h_0(t)$, different kinds of proportional hazard models can be specified.

6.2 Semiparametric Models

6.2.1 Piecewise Constant Baseline Hazard Models

In Bayesian survival analysis, the semiparametric proportional hazard model assigns a nonparametric prior to the baseline hazard function. One of the most convenient methods is to construct a piecewise constant baseline hazard model for $h_0(t)$ (Breslow, 1972).

Consider the proportional hazards model of the form (6.1). Let us construct the semiparametric model starting from a finite partition of the time axis. We partition the time axis into K intervals with cutpoints $0 = s_0 < s_1 < s_2 < ... < s_K$ with $s_K < \max\{t_i, i = 1,...,n\}$. Then we assume that the baseline hazard is constant within each interval,

$$h_0(t) = \lambda_k, \qquad t \in (s_{k-1}, s_k),\ k = 1,...,K.$$

Thus, the hazard rate for subject i with time $t_i \in (s_{k-1}, s_k]$ is:

$$\begin{aligned} h(t_i) &= h_0(t_i)\exp(\beta_1 x_{i1} + ... + \beta_p x_{ip}) \\ &= \exp(\log(\lambda_k) + \beta_1 x_{i1} + ... + \beta_p x_{ip}), \qquad t_i \in (s_{k-1}, s_k]. \end{aligned}$$

Denote $\eta_{ik} = \log(\lambda_k) + \beta_1 x_{i1} + ... + \beta_p x_{ip}$. The log-likelihood function for the ith observation can be written as

$$\begin{aligned} \log\left[h(t_i)^{\delta_i} S(t_i)\right] &= \delta_i \log h(t_i) - \int_0^{t_i} h(u)du \\ &= \delta_i \eta_{ik} - (t_i - s_k)\exp(\eta_{ik}) - \sum_{j=1}^{k-1}(s_{j+1} - s_j)\exp(\eta_{ij}). \qquad (6.2) \end{aligned}$$

The INLA approach is not directly applicable to such a model. However, Martino et al. (2011) pointed out that this semiparametric model still can be fit into the INLA framework after rewriting it. Note that (6.2) is equivalent to the log-likelihood of k Poisson-distributed "augmented data points," where the first two terms on the right-hand side of (6.2) can be seen as the log-likelihood from a Poisson distribution with mean $(t_i - s_k)\exp(\eta_{ik})$ observed to be 0 or 1 according to δ_i, and the third term in the right-hand side of (6.2) can be seen as the log-likelihood from Poisson distribution with mean $(s_{j+1} - s_j)\exp(\eta_{ij})$ observed to be zero.

In INLA, each original data point (t_i, δ_i) with $t_i \in (s_{k-1}, s_k]$ is reconstructed by k Poisson-distributed data points in an augmented dataset at its background process.

Such data augmentation brings us back to the latent Gaussian models with INLA algorithms.

To perform a regression analysis for time-to-event data in INLA, we need to define an object of class "`inla.surv`". It creates a survival object, to be used as a response variable in a model formula for the `inla` function for a variety of survival models.

Let us take a look at an example of analyzing the data of 1151 subjects of the AIDS Clinical Trials study (ACTG 320). The dataset has been distributed and illustrated in detail by Hosmer et al. (2008). The data come from a double-blind, placebo-controlled trial that compared the three-drug regimen of *Indinavir*, open label *Zidovudine or Stavudine* and *Lamivudine* (group with IDV) with the two-drug regimen of *Zidovudine or Stavudine* and *Lamivudine* (group without IDV) in HIV-infected patients (Hammer et al., 1997). Patients were eligible for the trial if they had no more than 200 CD4 cells per cubic millimeter and at least three months of prior zidovudine therapy. This study examined several factors, such as treatment, age, sex and CD4 cell counts, which may influence survival time to AIDS defining event or death. The specific variables we will use here and their codes for the data are provided in Table 6.1.

TABLE 6.1
Description of variables in the AIDS clinical trials group study (ACTG 320).

Variable Name	Description	Codes/Values
id	identification code	1-1156
time	time to AIDS diagnosis or death	days
censor	event indicator for AIDS defining diagnosis or death	1 = AIDS defining diagnosis or death 0 = otherwise
tx	treatment indicator	1 = treatment includes IDV 0 = control group (without IDV)
sex	sex	1 = male 2 = female
cd4	baseline CD4 count	cells/milliliter
priorzdv	months of prior ZDV use	months
age	age at enrollment	years

The primary goal of this study is to examine the effectiveness of the new three-drug treatment regimen when compared to the standard two-drug regimen in improving survival among HIV-infected patients. For comparison purposes, we begin the analysis with the conventional partial likelihood method using R `survival` package. We fit a model that contains five variables: treament (`tx`), age (`age`), sex (`sex`), and prior months use of ZDV (`priorzdv`):

```
data(ACTG320, package = "brinla")
library(survival)
ACTG320.coxph <- coxph(Surv(time, censor) ~ tx + age + sex + priorzdv,
    ↪  data = ACTG320)
```

```
round(coef(summary(ACTG320.coxph)), 4)
```

```
            coef exp(coef) se(coef)       z Pr(>|z|)
tx1      -0.6807    0.5063   0.2151 -3.1652   0.0015
age       0.0218    1.0220   0.0110  1.9843   0.0472
sex2      0.0144    1.0145   0.2837  0.0507   0.9595
priorzdv -0.0033    0.9967   0.0039 -0.8497   0.3955
```

The results show that tx and age are significant, while sex and priorzdx are not. Note that age is just marginally significant at level 0.05 ($p = 0.047$). We now fit the model using INLA with the default priors:

```
ACTG320.formula = inla.surv(time, censor) ~ tx + age + sex + priorzdv
ACTG320.inla1 <- inla(ACTG320.formula,  family = "coxph", data =
    ↪ ACTG320, control.hazard = list(model = "rw1", n.intervals = 20)
    ↪ , control.compute = list(dic = TRUE))
```

In the above commands, inla.surv created a survival object, where the first argument is the follow-up time for the right censored data and the second argument is the event indicator (1=observed event, 0=right censored event). To implement the semiparametric Cox regression, we need to specify the argument family = "coxph" in the inla function. The model for the piecewise constant baseline hazard is specified through control.hazard. Here we partition the time axis into $K = 20$ intervals. For $(\log(\lambda_1), ..., \log(\lambda_{20}))$, we assign a Gaussian prior with an intrinsic first-order random walk (RW1) model (Rue and Held, 2005, chp3). RW1 models are built by assuming the increments

$$\log(\lambda_{k+1}) - \log(\lambda_k) \sim N(0, \tau^{-1}), \quad k = 1, ..., K-1.$$

A diffuse gamma prior is assigned to τ. The INLA package also allows us to specify a random walk model of order 2 (RW2) for the baseline hazard in the program. More discussions about the random walk models can be found in Chapter 7 of this book.

Let us display the fitted result of the model:

```
round(ACTG320.inla1$summary.fixed, 4)
```

```
               mean     sd 0.025quant 0.5quant 0.975quant    mode kld
(Intercept) -8.4711 0.4797    -9.4193  -8.4691    -7.5354 -8.4649   0
tx1         -0.6879 0.2151    -1.1181  -0.6852    -0.2730 -0.6798   0
age          0.0214 0.0110    -0.0006   0.0215     0.0426  0.0218   0
sex2         0.0187 0.2836    -0.5664   0.0286     0.5483  0.0487   0
priorzdv    -0.0033 0.0039    -0.0115  -0.0031     0.0039 -0.0028   0
```

The results using INLA show that the posterior mean of the treatment effect is -0.6879, and its 95% credible interval is $(-1.1181, -0.2730)$. The posterior hazard ratio and its 95% credible interval, thus, are $\widehat{HR} = \exp(-0.6879) = 0.5026$ and $(\exp(-1.1181), \exp(-0.2730)) = (0.3269, 0.7611)$, respectively. The interpretation for the results is that the rate of progression to AIDS or death among HIV-infected patients on the three-drug regimen is 0.5026 times as much as that of patients on the two-drug regimen; this could be as little as 0.3269 times or as much as 0.7611 times with 95% credibility, assuming other covariates are fixed. The results using INLA are simliar to those using the conventional partial likelihood analysis. However, one difference in the results is the age effect. Its 95% credible interval is $(-0.0006, 0.0426)$, which indicates that age becomes not significant (marginally, in the Bayesian sense, 95% credible set includes 0). There is an estimate for "`Intercept`" in the INLA output, while there is not from the `coxph` function. This is due to the fact that the random walk model is used for the baseline hazard function in INLA.

6.2.2 Stratified Proportional Hazards Models

When modeling the Cox proportional hazard model the most important assumption is the proportional hazards. It means that, for any two individuals with certain covariates, the hazard ratio does not depend on time. There are a number of methods for testing proportionality, including graphical methods and testing procedures (Hosmer et al., 2008). In many real studies, the proportional hazards assumption is violated for some covariates. In such a case, one may stratify on that variable and apply a stratified proportional hazards model. Specifically, assume that the subjects are stratified into S disjoint groups. Each of the groups has a distinct baseline hazard function but common effects of the covariates. The baseline hazards h_{0j}, again, are assumed to be piecewise constant in each time interval $t \in (s_{k-1}, s_k)$, $k = 1, ..., K$:

$$h_{0j}(t) = \lambda_{kj}, \qquad \text{for } t \in (s_{k-1}, s_k),\ k = 1, ..., K,\ j = 1, ..., S.$$

We assume RW1 priors with common precision for $(\log(\lambda_{1j}), ..., \log(\lambda_{Kj}))$, $j = 1, ..., S$:

$$\log(\lambda_{k+1,j}) - \log(\lambda_{k,j}) \sim N(0, \tau^{-1}), \quad k = 1, ..., K-1, j = 1, ..., S.$$

Thus, the hazard of subject i with time $t_i \in (s_{k-1}, s_k]$ that belongs to stratum $j = 1, ..., S$ is:

$$h(t_i) = h_{0j}(t_i) \exp(\beta_1 x_{i1} + ... + \beta_p x_{ip}).$$

Such a stratified model can be easily implemented in INLA. In the ACTG 320 study, the effect of baseline CD4 counts on survival among HIV-infected patients is well-known. Different patients who have different baseline CD4 counts often have different baseline hazards (Hammer et al., 1997). Hence, we create a group variable, based on the observed CD4 quartiles, and stratify on it:

```
ACTG320$cd4group <- as.numeric(cut(ACTG320$cd4, breaks=c(-Inf,
    ↪ quantile(ACTG320$cd4, probs = c(.25,.5,.75)), Inf), labels=c("1
    ↪ ","2","3","4")))
```

We then fit a stratified proportional hazards models in INLA by specifying `strata.name = "cd4group"` in `control.hazard` argument:

```
ACTG320.formula2 = inla.surv(time, censor) ~ tx + age + sex + priorzdv
ACTG320.inla2 <- inla(ACTG320.formula2,  family = "coxph", data =
    ↪ ACTG320, control.hazard = list(model = "rw1", n.intervals = 20,
    ↪  strata.name = "cd4group"))
round(ACTG320.inla1$summary.fixed, 4)
```

```
               mean     sd 0.025quant 0.5quant 0.975quant    mode kld
(Intercept) -8.9584 0.5067    -9.9625  -8.9553    -7.9725 -8.9491   0
tx1         -0.6625 0.2152    -1.0930  -0.6598    -0.2474 -0.6544   0
```

```
age          0.0242 0.0110    0.0021   0.0243    0.0455  0.0245   0
sex2         0.0443 0.2838   -0.5412   0.0542    0.5743  0.0743   0
priorzdv    -0.0019 0.0038   -0.0099  -0.0017    0.0051 -0.0013   0
```

The posterior estimates show slight differences compared with the previous model. We could generate the plots for the estimated baseline hazard functions. We have written a convenience function `bri.basehaz.plot` to produce the baseline hazard plots in our `brinla` package:

```
library(brinla)
bri.basehaz.plot(ACTG320.inla2)
```

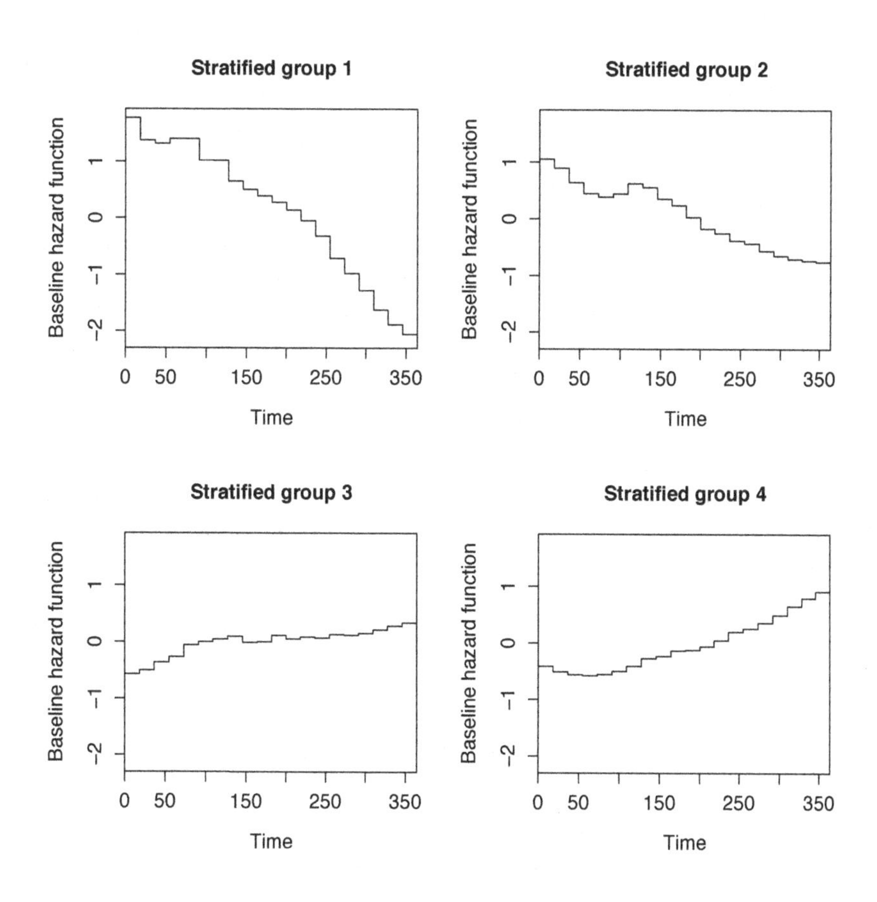

FIGURE 6.1
The estimated baseline hazard functions in the stratified proportional hazards models for the ACTG 320 data.

Figure 6.1 shows that the baseline functions are very different for each of the stratified groups. Baseline hazard functions for group 1 and group 2 decrease over time, while those for group 3 and group 4 increase. These findings indicate that stratifying the CD4 variable seems necessary. Checking the DICs for the two models also shows the stratified model fits better than the unstratified one:

```
c(ACTG320.inla1$dic$dic, ACTG320.inla2$dic$dic)
```

```
[1] 1316.253 1274.168
```

6.3 Accelerated Failure Time Models

Parametric models play an important rôle in Bayesian survival analysis. All of the parametric models we shall consider in this section have an *accelerated failure-time* (AFT) model representation and a linear model representation in log time.

The AFT model is defined by

$$S(t|\mathbf{x},\beta_1,...,\beta_p) = S_0\big(t\cdot\exp(\beta_1x_1+...+\beta_px_p)\big), \tag{6.3}$$

where the factor $\exp[\beta_1x_1+...+\beta_px_p]$ is known as the *acceleration factor.* This model states that the survival function of a subject with covariate $\mathbf{x}$ at time t is the same as the survival function of a subject with a baseline survival function at the time $t\cdot\exp[\beta_1x_1+...+\beta_px_p]$. The model implies that the hazard rate for a subject with covariate $\mathbf{x}$ is related to a baseline hazard rate by

$$h(t|\mathbf{x}) = h_0\big(t\cdot\exp(\beta_1x_1+...+\beta_px_p)\big)\times\exp(\beta_1x_1+...+\beta_px_p).$$

If we let $S_0(t)$ be the survival function of the random variable $\exp(\gamma_0+\sigma\varepsilon)$, then the accelerated failure-time model can be rewritten as a linear model for the log of the time variable, T, that is,

$$\log(T) = \gamma_0+\gamma_1x_1+...+\gamma_px_p+\sigma\varepsilon, \tag{6.4}$$

where $\gamma = (\gamma_0,\gamma_1,...,\gamma_p)^T$ is a vector of regression coefficients with $\gamma_j = -\beta_j$, $j = 1,...,p$, ε is the random error term, and σ is a scale parameter. A variety of distributions can be used for T or, equivalently for ε. Table 6.2 gives some popular parametric distributions.

TABLE 6.2
Common parametric distributions for the survival time T (equivalently for ε) in accelerated failure-time models.

Distribution of T	Distribution of ε
Exponential	Extreme value
Weibull	Extreme value
Log-normal	Normal
Log-logistic	Logistic

Here we shall discuss Weibull regression in detail. The Weibull distribution is a very flexible model for lifetime data, which gives a hazard rate that is either monotone increasing, decreasing, or constant. It is a parametric model which has both a proportional hazards representation and an accelerated failure-time representation. The survival function for the Weibull distribution is given by

$$S(t|\alpha,\lambda) = \exp(-\lambda t^{\alpha}), \qquad \alpha > 0, \lambda > 0,$$

its density function is

$$f(t|\alpha,\lambda) = \alpha t^{\alpha-1}\lambda \exp(-\lambda t^{\alpha}).$$

The hazard rate is expressed by

$$h(t) = \lambda \alpha t^{\alpha-1},$$

where the parameter α is a shape parameter. Note that when $\alpha = 1$, the Weibull distribution reduces to an exponential distribution.

If we redefine the parameters by letting $\sigma = 1/\alpha$, and $\mu = -\frac{1}{\alpha}\log\lambda$, then the log transform of survival time, $\log(T)$ follows a log linear model,

$$\log(T) = \mu + \sigma\varepsilon,$$

where ε is the extreme value distribution with density function $f_{\varepsilon}(x) = \exp(x - \exp(x))$. To build the Weibull regression model with covariates, we let

$$\mu = \gamma_0 + \gamma_1 x_1 + \ldots + \gamma_p x_p, \tag{6.5}$$

which results in the linear model (6.4) for log time.

Note that this is equivalent to introducing covariates through λ with a log link function in the Weibull distribution, that is,

$$\log(\lambda) = \theta_0 + \theta_1 x_1 + \ldots + \theta_p x_p,$$

where $\theta_0 = -\gamma_0/\sigma$, $\theta_j = -\gamma_j/\sigma$, $j = 1,\ldots,p$. This leads to a proportional hazards model for the survival time with a Weibull baseline hazard. The hazard rate given the covariates $\mathbf{x} = (x_1,\ldots,x_p)^T$ is

$$h(t|\mathbf{x}) = \alpha t^{\alpha-1} \exp\left(\theta_0 + \sum_{j=1}^{p} \theta_j x_j\right) = h_0(t)\exp\left(\sum_{j=1}^{p} \theta_j x_j\right),$$

where $h_0(t) = \alpha\lambda_0 t^{\alpha-1}$, and $\lambda_0 = \exp(-\gamma_0/\sigma)$.

Using the accelerated failure-time representation of the Weibull model, the hazard rate is given by

$$h(t|\mathbf{x}) = \exp\left(\sum_{j=1}^{p} \beta_j x_j\right) \cdot h_0\left(t\exp\left(\sum_{j=1}^{p} \beta_j x_j\right)\right), \tag{6.6}$$

where $\beta_0 = -\gamma_0 = \theta_0/\alpha$, $\beta_j = -\gamma_j = \theta_j/\alpha$, $j = 1, ..., p$. The factor $\exp\left(\sum_{j=1}^{p} \beta_j x_j\right)$ is the acceleration factor.

Suppose we observe independent survival times $\mathbf{t} = (t_1, t_2, ..., t_n)^T$, each having a Weibull distribution, and $D = (\mathbf{t}, \mathbf{X}, \boldsymbol{\delta})$ denotes the observed data for the model, $\mathbf{X}$ is the $n \times (p+1)$ design matix and $\boldsymbol{\delta} = (\delta_1, ..., \delta_n)^T$. The likelihood function of the unknown parameters $(\alpha, \boldsymbol{\beta}) = (\alpha, \beta_0, \beta_1, ..., \beta_p)$ is

$$L(\alpha, \boldsymbol{\beta}|D) = \prod_{i=1}^{n} f(t_i|\alpha, \boldsymbol{\beta})^{\delta_i} S(t_i|\alpha, \boldsymbol{\beta})^{1-\delta_i}. \tag{6.7}$$

In this basic survival model, we have a linear predictor $\eta_i = \beta_0 + \beta_1 x_{1i} + ... + \beta_p x_{pi}$. We assign Gaussian priors to all elements of $\boldsymbol{\beta} = (\beta_0, \beta_1, ...\beta_p)$, so that the model can easily be seen as a latent Gaussian model with a latent field. The hyperparameter α may be assigned a diffuse gamma prior. The likelihood (6.7) depends on the latent field only through the predictor η_i, so the INLA approach can be directly applied to such a model.

Similarly, other parametric survival models can be solved using the INLA approach. The R INLA library currently supports four popular parametric survival regression models, exponential, Weibull, log-normal, and log-logistic models. They correspond to specifying `family = "exponentialsurv"`, `"weibullsurv"`, `"lognormalsurv"`, and `"loglogistic"` in `inla` call, respectively.

Let us use a data example of male laryngeal cancer as an illustrating example for AFT models, which has been investigated in Klein and Moeschberger (2005). The dataset was first reported by Kardaun (1983), including 90 males diagnosed with cancer of the larynx during the period 1970 – 1978 at a Dutch hospital. Patient survival times were recorded between first treatment and either death or the end of the study. Patients were classified into one of four stages using the American Joint Committee for Cancer Staging. Other variables also recorded include the patient's age at the time of diagnosis, and the year of diagnosis. Table 6.3 shows the description of variables in the `larynx` data.

TABLE 6.3
Description of variables in the `larynx` data.

Variable Name	Description	Codes/Values
stage	Stage of disease	1=stage 1, 2=stage2, 3=stage 3, 4=stage 4
time	Time to death or on-study time	months
age	Age at diagnosis of larynx cancer	years
diagyr	Year of diagnosis of larynx cancer	year
delta	Death status	0=alive, 1=dead

We first consider a Weibull model with the independent variables, `age` and `stage`. Before using INLA to fit the model, let us apply the corresponding frequentist approach:

```
data(larynx, package = "brinla")
```

```
larynx.wreg <- survreg(Surv(time, delta)~ as.factor(stage) + age, data
    ↪ =larynx, dist="weibull")
round(summary(larynx.wreg)$table, 4)
```

```
                     Value Std. Error       z      p
(Intercept)         3.5288     0.9041  3.9030 0.0001
as.factor(stage)2 -0.1477     0.4076 -0.3624 0.7171
as.factor(stage)3 -0.5866     0.3199 -1.8333 0.0668
as.factor(stage)4 -1.5441     0.3633 -4.2505 0.0000
age               -0.0175     0.0128 -1.3667 0.1717
Log(scale)        -0.1223     0.1225 -0.9987 0.3179
```

Now we fit the AFT model using INLA with the default priors.

```
formula = inla.surv(time, delta) ~ as.factor(stage) + age
larynx.inla1 <- inla(formula, control.compute = list(dic = TRUE),
    ↪ family = "weibullsurv", data = larynx)
round(larynx.inla1$summary.fixed, 4)
```

```
                     mean     sd 0.025quant 0.5quant 0.975quant    mode kld
(Intercept)       -3.9348 1.0129    -5.9852  -3.9131    -2.0045 -3.8692   0
as.factor(stage)2  0.1620 0.4608    -0.7827   0.1761     1.0285  0.2046   0
as.factor(stage)3  0.6598 0.3554    -0.0339   0.6584     1.3607  0.6556   0
as.factor(stage)4  1.7059 0.4094     0.8896   1.7101     2.4983  1.7185   0
age                0.0197 0.0142    -0.0077   0.0196     0.0481  0.0192   0
```

It is important to point out here that there is a sign difference between the estimates from the output of `survreg` object and `inla` object. The estimates of γ_j's in (6.5) are reported using `survreg` function, while the estimates of β_j's in (6.6) are reported using the `inla` function. So, in the INLA output, a positive value of the risk coefficient reflects poor survival for Weibull models, which is consistent with the Cox model using INLA.

From the results, we see that patients with stage 4 disease do significantly (in the Bayesian sense, 95% credible interval excludes 0) worse than patients with stage 1 disease. The acceleration factor for stage 4 disease compared to stage 1 disease is $\exp(1.7059) = 5.5063$ with the credible interval $(\exp(0.8897), \exp(2.4983)) = (2.4344, 12.1618)$. So, we interpret the result as meaning that the median lifetime for a stage 1 patient is estimated to be 5.5063 as much as that of a stage 4 patient.

6.4 Model Diagnosis

Checking the adequacy of a fitted model is very important in survival regression analysis. Examining the residuals of a model is a common way of regression diagnostics. In the standard linear regression setup, it is quite easy to define a residual for the fitted model (see Chapter 3). The definition of the residual in survival models is not as clear-cut. A number of residuals have been proposed to examine different aspects of the model in survival analysis literature.

The first type of residuals we introduce here is the so-called *Cox–Snell residuals*. Assume that Θ is the vector of all unknown parameters in the model. Cox and Snell

(1968) define the residuals as

$$r^*_{C_i} = H_i(t_i, \hat{\Theta}|\mathbf{x}_i) = -\log(S(t_i, \hat{\Theta}|\mathbf{x}_i)), \quad i = 1, ..., n,$$

where $\hat{\Theta}$ are the maximum likelihood estimates of Θ. That is, the residuals are the maximum likelihood estimates of the cumulative hazards for the observed survival time given the covariates.

In Bayesian analysis, Chaloner (1991) defined the Bayesian version of the residuals:

$$r_{C_i} = H_i(t_i, \Theta|\mathbf{x}_i) = -\log(S(t_i, \Theta|\mathbf{x}_i)), \quad i = 1, ..., n. \tag{6.8}$$

Each r_{C_i} is just a function of unknown parameters, and posterior distribution is therefore straightforward to calculate. For example, one can draw samples from the posterior distribution of Θ and then substitute these samples into (6.8) to produce samples from the posterior distribution of the residuals. The posterior mean or median of the r_{C_i}'s can be calculated and evaluated. More simply, Wakefield (2013) suggested that one could substitute the posterior mean or median of Θ directly to obtain the approximate Bayesian residuals.

For the Cox semiparametric model, the Bayesian Cox–Snell residuals are

$$r_{C_i} = H_0(t_i) \exp\left(\sum_{j=1}^{p} \beta_j x_j\right), \quad i = 1, ..., n,$$

where $H_0(t_i) = \int_0^{t_i} h(u)du$ is the estimator of the baseline cumulative hazard rate at time t_i. For the Weibull model, the Bayesian Cox–Snell residuals are

$$r_{C_i} = \lambda_0 t_i^{\alpha} \exp\left(\sum_{j=1}^{p} \theta_j x_j\right) = \lambda_0 t_i^{\alpha} \exp\left(\alpha \sum_{j=1}^{p} \beta_j x_j\right), \quad i = 1, ..., n.$$

For the exponential model, the Bayesian Cox–Snell residuals are

$$r_{C_i} = \lambda_0 t_i \exp\left(\sum_{j=1}^{p} \theta_j x_j\right) = \lambda_0 t_i \exp\left(\sum_{j=1}^{p} \beta_j x_j\right), \quad i = 1, ..., n.$$

If the model fits well and the posterior mean $\hat{\beta}_j$ is close to the true value of β_j ($j = 1, ..., p$), then the posterior mean or median of r_{C_i}'s should look like a censored sample from a unit exponential distribution. In order to check whether the r_{C_i}'s behave as a sample from a unit exponential distribution, we could compute the Nelson–Aalen estimator of the cumulative hazard rate of r_{C_i}'s, which is defined as

$$\tilde{H}(t) = \sum_{r_{C_i} \leq t} \frac{d_i}{m_i},$$

with d_i the number of events at r_{C_i} and m_i the total individuals at risk (i.e., alive and not censored) just prior to time r_{C_i}. If the exponential distribution fits the residuals, the estimate should be very close to the true cumulative hazard rate of the unit exponential model, that is, $H(t) = t$. Hence, one could check the so-called *Cox–Snell residual plot*, a plot of the residual r_{C_i} versus its Nelson–Aalan estimate $\tilde{H}(r_{C_i})$. If a model fits well, this plot should follow a straight line through the origin with a slope of 1.

Let us go back to the example of the larynx data. We want to examine whether the Weibull model fits the data or not and compare it with other models. We refit the data with an exponential model and a Cox semiparametric model:

```
larynx.inla2 <- inla(formula, control.compute = list(dic = TRUE),
    ↪ family = "exponential.surv", data = larynx)
round(larynx.inla2$summary.fixed, 4)
```

```
                      mean     sd 0.025quant 0.5quant 0.975quant    mode kld
(Intercept)        -3.7700 0.9901    -5.7761  -3.7483    -1.8845 -3.7043   0
as.factor(stage)2   0.1475 0.4601    -0.7960   0.1617     1.0125  0.1903   0
as.factor(stage)3   0.6505 0.3551    -0.0427   0.6491     1.3509  0.6463   0
as.factor(stage)4   1.6301 0.3985     0.8343   1.6346     2.4001  1.6438   0
age                 0.0197 0.0142    -0.0077   0.0196     0.0481  0.0192   0
```

```
larynx.inla3 <- inla(formula, control.compute = list(dic = TRUE),
    ↪ family = "coxph", data = larynx, control.hazard=list(model="rw1
    ↪ ", n.intervals=20))
round(larynx.inla3$summary.fixed, 4)
```

```
                      mean     sd 0.025quant 0.5quant 0.975quant    mode kld
(Intercept)        -3.7693 0.9903    -5.7756  -3.7476    -1.8835 -3.7036   0
as.factor(stage)2   0.1477 0.4601    -0.7958   0.1619     1.0127  0.1905   0
as.factor(stage)3   0.6506 0.3551    -0.0426   0.6491     1.3509  0.6463   0
as.factor(stage)4   1.6310 0.3986     0.8350   1.6355     2.4013  1.6447   0
age                 0.0197 0.0142    -0.0077   0.0196     0.0481  0.0192   0
```

It appears that the coefficient estimates for the covariates are quite close in comparison to the three models. Our `brinla` library provides a convenient R function, `bri.surv.resid`, to compute Bayesian Cox–Snell residuals. We now calculate the residuals for each of the three models:

```
larynx.inla1.res <- bri.surv.resid(larynx.inla1, larynx$time, larynx$
    ↪ delta)
larynx.inla2.res <- bri.surv.resid(larynx.inla2, larynx$time, larynx$
    ↪ delta)
larynx.inla3.res <- bri.surv.resid(larynx.inla3, larynx$time, larynx$
    ↪ delta)
```

The arguments are needed in the `bri.surv.resid` function: the fitted INLA survival model object, the follow-up time for the right censored data, and the status indicator. Then we can generate the Cox–Snell residual plots using the `bri.csresid.plot` function in `brinla` library:

```
bri.csresid.plot(larynx.inla1.res, main = "Weibull model")
bri.csresid.plot(larynx.inla2.res, main = "Exponential model")
bri.csresid.plot(larynx.inla3.res, main = "Cox model")
```

The resulting plots are displayed in Figure 6.2. These plots suggest that all three models do not fit badly. However, for the Weibull model, the estimated cumulative hazard rate lies above the 45° line, except in the tail where the variability of the

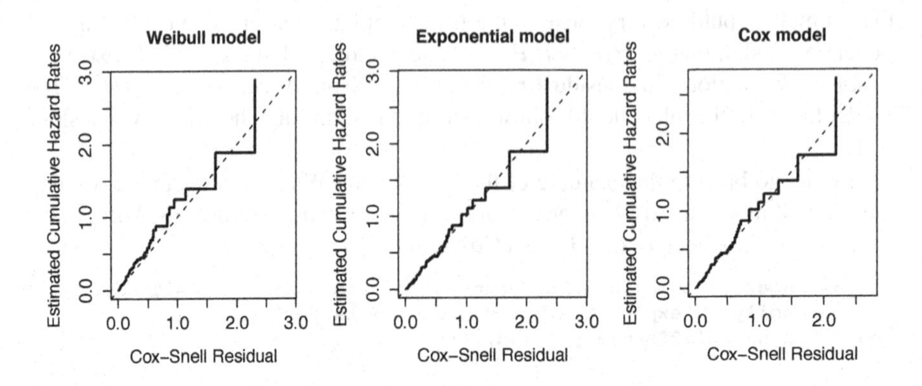

FIGURE 6.2
Cox–Snell residual plots for the different models that are used to fit the `larynx` data.

cumulative hazard estimate is large. It seems that exponential model has the best fit among the three models.

Based on the Bayesian Cox–Snell residuals, two other types of residuals can be computed. The Bayesian version of the *martingale residuals* (Barlow and Prentice, 1988) are defined by

$$r_{M_i} = \delta_i - r_{C_i}, \quad j = 1, \ldots, n.$$

This quantity provides us a measure of the difference between the indicator of whether a given individual experiences the event of interest and the expected number of events the individual would experience.

Martingale residuals are a reallocation of the Cox–Snell residuals to a mean of zero for uncensored observations, so that they follow similar properties as the residuals in linear regression analysis. To assess the model fitting, one could construct a plot of the martingale residuals versus a certain covariate or a plot of the residuals by case (index plot). However, martingale residuals for censored observations take negative values, and the maximum possible value of the residuals is $+1$ but the minimum possible value could go to $-\infty$. So, in practice, martingale residuals are not often symmetrically distributed around zero which makes the martingale residual plots difficult to interpret.

The Bayesian version of the *deviance residuals* (Therneau et al., 1990) are used to obtain residuals that have a distribution more normally shaped than the martingale residuals. They are defined by

$$r_{D_i} = \text{sgn}(r_{M_i})[-2\{r_{M_i} + \delta_i \log(\delta_i - r_{M_i})\}]^{1/2}, \quad i = 1, \ldots, n,$$

where $\text{sgn}(\cdot)$ is the sign function. Note that r_{D_i} has a value of 0 when r_{M_i} is zero. The logarithm tends to inflate the values of the residuals when r_{M_i} is close to 1 and to shrink large negative values of r_{M_i}.

The `bri.surv.resid` function also outputs Bayesian martingale residuals and

Bayesian deviance residuals for a fitted INLA survival model object. We now construct a plot of the deviance residuals versus the covariate, `age`, for each of the three models. The `bri.dresid.plot` function in the `brinla` library generates the deviance residual plots:

```
par(mfrow=c(1,3))
bri.dresid.plot(larynx.inla1.res, larynx$age, xlab = "age", main = "
    ↪ Weibull model")
bri.dresid.plot(larynx.inla2.res, larynx$age, xlab = "age", main = "
    ↪ Exponential model")
bri.dresid.plot(larynx.inla3.res, larynx$age, xlab = "age", main = "
    ↪ Cox model")
```

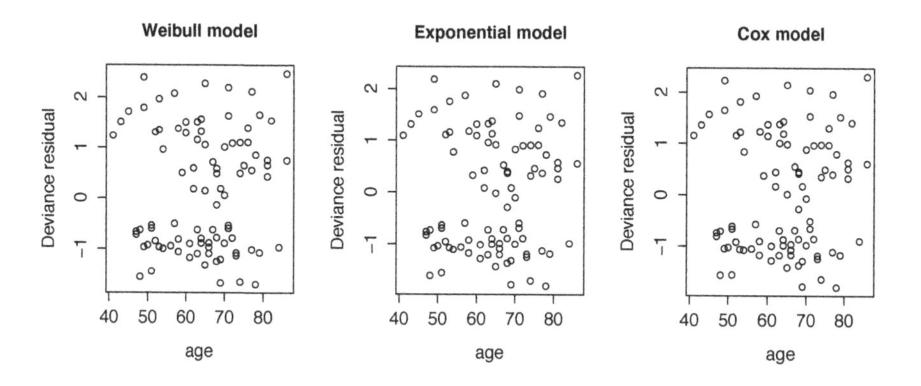

FIGURE 6.3
Bayesian deviance residual plots for the different models that are used to fit `larynx` data.

All residual plots show random patterns (Figure 6.3), which indicate that the three models all have good fits. The second argument in the `bri.dresid.plot` function is the covariate that you want to plot against the residuals. If the argument is missing, a residual index plot will be generated.

In practice, the martingale residuals are often used to determine the functional form of a given covariate to best explain its effect on survival through a Cox, Weibull, or exponential model. Suppose that the covariate vector $\mathbf{x} = (x_1, x_2, ..., x_p)^T$ can be partitioned into a $p-1$ vector $\mathbf{x}^* = (x_2, ..., x_p)^T$, for which we have a known proper functional form of the Cox model, Weibull, or exponential model, and a single covariate x_1, for which we are not sure what kind of functional form of x_1 to use. Assume that $g(x_1)$ is the best form of x_1, and x_1 and $\mathbf{x}^*$ are independent. The optimal Cox, Weibull, or exponential model is

$$H(t|\mathbf{x}^*, x_1) = H_0(t) \exp\left(\sum_{j=2}^{p} \beta_j x_j\right) \exp\{g(x_1)\}.$$

In order to find an appropriate g function, we construct martingale residual plot as follows: First, we fit a Cox, Weibull, or exponential model to the data based on $\mathbf{x}^*$ and compute the Bayesian martingale residuals, r_{M_i}, $i = 1,...,n$. Then, we draw the scatterplot of r_{M_i} versus the value of x_1 for the ith observation, and overlay it with a smoothed-fitted curve. Many smoothing techniques can be applied here, such as the random walk model of second order that we will discuss in Chapter 7, local polynomial regression, splines etc. The curve gives an indication of the function g. For example, if the curve is linear, we use x_1 directly; if there are some change points or thresholds, a discretized version of the covariate is indicated. If there is some nonlinear pattern, a transformation of the covariate, such as $\log(x_1)$, x_1^2 could be used. Note that the above approach cannot be used for the AFT model with the log-normal or log-logistic distribution, since the proportional hazards representation does not exist in those two models.

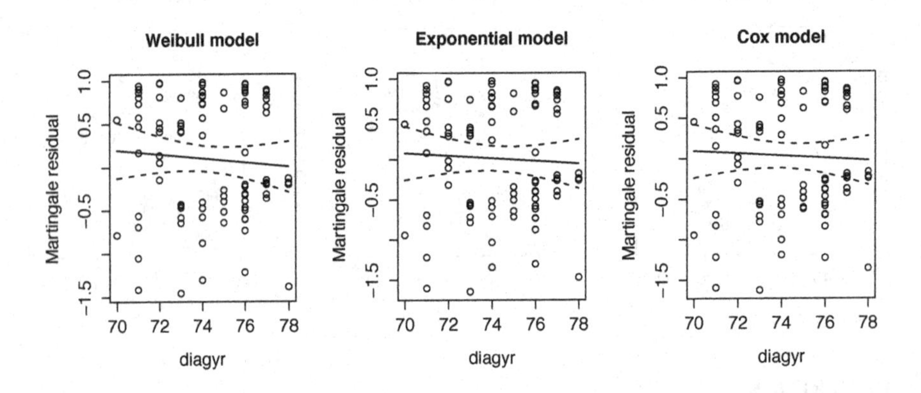

FIGURE 6.4
Bayesian martingale residual plots for the different models that are used to fit the `larynx` data. The solid curves are the smoothed-fitted curves and the dashed curves are their 95% credible intervals.

In the larynx data example, the above three models did not include the variable, `diagyr`, as a covariate. Let us examine the best function of `diagyr` if we want to include it in the models. The `bri.mresid.plot` function in the `brinla` library generates the Bayesian martingale residual plots:

```
bri.mresid.plot(larynx.inla1.res, larynx$diagyr, smooth = TRUE, xlab =
    ↪ "diagyr", main = "Weibull model")
bri.mresid.plot(larynx.inla2.res, larynx$diagyr, smooth = TRUE, xlab =
    ↪ "diagyr", main = "Exponential model")
bri.mresid.plot(larynx.inla3.res, larynx$diagyr, smooth = TRUE, xlab =
    ↪ "diagyr", main = "Cox model")
```

The third argument in the `bri.mresid.plot` function, `smooth = TRUE` is to add the smooth curve with its 95% credible interval on the residual plot. Figure 6.4 displays Bayesian martingale residual plots for the three survival models that are used to fit larynx data. The curves are the smoothed-fitted curves (solid lines) with their 95% credible intervals (dashed lines) using the random walk model of order 2 with INLA.

The plots indicate that using a linear form for the `diagyr` variable in the regression equation is sufficient for the three models.

6.5 Interval Censored Data

Interval censoring arises in many practical settings. For instance, subjects are periodically assessed for disease progression in a clinical study. In case of interval censoring, the survival time T is observed only to lie in an interval $[t^{lo}, t^{up}]$. Another feature of lifetime data is that of left truncation. Left truncation in survival analysis means that some subjects do not enter the risk set until a known period after the time origin. For example, in a study of disease mortality where the outcome of interest is survival from the time of diagnosis, many patients may not have been enrolled in the study until several months or years after their diagnosis. We say, an observation is left truncated if only values above a known truncation time t^{tr} are reported.

In a more general framework for survival data, an observation can be described by a quintuple $(t^{lo}, t^{up}, t^{tr}, \delta, \mathbf{x})$. The general log-likelihood contribution for the ith observation is given by:

$$l_i = \delta_i \log(h_i(t_i^{up}|\mathbf{x})) - \int_{t_i^{tr}}^{t_i^{up}} h_i(u|\mathbf{x})du + \log\left\{1 - \exp\left(-\int_{t_i^{lo}}^{t_i^{up}} h_i(u|\mathbf{x})du\right)\right\}. \quad (6.9)$$

We assign Gaussian priors to all unknown fixed effects. So, the likelihood in (6.9) depends on a latent Gaussian field only through the linear predictor $\mathbf{x}_i^T\beta$, just as with right-censored data. Then, it is straightforward to apply INLA algorithms on the accelerated failure time models with interval-censored data.

However, for the semiparametric model with piecewise constant baseline hazard, the log-likelihood in the case of interval censoring does not allow us to use the same trick of data augmentation discussed in Section 6.2.1. Therefore, INLA cannot handle the case of the semiparametric model with interval censoring.

Let us look at the Signal Tandmobiel study (`tooth24`), a prospective oral health study conducted in Belgium from 1996 to 2001. The study contains a cohort of randomly sampled schoolchildren who attended the first year of the primary school at the beginning of the study. The original dataset has been presented and studied by Bogaerts and Lesaffre (2004) and Gomez et al. (2009). Here we restrict the analysis to the age of the emergence time of the permanent upper left first premolars (tooth 24 in European dental notation). Since permanent teeth do not emerge before the age of 5, the origin time for all analyses is set at 5 years (Gomez et al., 2009). The covariates of interest include `GENDER` (0 = boy; 1 = girl) and `DMF` (the status of the primary predecessor of this tooth: 0 if the primary predecessor was sound, 1 if it was decayed, missing due to caries or filled). The variables and their codes for the data are shown in Table 6.4.

We load the data from the `brinla` package and compactly display the structure of the R data frame.

TABLE 6.4
Description of variables in the Signal Tandmobiel study (tooth24).

Variable Name	Description	Codes/Values
ID	subject's identification code	unique integers
LEFT	lower limit of tooth emergence	years since age 5
RIGHT	upper limit of tooth emergence	years since age 5
SEX	gender of the child	0 = boy 1 = girl
DMF	status of primary predecessor	0 = sound 1 = decayed, missing, or filled

```
data(tooth24, package = "brinla")
str(tooth24)
```

```
'data.frame':        4386 obs. of  5 variables:
 $ ID   : int  1 2 3 4 5 6 7 8 9 10 ...
 $ LEFT : num  2.7 2.4 4.5 5.9 4.1 3.7 4.9 5.4 4 5.9 ...
 $ RIGHT: num  3.5 3.4 5.5 999 5 4.5 5.8 6.5 4.9 6.7 ...
 $ SEX  : Factor w/ 2 levels "0","1": 2 1 2 2 2 1 1 2 2 2 ...
 $ DMF  : Factor w/ 2 levels "0","1": 2 2 1 1 2 2 2 2 2 1 ...
```

The dataset contains 4386 observations of 5 variables. First, a new censoring variable needs to be added to the dataframe tooth24:

```
tooth24$cens <- with(tooth24, ifelse(RIGHT == 999, 0, 3))
```

The variable cens indicating the type of censoring has to be defined as being 0 for the indicator for right-censoring and 3 for the indicator for interval-censoring.

Let us consider an AFT model with a log-logistic distribution for the interval censored data. The log-logistic distribution has cumulative distribution function

$$F(t|\alpha,\beta) = \frac{1}{1+(t/\gamma)^{-\alpha}},$$

where $\alpha, \gamma > 0$. The parameter γ is linked to the covariates $(x_1,...,x_p)$ as:

$$\log(\gamma) = \beta_0 + \beta_1 x_1 + ... + \beta_p x_p.$$

The parameter α is a shape parameter. Sometimes, an alternative parameterization for the log-logistic distribution is given by $s = 1/\alpha$.

To fit the model with the conventional maximum likelihood approach, we need to use the R function survreg in the survival library. The survreg function requires to create a so-called Surv object, which combines all those vectors containing information on the survival times and its censoring status. With our data, the Surv object can be defined as follows:

```
library(survival)
sur24 <- with(tooth24, Surv(LEFT, RIGHT, cens, type = "interval"))
sur24[1:5]
```

```
[1] [2.7, 3.5] [2.4, 3.4] [4.5, 5.5] 5.9+       [4.1, 5.0]
```

The object `sur24` contains the observed intervals of emergence times of permanent tooth24 of 4386 children, five of which are shown above. Note that we model $T_i - 5$ instead of T_i. So, the values `[2.7, 3.5]` in `sur24` correspond to 7.7 and 8.5 years of age. The sign + indicates a right-censored observation. For example, the value `5.9+` means that the child #4 had not emerged yet by its last dental examination at the age of $5 + 5.9 = 10.9$ years.

We consider two covariates in the model, `SEX` and `DMF`, and fit the model with the R function `survreg`:

```
tooth24.survreg <- survreg(sur24 ~ SEX + DMF, data = tooth24, dist="
    ↪ loglogistic")
```

The model fit is stored in the object `tooth24.survreg`. Let us have a close look at the estimated parameters:

```
round(summary(tooth24.survreg)$table, 4)
```

```
              Value Std. Error         z p
(Intercept)  1.7718     0.0069  257.2276 0
SEX1        -0.0719     0.0082   -8.7980 0
DMF1        -0.0896     0.0082  -10.8993 0
Log(scale)  -1.9913     0.0164 -121.1760 0
```

According to the results, both `SEX` and `DMF` are highly significant. The negative sign of both parameter estimates indicates, on average, shorter times until tooth emergence for girls (the categories with label 1) and children with a decayed, missing or filled primary predecessor of permanent tooth24. The `Log(scale)` in the output reports the estimate for the logarithm of s. We can extract the estimate of s from the fit directly:

```
round(summary(tooth24.survreg)$scale, 4)
```

```
[1] 0.1365
```

Now we fit the model using INLA. We first define the model formula:

```
tooth24.formula = inla.surv(LEFT, cens, RIGHT) ~ SEX + DMF
```

Note that, in the object `inla.surv`, we need to specify the starting time for the interval, `LEFT`, the status indicator, `cens`, and the ending time for the interval, `RIGHT`, sequentially. The model fit is stored in the object `tooth24.inla` and we output the estimates:

```
tooth24.inla <- inla(tooth24.formula,  family = "loglogistic", data =
    ↪ tooth24)
round(tooth24.inla$summary.fixed, 4)
```

```
               mean     sd 0.025quant 0.5quant 0.975quant    mode kld
(Intercept)  1.7732 0.0071     1.7592   1.7732     1.7872  1.7731   0
SEX1        -0.0726 0.0084    -0.0892  -0.0726    -0.0560 -0.0726   0
DMF1        -0.0903 0.0085    -0.1070  -0.0903    -0.0737 -0.0903   0
```

Compared with the maximum likelihood approach, the results from INLA are very close. All values in the 95% credible intervals are on the same side of zero (all negative) for both covariates.

We could interpret the results in terms of the acceleration factor. For instance, comparing girls and boys with the same value of `DMF`, the survival time of tooth emergence in girls is "accelerated" by a factor of $\exp(-0.073) = 0.930$ compared to the boys, that is, the median time (from age 5) until tooth emergence in girls is 0.930 times the median time in boys.

We now extract the result for the hyperparameter:

```
round(tooth24.inla$summary.hyper, 4)
```

```
                                  mean     sd 0.025quant 0.5quant 0.975quant  mode
alpha parameter for loglogistic 7.0231 0.1063      6.819   7.0224     7.2307 7.021
```

Note that `inla` reports the estimate of the parameter α in the log-logistic model, while `survreg` outputs the estimate of the scale parameter, the reciprocal of α.

We have applied the log-logistic model here. This model is the only one that has the *proportional odds* form among all accelerated failure-time models. And it is not a proportional hazards model any more. In order to develop the odds-ratio interpretation, we begin by expressing the survivorship function for the model as

$$S(t_i|\mathbf{x}_i) = \frac{1}{1 + \left(t \exp(-\mathbf{x}_i^T \beta)\right)^{1/\sigma}}, \quad i = 1, \ldots, n.$$

With some simple algebra, it can be shown that

$$\log\left(\frac{S(t_i|\mathbf{x}_i)}{1 - S(t_i|\mathbf{x}_i)}\right) = \beta_0^* + \beta_1^* x_{i1} + \ldots + \beta_p^* x_{ip} - \sigma^{-1} \log(t_i),$$

where $\beta_j^* = \beta_j/\sigma$ for $j = 0, 1, \ldots, p$. This is nothing but a logistic regression model with the intercept depending on time. Since $S(t)$ is the probability of surviving to time t, the ratio $S(t)/(1 - S(t))$ is often called the *survival odds*, i.e., the odds of surviving beyond time t; the ratio $(1 - S(t))/S(t)$ is often called the *failure odds*, i.e., the odds of getting the event by time t.

With a unit increase in x_k while other covariates are being held fixed, the *survival odds ratio* is given by

$$\frac{S(t|x_k + 1)/(1 - S(t|x_k + 1))}{S(t|x_k)/(1 - S(t|x_k))} = \exp(\beta_k^*), \quad \forall t \geq 0,$$

which is a constant over time. Therefore, we have a *proportional odds model*, and $\exp(\beta_k^*)$ can be interpreted as the odds ratio for surviving with a unit increase in x_k and $\exp(-\beta_k^*)$ can be interpreted as the odds ratio for getting the event with a unit increase in x_k.

Back to our example, we can calculate the estimated failure odds ratio for `SEX`, that is, $\exp(-1 \times (-0.073)/(1/7.024)) = 1.67$. Hence, The odds of girls to have tooth emergence is 1.67 higher than boys with the same value of `DMF`.

6.6 Frailty Models

The concept of frailty provides a suitable way to introduce random effects in survival models to account for association and unobserved heterogeneity. In its simplest form, a frailty is an unobserved random factor that modifies multiplicatively the hazard function of an individual or a group or cluster of individuals.

The standard situation of the application of survival methods to a study assumes that a homogeneous population is investigated when subject to different conditions (e.g., treatment vs. control). The appropriate survival model then assumes that the survival data of the different patients are independent from each other and that each patient's individual survival time distribution is the same (independent and identically distributed failure times) conditional on certain fixed covariates. However, heterogeneity in survival data often occurs in practice. For example, in a clinical trial study, the effect of a drug, a treatment or the influence of various explanatory variables may differ greatly between subgroups of patients. Vaupel et al. (1979) introduced univariate frailty models into survival analysis for heterogeneous data, where the random effect (the frailty) has a multiplicative effect on the hazard. The frailty takes into account the effects of unobserved or unobservable heterogeneity, caused by different sources.

To be specific, let t_{ij} be the survival time for the jth individual in the ith cluster, $i = 1, \ldots, n$, and $k = 1, \ldots, m_i$. Here the m_i's represent the number of individuals in the ith cluster. Consider an AFT frailty model by assuming that t_{ik} follows a Weibull distribution,

$$t_{ik}|\lambda_{ik} \sim \text{Weibull}(\alpha, \lambda_{ik}), \quad \alpha > 0,$$

where

$$\log(\lambda_{ik}) = \exp(\mathbf{x}_{ik}^T \boldsymbol{\beta} + b_i).$$

The b_i is the unobserved frailty for the ith cluster, the fixed-effect vector $\mathbf{x}_{ik} = (1, x_{ik1}, \ldots, x_{ikp})^T$ and the vector of regression coefficients $\boldsymbol{\beta} = (\beta_0, \beta_1, \ldots, \beta_p)^T$. Hence, the hazard function for the model is given as

$$h(t_{ik}) = \alpha t_{ik}^{\alpha-1} \exp(\mathbf{x}_{ik}^T \boldsymbol{\beta} + b_i), \tag{6.10}$$

which reduces to the exponential hazard if $\alpha = 1$.

If we want to consider the semiparametric proportional frailty model, the hazard in (6.10) is replaced by

$$h(t_{ik}) = h_0(t_{ik}) \exp(\mathbf{x}_{ik}^T \boldsymbol{\beta} + b_i),$$

where the baseline hazard $h_0(t)$ can be modeled by the piecewise constant model.

Typically, the frailty term b_i is assumed to have a Gaussian distribution $N(0, \tau_b^{-1})$. Diffuse Gaussian priors are assigned to $\boldsymbol{\beta}$, while diffuse gamma priors are assigned to τ_b and α. Hence, the frailty models reduce to latent Gaussian models, and INLA methodology can be applied to solve the models.

Let us look at an example of the kidney infection data (McGilchrist and Aisbett, 1991). The study concerns the recurrence times to infection, at the point of insertion of the catheter, for kidney patients using portable dialysis equipment. The data consist of times until the first and second recurrences of kidney infection in 38 patients. Each patient has exactly 2 observations. Each survival time is the time until infection since the insertion of the catheter. The survival times for the same patient are likely to be related because of a shared frailty describing the common patient's effect. Catheters may be removed for reasons other than infection, in which case the observation is censored. There are about 24% censored observations in the dataset. This dataset can be found in the R `survival` package.

TABLE 6.5
Description of variables in the `kidney` catheter data.

Variable Name	Description	Codes/Values
id	Patient's identification code	integers
time	Time to infection	days
status	Event status	1 = infection occurs; 0 = censored
age	Age	years
sex	Sex	1 = male; 2 = female
disease	Disease type	0=GN; 1=AN; 2=PKD; 3=Other

Table 6.5 displays the description of variables in the `kidney` catheter data. The risk variables include `age`, `sex`, and `disease`. We start with fitting a Weibull frailty model using conventional maximum likelihood method:

```
data(kidney, package = "survival")
kidney.weib <- survreg(Surv(time,status) ~ age + sex + disease +
    ↪ frailty.gaussian(id), dist='weibull', data = kidney)
round(summary(kidney.weib)$table, 4)
```

```
              Value Std. Error       z      p
(Intercept)  2.1153     0.7612  2.7787 0.0055
age         -0.0030     0.0127 -0.2368 0.8128
sex          1.5620     0.3518  4.4393 0.0000
diseaseGN   -0.1571     0.4563 -0.3441 0.7307
diseaseAN   -0.5586     0.4565 -1.2237 0.2211
diseasePKD   0.6378     0.6346  1.0050 0.3149
Log(scale)  -0.5987     0.1278 -4.6842 0.0000
```

The `frailty.gaussian` function allows one to add a simple Gaussian random effects term to a `survreg` model. We could extract the estimate of standard deviation of the random effect:

```
round(kidney.weib$history$`frailty.gaussian(id)`$theta, 4)
```

```
[1] 0.6279
```

We now fit the model using the INLA approach:

```
formula = inla.surv(time, status) ~ age + sex + disease + f(id, model
    ↪ = "iid", hyper = list(prec = list(param=c(0.1, 0.1))))
kidney.inla <- inla(formula, family = "weibullsurv", data = kidney)
round(kidney.inla$summary.fixed, 4)
```

```
               mean     sd 0.025quant 0.5quant 0.975quant    mode kld
(Intercept) -2.4536 0.9511    -4.4148  -2.4222    -0.6680 -2.3593   0
age          0.0028 0.0146    -0.0256   0.0027     0.0320  0.0024   0
sex         -1.8765 0.4616    -2.8145  -1.8672    -0.9920 -1.8506   0
diseaseGN    0.1308 0.5305    -0.9146   0.1298     1.1828  0.1296   0
diseaseAN    0.6031 0.5293    -0.4254   0.5971     1.6672  0.5867   0
diseasePKD  -1.0839 0.7883    -2.6456  -1.0822     0.4682 -1.0773   0
```

Here the function `f()` with the argument `model="iid"` is used to specify the subject level frailty term. We assign the hyperparameter a diffuse gamma prior, $Gamma(0.1, 0.1)$, by using the argument `hyper = list(prec = list(param=c(1, 0.1)))` in `f()`. The result for the fixed effects shows that the only covariate whose 2.5% and 97.5% posterior quantiles are on the same side of zero is `sex`, which is concordant with the result using the MLE method. This indicates a lower infection rate for female patients, with high probability. The estimates of the hyperparameters in the model can be obtained as follows:

```
round(kidney.inla$summary.hyper, 4)
```

```
                                 mean      sd 0.025quant 0.5quant 0.975quant   mode
alpha parameter for weibullsurv 1.1691  0.1582     0.8491   1.1771     1.4570 1.2229
Precision for id                5.9519 11.9417     0.7387   2.8416    30.3126 1.2416
```

We could calculate the improved estimates of the posterior marginals for the hyperparameters in the model. The function `inla.hyperpar` uses the grid integration strategy to compute more accurate posterior estimates for the hyperparameters.

```
kidney.inla.hp <- inla.hyperpar(kidney.inla)
round(kidney.inla.hp$summary.hyper, 4)
```

```
                                 mean     sd 0.025quant 0.5quant 0.975quant   mode
alpha parameter for weibullsurv 1.1630 0.1182     0.9562   1.1548     1.4169 1.1416
Precision for id                3.5552 3.3492     0.7266   2.4378    13.5488 1.4124
```

Oftentimes, we are interested in the posterior estimate of the standard deviation parameter of the frailty term $\sigma_b = \tau_b^{-1/2}$, instead of the precision τ_b which is given by default in the summary of the hyperparameters. The R function `bri.hyperpar.summary` in our `brinla` library produces the summary statistics of hyperparameters in terms of σ_b:

```
round(bri.hyperpar.summary(kidney.inla.hp),4)
```

```
                                 mean     sd q0.025   q0.5 q0.975   mode
alpha parameter for weibullsurv 1.1630 0.1182 0.9557 1.1540 1.4161 1.1415
SD for id                       0.6594 0.2323 0.2724 0.6398 1.1674 0.6105
```

The posterior mean of σ_b appears close to the estimate using the conventional MLE method. We could further check the plots of the approximate posterior distribution of the alpha parameter for Weibull distribution and the standard deviation parameter for the frailty term using the function `bri.hyperpar.plot` in our `brinla` library:

```
bri.hyperpar.plot(kidney.inla.hp, together = F)
```

Figure 6.5 shows the posterior densities of the parameter α and the parameter σ_b for frailty term. The distribution of α appears relatively symmetric, while the distribution of σ_b is right skewed.

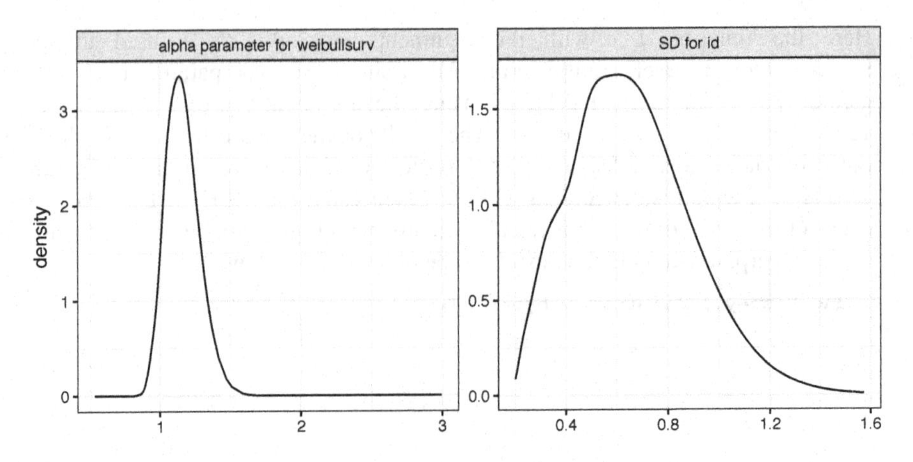

FIGURE 6.5
Approximate marginal posterior distributions of the alpha parameter for Weibull distribution and the standard deviation parameter for the frailty term.

6.7 Joint Modeling of Longitudinal and Time-to-Event Data

Many clinical studies produce two types of outcomes, a set of longitudinal response measurements as well as the time to an event of interest, such as death, or development of a disease. A common objective in these studies is to characterize the relationship between a longitudinal response process and a time-to-event. A common example of this setting is an AIDS clinical study, where infected patients are monitored until they develop AIDS or die and they are regularly measured for the condition of the immune system using markers such as the CD4 lymphocyte count (Abrams et al., 1994).

Many models exist for analyzing such data separately, including linear mixed-effects models in Chapter 5 for longitudinal data, and semiparametric proportional hazards models and AFT models in this chapter for survival data. However, separate modeling of longitudinal and time-to-event data may be inappropriate when the longitudinal variable is correlated with the patient's survival endpoint. A joint modeling approach is often preferable in such data (Tsiatis and Davidian, 2004).

Henderson et al. (2000) proposed a very flexible joint model that allows a very broad range of dependencies between the longitudinal responses and the survival endpoints. The model includes fixed effects, random effects, serial correlation, and pure measurement error for the longitudinal data, and use a semiparametric proportional hazards model with frailty terms for the survival data. The key idea of the method is to connect the longitudinal and survival processes with a latent bivariate Gaussian process. The longitudinal and time-to-event data are then assumed independent given the linking latent process and the covariates. Guo and Carlin (2004) developed a fully Bayesian version of this approach, implemented via MCMC methods. Martino et al. (2011) showed how INLA can be adapted and applied to this complex model. Following their work, in this section, we demonstrate the joint analysis of longitudinal and time-to-event data using INLA.

Let us assume that a set of m subjects is followed over a time interval $[0,\omega)$. The ith subject provides a set of (possibly partly missing) longitudinal measurements $\{y_{ij}, j = 1,...,n_i$ at times $\{s_{ij}, j = 1,...,n_i\}$, and a (possibly censored) survival time t_i to a certain endpoint. The joint model is composed of two submodels, one for each type of data. The longitudinal data y_{ij} are modeled as

$$\begin{cases} y_{ij}|\eta_{ij},\sigma^2 \sim N(\eta_{ij},\sigma^2)), \\ \eta_{ij} = \mu_i(s_{ij}) + W_{1i}(s_{ij}), \end{cases} \tag{6.11}$$

where $\mu_i(s) = \mathbf{x}_{1i}^T\boldsymbol{\beta_1}$ is the mean response, and $W_{1i}(s)$ incorporates subject-specific random effects. The vector $\mathbf{x}_{1i}$ and $\boldsymbol{\beta}$ represent possibly time-varying explanatory variables and their corresponding regression coefficients, respectively.

The survival data can be modeled by an AFT model or semiparametric proportional model. In an AFT model, for example, we assume that the survival time for the ith subject follows a Weibull distribution,

$$\begin{cases} t_i|\lambda_i(t) \sim \text{Weibull}(\alpha,\lambda_i(t)), \\ \log(\lambda_i(t)) = \mathbf{x}_{2i}^T(t)\boldsymbol{\beta_2} + W_{2i}(t), \end{cases} \tag{6.12}$$

where the vectors $\mathbf{x}_{2i}(t)$ and $\boldsymbol{\beta_2}$ represent (possibly time-dependent) explanatory variables and their corresponding regression coefficients. The $\mathbf{x}_{2i}$ may or may not have variables in common with $\mathbf{x}_{1i}$ in the longitudinal model. The form of $W_{2i}(t)$ is similar to $W_{1i}(s)$, including subject-specific covariate effects and an intercept (i.e., a frailty).

Henderson et al. (2000) proposed to jointly model the longitudinal and survival processes via a latent zero-mean bivariate Gaussian process on $(W_{1i}, W_{2i})^T$, which is independent across different subjects. Specifically, the joint model links (6.11) and (6.12) by taking

$$W_{1i}(s) = U_{1i} + U_{2i}s, \tag{6.13}$$

and

$$W_{2i}(t) = \gamma_1 U_{1i} + \gamma_2 U_{2i} + \gamma_3(U_{1i} + U_{2i}t) + U_{3i}. \tag{6.14}$$

The parameters γ_1, γ_2 and γ_3 measure the association between the two submodels induced by the random intercepts, slopes, and fitted longitudinal value at the event time $W_{1i}(t)$, respectively. The pair of latent variables $(U_{1i}, U_{2i})^T$ has a mean-zero bivariate Gaussian distribution $N(0,\Sigma_U)$, while the U_{3i} are independent frailty terms, assumed to have a Gaussian $N(0,\sigma_{U_3}^2)$, independent of the $(U_{1i}, U_{2i})^T$.

Denote $\tau = \sigma^{-2}$, $\tau_{U_3} = \sigma_{U_3}^{-2}$, and $Q_U = \Sigma_U^{-1}$. If we assign gamma priors to τ, τ_{U_3}, α; a Wishart prior to Q_U; and diffuse Gaussian priors to $\boldsymbol{\beta}_1$, $\boldsymbol{\beta}_2$, γ_1, γ_2, and γ_3, the above complex joint model reduces to a latent Gaussian field. INLA methodology can be again applied to solve the model. Note that, in this complex model, not all

TABLE 6.6
Description of variables in the AIDS clinical trial data.

Variable Name	Description	Codes/Values
y	the square root of the CD4 count	numbers
SURVTIME	time to death	months
CENSOR	death status	1 = death; 0 = censored
TIME	time that CD4 counts were recorded	months
DRUG	receive either didanosine (ddI) or zalcitabine (ddC)	1 = ddI; 0 = ddC
TIMEDRUG	time and drug interaction	numbers ranged from 0 to 12
SEX	Sex	1 = male; -1 = female
PREVOI	previous opportunistic infection (AIDS diagnosis) at study entry	1 = yes; -1 = no
STRATUM	failure or intolerance of zidovudine (AZT) therapy	1 = failure; -1 = intolerance
ID	patient identification	integers

data points have the same likelihood. Some coding tricks are needed to manipulate the likelihoods in the `inla` function.

We use the AIDS clinical trial data that has been presented in Guo and Carlin (2004) and Martino et al. (2011) as the illustrating example. In this AIDS study, both longitudinal and survival data were collected to compare the efficacy and safety of two antiretroviral drugs in treating patients who had failed or were intolerant of zidovudine (AZT) therapy. There were 467 HIV-infected patients who met entry conditions (either an AIDS diagnosis or two CD4 counts of 300 or fewer, and fulfilling specific criteria for AZT intolerance or failure). The patients were randomly assigned to receive either didanosine (ddI) or zalcitabine (ddC). CD4 counts were recorded at study entry, and again at the 2-, 6-, 12-, and 18-month visits. The times to death were also recorded. Four explanatory variables are also recorded: `DRUG` (ddI = 1, ddC = 0), `SEX` (male = 1, female = -1), `PREVOI` (previous opportunistic infection (AIDS diagnosis) at study entry = 1, no AIDS diagnosis = -1), and `STRATUM` (AZT failure = 1, AZT intolerance = -1). Table 6.6 displays the variables and their descriptions in the dataset.

Following Guo and Carlin (2004)'s suggestion, the longitudinal submodel assumes a Gaussian model with mean

$$\begin{aligned}\eta_{ij} = \beta_{11} + \beta_{12} TIME_{ij} + \beta_{13} TIMEDRUG_{ij} + \beta_{14} SEX_i \\ + \beta_{15} PREVOI_i + \beta_{16} STRATUM_i + W_{1i}(TIME_{ij}),\end{aligned}$$

and the survival submodel assumes an exponential model for survival time with the log hazard

$$\begin{aligned}\log(\lambda_{ij}) = \beta_{21} + \beta_{22} DRUG_i + \beta_{23} SEX_i \\ + \beta_{24} PREVOI_i + \beta_{25} STRATUM_i + W_{2i}(SURTIME_{ij}).\end{aligned}$$

Guo and Carlin proposed a variety of joint models with different forms of the latent processes $W_1(s)$ and $W_2(t)$ and compare them using DIC. Here we only demonstrate the code using INLA for the model with the smallest DIC. That is, the case where $W_{1i}(s) = U_{1i} + U_{2i}s$ and $W_{2i}(t) = \gamma_1 U_{1i} + \gamma_2 U_{2i}$ is considered here. More detailed model comparisons for the study can be found in Guo and Carlin (2004) and Martino et al. (2011).

In order to fit the joint model in INLA, we need to manipulate the data and prepare the fixed and random covariates before using the `inla` function. Let us first read the dataset:

```
data(joint, package = "brinla")
longdat <- joint$longitudinal
survdat <- joint$survival
n1 <- nrow(longdat)
n2 <- nrow(survdat)
```

Now we need to prepare the response variables:

```
y.long <- c(longdat$y, rep(NA, n2))
y.surv <- inla.surv(time = c(rep(NA, n1), survdat$SURVTIME), event = c
    ↪ (rep(NA, n1), survdat$CENSOR))
Yjoint <- list(y.long, y.surv)
```

Then we prepare the fixed covariates and random covariates for the model:

```
linear.covariate <- data.frame(mu = as.factor(c(rep(1, n1), rep(2, n2)
    ↪ )), l.TIME = c(longdat$TIME, rep(0, n2)), l.TIMEDRUG = c(
    ↪ longdat$TIMEDRUG, rep(0, n2)), l.SEX = c(longdat$SEX, rep(0, n2
    ↪ )), l.PREVOI = c(longdat$PREVOI, rep(0, n2)), l.STRATUM = c(
    ↪ longdat$STRATUM, rep(0, n2)), s.DRUG = c(rep(0, n1), survdat$
    ↪ DRUG), s.SEX = c(rep(0, n1), survdat$SEX), s.PREVOI = c(rep(0,
    ↪ n1), survdat$PREVOI), s.STRATUM = c(rep(0, n1), survdat$STRATUM
    ↪ ))
ntime <- length(unique(longdat$TIME))

random.covariate <- list(U11 = c(rep(1:n2, each = ntime),rep(NA, n2)),
    ↪ U21 = c(rep(n2+(1:n2), each = ntime),rep(NA, n2)), U12 = c(rep
    ↪ (NA, n1), 1:n2), U22 = c(rep(NA, n1), n2+(1:n2)), U3 = c(rep(NA
    ↪ , n1), 1:n2))
```

We can finalize the joint dataset for the `INLA` program now:

```
joint.data <- c(linear.covariate,random.covariate)
joint.data$Y <- Yjoint
```

INLA allows different likelihoods for different observations. After implementing the commands of manipulating the orginal data in R, we are ready to fit the joint model as usual:

```
formula = Y ~ mu + l.TIME + l.TIMEDRUG + l.SEX + l.PREVOI + l.STRATUM
    ↪ + s.DRUG + s.SEX + s.PREVOI + s.STRATUM - 1 + f(U11 , model="
    ↪ iid2d", param = c(23,100,100,0), initial = c(-2.7,0.9,-0.22), n
    ↪ =2*n2) + f(U21, l.TIME, copy="U11") + f(U12, copy="U11", fixed
    ↪ = FALSE, param=c(0,0.01), initial = -0.2) + f(U22, copy="U11",
    ↪ fixed = FALSE, param = c(0,0.01), initial = -1.6)
joint.inla <- inla(formula, family = c("gaussian","exponentialsurv"),
    ↪ data = joint.data, control.compute=list(dic=TRUE))
round(joint.inla$summary.fixed, 4)
```

```
                 mean     sd 0.025quant 0.5quant 0.975quant    mode kld
mu1            8.0493 0.3516     7.3585   8.0494     8.7392  8.0495   0
mu2           -4.0592 0.2081    -4.4830  -4.0537    -3.6663 -4.0424   0
l.TIME        -0.2653 0.0487    -0.3612  -0.2653    -0.1699 -0.2652   0
l.TIMEDRUG     0.0288 0.0692    -0.1071   0.0288     0.1648  0.0288   0
l.SEX         -0.1060 0.3273    -0.7487  -0.1060     0.5364 -0.1061   0
l.PREVOI      -2.3492 0.2412    -2.8230  -2.3492    -1.8758 -2.3491   0
l.STRATUM     -0.1089 0.2375    -0.5754  -0.1089     0.3571 -0.1089   0
s.DRUG         0.2582 0.1777    -0.0900   0.2578     0.6080  0.2571   0
s.SEX         -0.1310 0.1469    -0.4098  -0.1346     0.1678 -0.1418   0
s.PREVOI       0.7544 0.1302     0.5055   0.7521     1.0165  0.7475   0
s.STRATUM      0.0730 0.0979    -0.1185   0.0727     0.2661  0.0720   0
```

```
round(joint.inla$summary.hyper, 4)
```

```
                                 mean     sd 0.025quant 0.5quant 0.975quant    mode
Precision for
the Gaussian observations      0.3480 0.0193     0.3114   0.3476     0.3871  0.3469
Precision for U11
(component 1)                  0.0648 0.0047     0.0561   0.0647     0.0745  0.0643
Precision for U11
(component 2)                  2.5754 0.1987     2.2010   2.5706     2.9804  2.5635
Rho1:2 for U11                -0.0630 0.0541    -0.1705  -0.0624     0.0414 -0.0602
Beta for U12                  -0.1933 0.0289    -0.2514  -0.1928    -0.1381 -0.1910
Beta for U22                  -1.6047 0.2504    -2.0888  -1.6082    -1.1050 -1.6187
```

From the above results of the longitudinal submodel, the estimated posterior mean of regression coefficient of `TIME` is -0.2653 with 95% credible interval of (-0.3612, -0.1699), suggesting a significant decrease (in the Bayesian sense) in CD4 count over the study period. `PREVOI` is also significant with the estimate -2.3492 and 95% credible interval of (-2.8230, -1.8758). In the survival submodel, only `PREVOI` is associated with the survival time with high probability (95% credible interval excludes 0). Note that, through the joint modeling, the posterior estimates of the association parameters are negative and significantly different from zero, which provides strong evidence of association between the two submodels. It indicates that a patient's survival is related to two characteristics driving the patient's longitudinal data pattern, the initial CD4 level and the rate of CD4 decrease. The finding is clinically reasonable, since high CD4 count represents better health status and patients with CD4 counts that are in more rapid decline would be expected to have poorer survival.

7

Random Walk Models for Smoothing Methods

Smoothing methods have been playing an important role in the *nonparametric* approach to regression. In this chapter, we introduce a few smoothing models that have been extensively used in statistical fields, e.g., smoothing splines, thin-plate splines and penalized regression splines (P-splines). We demonstrate how these models are linked to random walk (RW) priors under the Bayesian framework, and how to make Bayesian inference on those models using INLA in simulated and real data examples.

7.1 Introduction

Let's begin with the general regression problem. Given fixed $x_1, \ldots, x_n \in \mathbf{R}$, we observe $y_1, \ldots, y_n$ and assume the following model

$$y_i = f(x_i) + \varepsilon_i, \tag{7.1}$$

where ε_i are identically independently distributed with mean zero and unknown variance σ_ε^2. The problem is to estimate function f. There are two main approaches: parametric and nonparametric modeling.

The parametric approach is to assume that $f(x)$ belongs to a parametric family of functions $f(x \mid \beta)$ with a finite number of parameters. For example, we can have the linear function: $f(x \mid \beta) = \beta_0 + \beta_1 x$, polynomial function: $f(x \mid \beta) = \beta_0 + \beta_1 x + \beta_2 x^2$ or nonlinear function: $f(x \mid \beta) = \beta_0 \exp(\beta_1 x)$. The parametric modeling approach has several advantages. It is usually efficient when it is the correct model. It reduces the information necessary for prediction. The parameters may have intuitive interpretations. In short, you should prefer a parametric model if you have good information about an appropriate model family. However, the parametric approach will always exclude many plausible functions.

The nonparametric approach is to choose f from a specified family of *smooth* functions, resulting in a much larger range of potential fits to the data than the parametric approach. This is the process also called *smoothing* data, which aims at capturing important patterns (signals) in the data and leaving out noise. Unfortunately, such nonparametric regression models do not have a formulaic way of describing the relationship between the predictors and the response. This often needs to be done graphically, and therefore more data information is necessary for nonparametric prediction. However, the high flexibility of the nonparametric approach makes you less liable to make bad mistakes by using incorrect models, and thus particularly useful when little information about appropriate models is available.

7.2 Smoothing Splines

Assuming model (7.1), the smoothing spline of degree $2m-1$ is given by choosing f to minimize the *penalized least squares* criterion:

$$\sum_{i=1}^{n} [y_i - f(x_i)]^2 + \lambda \int \left(f^{(m)}(x)\right)^2 dx, \tag{7.2}$$

where $f^{(m)}$ is the mth derivative of f. The value of m is often chosen to be 1 or 2. This results in linear (degree = 1) and cubic (degree = 3) smoothing splines, respectively. The first term in (7.2) measures closeness to the data, while the second term, called the *penalty function*, penalizes roughness in the function. The smoothing parameter λ establishes a tradeoff between the two. When $\lambda = 0$, f is any function interpolating the data. When $\lambda = \infty$ it is the least squares line fit. From the frequentist point of view, the solution to minimizing (7.2) can be explicitly derived within a reproducing kernel Hilbert space, and λ is usually estimated via *cross validation* procedures (see, e.g., Wahba, 1990).

7.2.1 Random Walk (RW) Priors for Equally-Spaced Locations

We assume that each y_i independently follows a normal distribution with mean $f(x_i)$ and variance σ_ε^2. Without loss of generality, we assume observed x_i's are ordered as $x_1 < \cdots < x_n$ and define $d_i = x_i - x_{i-1}$ for $i = 1, \ldots, n-1$. Let's begin with the situation where x_i's are equally spaced, that is $d_i = d$, where d is some constant.

Under the Bayesian framework, we need to take a prior on f in order to estimate it. Following Speckman and Sun (2003) among others, we approximate the penalty function in (7.2) as follows:

$$\int \left(f^{(m)}(x)\right)^2 dx \approx d^{-(2m-1)} \sum_{i=m+1}^{n} [\nabla^m f(x_i)]^2,$$

given d is small and the mth derivative of f is continuous. The ∇^m is the notation for m^{th} order backward difference operator. For example, $\nabla^1 f(x) = f(x_i) - f(x_{i-1})$ and $\nabla^2 f(x) = f(x_i) - 2f(x_{i-1}) + f(x_{i-2})$ for $m = 1$ and 2, respectively. We then assume each difference independently follows a normal distribution, i.e.,

$$\nabla^m f(x_i) \overset{iid}{\sim} N\left(0, \sigma_f^2\right), \quad i = m+1, \ldots, n.$$

Letting $\boldsymbol{f} = (f(x_1), \ldots, f(x_n))^T$ be the vector of all function realizations, we can show that $\boldsymbol{f}$ follows a *singular* multivariate normal distribution with density function

$$\sqrt{\frac{|\boldsymbol{Q}_m|_+}{2\pi\sigma_f^2}} \exp\left(-\frac{1}{2\sigma_f^2}\boldsymbol{f}^T\boldsymbol{Q}_m\boldsymbol{f}\right),$$

where $\boldsymbol{Q}_m = \boldsymbol{D}_m^T\boldsymbol{D}_m$ and for $m = 1$ and 2,

$$\boldsymbol{D}_1 = \begin{pmatrix} -1 & 1 & & & \\ & -1 & 1 & & \\ & & \ddots & \ddots & \\ & & & -1 & 1 \end{pmatrix}, \quad \boldsymbol{D}_2 = \begin{pmatrix} 1 & -2 & 1 & & & \\ & 1 & -2 & 1 & & \\ & & \ddots & \ddots & \ddots & \\ & & & 1 & -2 & 1 \end{pmatrix},$$

and $|\boldsymbol{Q}_m|_+$ denotes the product of nonzero eigenvalues of $\boldsymbol{Q}_m$. The distribution is singular because $\boldsymbol{Q}_m$ is singular with a rank of $n-m$ and its null space is spanned by m^{th} order polynomials. This distribution is one kind of random walk (RW) prior . It can be shown that with this prior and a Gaussian likelihood, the posterior mean of $\boldsymbol{f}$ is a (discretized) Bayesian version of the smoothing spline solution defined in (7.2) with $\lambda = \sigma_\varepsilon^2/\sigma_f^2$ (e.g., Speckman and Sun, 2003).

To understand how this RW prior brings smoothness to the fitted function, we should take a look at the full conditional distribution $p(f(x_i) \mid \boldsymbol{f}_{-i})$, where $\boldsymbol{f}_{-i}$ denotes all the elements in $\boldsymbol{f}$ except $f(x_i)$. The distribution turns out to be normal with mean that is a weighted average of function values that come from the neighbors of x_i. For example, the full conditionals of first- and second-order RW priors (abbreviated as RW1 and RW2, respectively) are

$$\text{RW1:} \quad N\left(\frac{1}{2}\left[f(x_{i-1}) + f(x_{i+1})\right], \frac{\sigma_f^2}{2}\right),$$

$$\text{RW2:} \quad N\left(\frac{4}{6}\left[f(x_{i-1}) + f(x_{i+1})\right] - \frac{1}{6}\left[f(x_{i-2}) + f(x_{i+2})\right], \frac{\sigma_f^2}{6}\right),$$

where we can see the conditional independence properties of the two models: mean of $f(x_i)$ only depends on its first- or second-order neighbors, and is conditionally independent of those outside the neighborhood. This local structure applies smoothness to the estimated function. Moreover, it also makes $\boldsymbol{Q}_m$ a highly sparse matrix, which facilitates Bayesian computation greatly because a fast decomposition can be done to compute the matrix inverse (see details in Rue and Held, 2005).

As mentioned, $\boldsymbol{Q}_m$ in the RW prior is singular and its null space is spanned by m^{th} order polynomials. As a result, there will be an issue of identifiability if the model includes both polynomials and RW prior. For example, the RW1 prior is not identifiable with an intercept term because the null space of $\boldsymbol{Q}_1$ consists of a vector of ones; the RW2 prior is not identifiable with both intercept and slope terms because the null space of $\boldsymbol{Q}_2$ is spanned by $\mathbf{1}$ (vector of ones) and the vector of $(x_1, \ldots, x_n)$. To solve the problem we need to add a few linear constraints to the RW priors. They can be written in general as $\boldsymbol{A}\boldsymbol{f} = \mathbf{0}$, where $\boldsymbol{A}$ is the $m \times n$ matrix of mth order polynomials and $\mathbf{0}$ is the vector of n zeroes. If intercept is included in the model we need $\boldsymbol{A} = \mathbf{1}^T$, so-called sum-to-zero constraint $f(x_1) + \cdots + f(x_n) = 0$, to make $\boldsymbol{f}$ identifiable. If slope is also considered we need one more constraint, making $\boldsymbol{A} = (\mathbf{1}, \boldsymbol{x})^T$. Such constrained RW priors can be constructed by making easy adjustments to their unconstrained counterparts (see Rue and Held, 2005, Section 2.3).

Example: Simulated Data

Here we test the performance of RW models when estimating a smooth curve. The data are simulated from the following model:

$$y_i = \sin^3(2\pi x_i^3) + \varepsilon_i, \quad \varepsilon_i \sim N(0, \sigma_\varepsilon^2),$$

where $x_i \in [0, 1]$ are equally spaced and $\sigma_\varepsilon^2 = 0.04$. We simulate $n = 100$ data points:

```
library(INLA); library(brinla)
set.seed(1)
n <- 100
x <- seq(0, 1,, n)
f.true <- (sin(2*pi*x^3))^3
y <- f.true + rnorm(n, sd = 0.2)
```

We set the seed of the random number generator, so you will get the same results if you repeat this. We fit both RW1 and RW2 models to the same data for comparison purposes:

```
data.inla <- list(y = y, x = x)
formula1 <- y ~ -1 + f(x, model = "rw1", constr = FALSE)
result1 <- inla(formula1, data = data.inla)
formula2 <- y ~ -1 + f(x, model = "rw2", constr = FALSE)
result2 <- inla(formula2, data = data.inla)
```

We use '-1' to exclude the intercept from the model. The `constr` argument is a logical option that decides whether or not a sum-to-zero constraint is added to the fitted function. This option is turned off here, but it should be turned on (which is the default choice) if the intercept is included in the model.

The posterior summary regarding function estimation using RW1 is saved in `result1$summary.random$x`, and it is partially presented below:

```
round(head(result1$summary.random$x), 4)
```

```
      ID    mean     sd 0.025quant 0.5quant 0.975quant    mode kld
1 0.0000 -0.0637 0.1147    -0.2883  -0.0641     0.1627 -0.0649   0
2 0.0101 -0.0177 0.0993    -0.2132  -0.0177     0.1775 -0.0177   0
3 0.0202 -0.0177 0.0969    -0.2082  -0.0178     0.1729 -0.0179   0
4 0.0303  0.1023 0.0994    -0.0924   0.1020     0.2982  0.1014   0
5 0.0404  0.0527 0.0953    -0.1348   0.0527     0.2400  0.0526   0
6 0.0505 -0.0061 0.0972    -0.1976  -0.0060     0.1845 -0.0057   0
```

It includes the value of x_i (`ID`) as well as the posterior mean, standard deviation, quantiles and mode of each $f(x_i)$. We then extract the posterior mean and the quantiles for building 95% credible band:

```
fhat <- result1$summary.random$x$mean
f.lb <- result1$summary.random$x$'0.025quant'
```

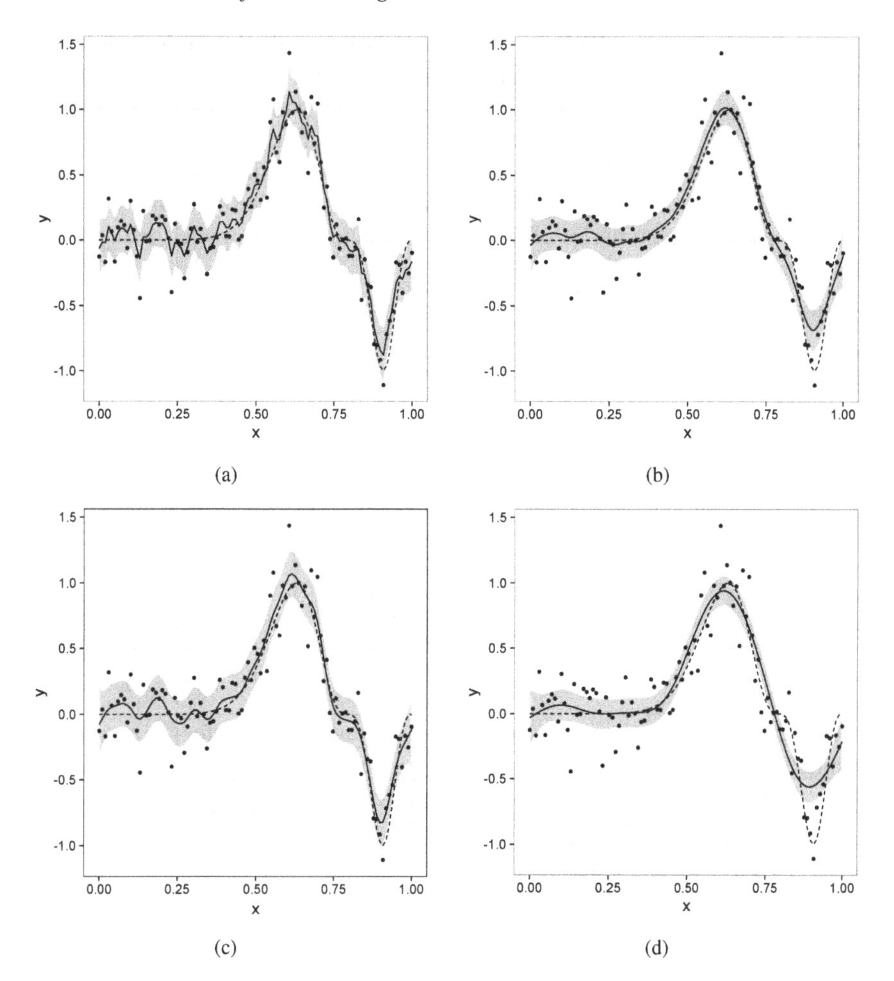

FIGURE 7.1
Simulated example: data points (dot), true mean function (dashed), posterior mean (solid) and 95% credible band (gray) using (a) RW1, (b) RW2, (c) cubic smoothing spline and (d) low-rank thin-plate spline.

```
f.ub <- result1$summary.random$x$'0.975quant'
```

We plot them in Figure 7.1(a), together with the true mean curve and data points:

```
library(ggplot2)
data.plot <- data.frame(y = y, x = x, f.true = f.true, fhat = fhat, f.
    ↪ lb = f.lb, f.ub = f.ub)
ggplot(data.plot, aes(x = x, y = y)) + geom_line(aes(y = fhat)) + geom
    ↪ _line(aes(y = f.true), linetype = 2) + geom_ribbon(aes(ymin = f
    ↪ .lb, ymax = f.ub), alpha = 0.2) + geom_point(aes(y = y)) +
    ↪ theme_bw(base_size = 20)
```

Similarly, we extract the function estimates using RW2 and plot them in Figure

7.1(b). We see that the RW1 fit, which corresponds to the linear smoothing spline, is too wiggly due to its lack of smoothness. The RW2 fit, corresponding to a cubic smoothing spline, seems to be much better, although it misses the point of inflection around $x = 0.75$ and fails to capture the minimum at about $x = 0.9$. The difficulty is that the true function has variable smoothness while the model assumes this is constant. This explains the underfitting for the more variable parts of the function.

The posterior summaries of the hyperparameters in the models using RW1 and RW2 priors are given by:

```
result1$summary.hyperpar
```

```
                                             mean         sd 0.025quant
Precision for the Gaussian observations 43.6093549 10.2886132 26.8688002
Precision for x                          0.6434155  0.2154812  0.3142227
                                         0.5quant 0.975quant       mode
Precision for the Gaussian observations 42.4460104  67.066242 40.219754
Precision for x                          0.6132653   1.149254  0.556862
```

```
result2$summary.hyperpar
```

```
                                               mean          sd   0.025quant
Precision for the Gaussian observations 28.262015420 4.683383372 2.006346e+01
Precision for x                          0.002233327 0.001396881 5.741318e-04
                                           0.5quant   0.975quant          mode
Precision for the Gaussian observations 27.927836236 38.410270375 27.307608331
Precision for x                          0.001905561  0.005837633  0.001348256
```

Note that they are the precision parameters, denoted by $1/\sigma_{\varepsilon}^2$ for "Precision for the Gaussian observations" and $1/\sigma_f^2$ for "Precision for x" in the output. We can use `bri.hyperpar.summary()` to obtain the posterior summaries of σ_{ε} and σ_f, the two standard deviations (SDs), which are usually easier to interpret:

```
round(bri.hyperpar.summary(result1), 4)
```

```
                                    mean     sd q0.025   q0.5 q0.975   mode
SD for the Gaussian observations 0.1545 0.0178 0.1224 0.1534 0.1924 0.1514
SD for x                         1.2982 0.2151 0.9350 1.2763 1.7784 1.2321
```

```
round(bri.hyperpar.summary(result2), 4)
```

```
                                   mean     sd  q0.025    q0.5  q0.975    mode
SD for the Gaussian observations  0.19 0.0156  0.1616  0.1892  0.2228  0.1875
SD for x                         24.06 7.2510 13.1515 22.8833 41.4569 20.7169
```

The two models give very different estimates on σ_f because they have different degrees of smoothness. The RW2 model yields a better estimate for σ_{ε} (recall its true value is 0.2), indicating that it is a more appropriate prior to use in this case. The low value for σ_{ε} in RW1 explains the rough fit as it thinks the errors are too small and that the true function is rougher than it really is.

We now compare INLA to other spline smoothing methods that can be implemented in R. We first try `smooth.spline()`, which fits a cubic smoothing spline using *generalized cross-validation (GCV)* to select the smoothing parameter λ as defined in (7.2):

```
fit.ss <- smooth.spline(x, y)
```

To account for the uncertainty in function estimation, we may build a 95% confidence band based on the following jackknife residuals (Green and Silverman, 1994):

```
res <- (fit.ss$yin - fit.ss$y)/(1 - fit.ss$lev)
```

Then we extract the fitted curve and compute the bounds of the confidence band:

```
fhat <- fit.ss$y
f.lb <- fhat - 2*sd(res)*sqrt(fit.ss$lev)
f.ub <- fhat + 2*sd(res)*sqrt(fit.ss$lev)
```

and plot them in Figure 7.1(c). Note that we use 2 (instead of 1.96) as a multiplier in forming in intervals. It is because the relevant quantity is not exactly normally distributed but might be more like a t distribution with a moderate number of degrees of freedom. Also, using 1.96 gives the impression that we assume exact normality, but using 2 makes it obvious that it is an approximation.

The fit given by `smooth.spline()` is smoother than the RW1 fit as we expect, but rougher than the RW2 fit. This can be explained by its estimated λ:

```
(fit.ss$lambda)
```

```
[1] 8.579964e-06
```

which is smaller than its counterpart in RW2:

```
result2$summary.hyperpar$mean[2]/result2$summary.hyperpar$mean[1]
```

```
[1] 7.902222e-05
```

It therefore outperforms RW2 on more variable parts of the function, while underperforms on the less variable parts.

Next we try `gam()` from the `mgcv` package:

```
library(mgcv)
fit.gam <- gam(y ~ s(x))
res.gam <- predict(fit.gam, se.fit = TRUE)
```

It fits a low-rank thin-plate spline, which is constructed by starting with the basis and penalty for a full thin-plate spline (see Section 7.3), and then truncating this basis in an optimal manner to obtain a low rank smoother (Wood, 2003). Note that `predict()` is used to compute standard errors based on the Bayesian posterior covariance matrix of the parameters in `fit.gam`. We then extract the fitted curve and compute the 95% credible bounds:

```
fhat <- res.gam$fit
f.lb <- res.gam$fit - 2*res.gam$se.fit
f.ub <- res.gam$fit + 2*res.gam$se.fit
```

and plot them in Figure 7.1(d). The fit given by `gam()` is the smoothest of all: it is very close to the true function on the flat part, but completely misses the point of inflection around $x = 0.75$ and is too far away from the minimum. It seems that the variable smoothness in the true function gives difficulty to all the methods we tried here, because they can only apply constant smoothing. In Section 7.6 we introduce a method to solve this issue by providing adaptive smoothing.

7.2.2 Choice of Priors on σ_ε^2 and σ_f^2

We want sensible priors on σ_ε^2 and σ_f^2 because these parameters together control how smooth the function fit will be. We need to be aware that INLA assigns the priors on the precisions $\delta = 1/\sigma_\varepsilon^2$ and $\tau = 1/\sigma_f^2$ instead of the original variances. Both precisions have the same default prior defined internally as a log gamma distribution. The corresponding gamma distribution gamma(a,b) has mean a/b and variance a/b^2. The values used for the default prior are $a = 1$ and $b = 5 \times 10^{-5}$, but these deserve a closer look.

Non-Informative Priors

The choice of values of a and b is not trivial because it is hard (if not impossible) to elicit any prior information on those parameters. To allow the data to speak for themselves, weakly informative or non-informative priors are often used in this situation. For example, setting $a = \varepsilon_1$ and $b = \varepsilon_2$, where ε_1 and ε_2 are small positive numbers, yields a gamma distribution with large variance (as the default prior); setting $a = -1$ and $b = 0$ corresponds to a flat prior, which is the restricted maximum likelihood REML estimation in an empirical Bayes approach; setting $a = -0.5$ and $b = 0$ is recommended as a standard choice in practical work by Gelman (2006); setting $a = b = 0$ results in the so-called Jeffreys prior, a popular choice for objective Bayesian inference.

Since some priors mentioned above have improper densities, the corresponding joint posterior distribution may be improper as well, which leads to invalid Bayesian inference. Sun et al. (1999), Speckman and Sun (2003), Sun and Speckman (2008) and Fahrmeir and Kneib (2009) investigate the posterior propriety in Bayesian nonparametric regression models and generalized additive models (see Chapter 9). To summarize their work, the Jeffreys prior can only be used on δ, otherwise the posterior distribution will be improper. The other priors can be used on both δ and τ, and will lead to a proper posterior distribution under mild conditions. The gamma$(\varepsilon_1, \varepsilon_2)$ is recommended because it provides a robust performance with respect to the choice of $(\varepsilon_1, \varepsilon_2)$, and yields more stable precision estimates than the other priors.

We can use priors other than the default ones for δ and τ in INLA. Suppose we want to take a Jeffreys prior ($a_1 = b_1 = 0$) on δ:

```
a1 <- 5e-5
b1 <- 5e-5
lgprior1 <- list(prec = list(param = c(a1, b1)))
```

and a Gelman recommended prior ($a_2 = -0.5, b_2 = 0$) on τ:

```
a2 <- -0.5
b2 <- 5e-5
lgprior2 <- list(prec = list(param = c(a2, b2)))
```

Note that we use a tiny number instead of 0 for a_1, b_1 and b_2, as required by INLA. We then apply these two priors to the RW2 model used in the simulated example in Section7.2.1:

```
formula <- y ~ -1 + f(x, model = "rw2", constr = FALSE, hyper =
    ↪ lgprior2)
result <- inla(formula, data = data.inla, control.family = list(hyper
    ↪ = lgprior1))
```

Here `control.family` is used to specify the prior on δ and `hyper` in `f()` on τ. We find out that the resulting posterior distributions do not differ much from those given by the default priors, although the prior specification on τ is relatively more important than that on δ.

Prior Scaling

We start this section with an example. Following the simulated example in Section 7.2.1, we simulate data as follows:

```
set.seed(1)
n <- 100
t <- 0.1
x <- seq(0, t,, n)
f.true <- (sin(2*pi*(x/t)^3))^3
y <- f.true + rnorm(n, sd = 0.2)
```

We have changed the range of x from [0, 1] to [0, 0.1], but everything else remains the same. We then fit the model in INLA:

```
data.inla <- list(y = y, x = x)
formula1 <- y ~ -1 + f(x, model = "rw2", constr = FALSE)
result1 <- inla(formula1, data = data.inla)
```

and plot the fitted curve and 95% credible band in Figure 7.2(a):

```
p <- bri.band.ggplot(result1, name = 'x', type = 'random')
p + geom_line(aes(y = f.true), linetype = 2)
```

where we see the RW2 model completely fails by providing a linear fit. The posterior summary of the SDs in the Gaussian noise and the RW2 model is given by:

```
round(bri.hyperpar.summary(result1), 4)
```

```
                                 mean     sd q0.025   q0.5 q0.975   mode
SD for the Gaussian observations 0.4662 0.0326 0.4062 0.4646 0.5342 0.4612
SD for x                         0.0100 0.0059 0.0037 0.0083 0.0261 0.0061
```

The noise SD is significantly overestimated, compared to its true value 0.2. This result is based on using the default prior on the precision τ (inverse of SD^2) in RW2, which is $\log\text{Gamma}(1, 5\text{e-}5)$.

We now try a much bigger scale with $t = 1{,}000$, but use the same data, same model and same priors. The resulting fit is shown in Figure 7.2(b), where we see it is more wiggly than it is supposed to be. Let us look at the posterior SDs (the R code is not displayed here):

```
                                 mean     sd q0.025   q0.5 q0.975   mode
SD for the Gaussian observations 0.1702 0.0145 0.1438 0.1693 0.2009 0.1676
SD for x                         0.0027 0.0004 0.0020 0.0027 0.0036 0.0026
```

and we see the noise SD is now underestimated, and the model SD (0.0027) is much smaller than the one (0.009) in the case when $t = 0.1$. We finally try $t = 10$, a closer

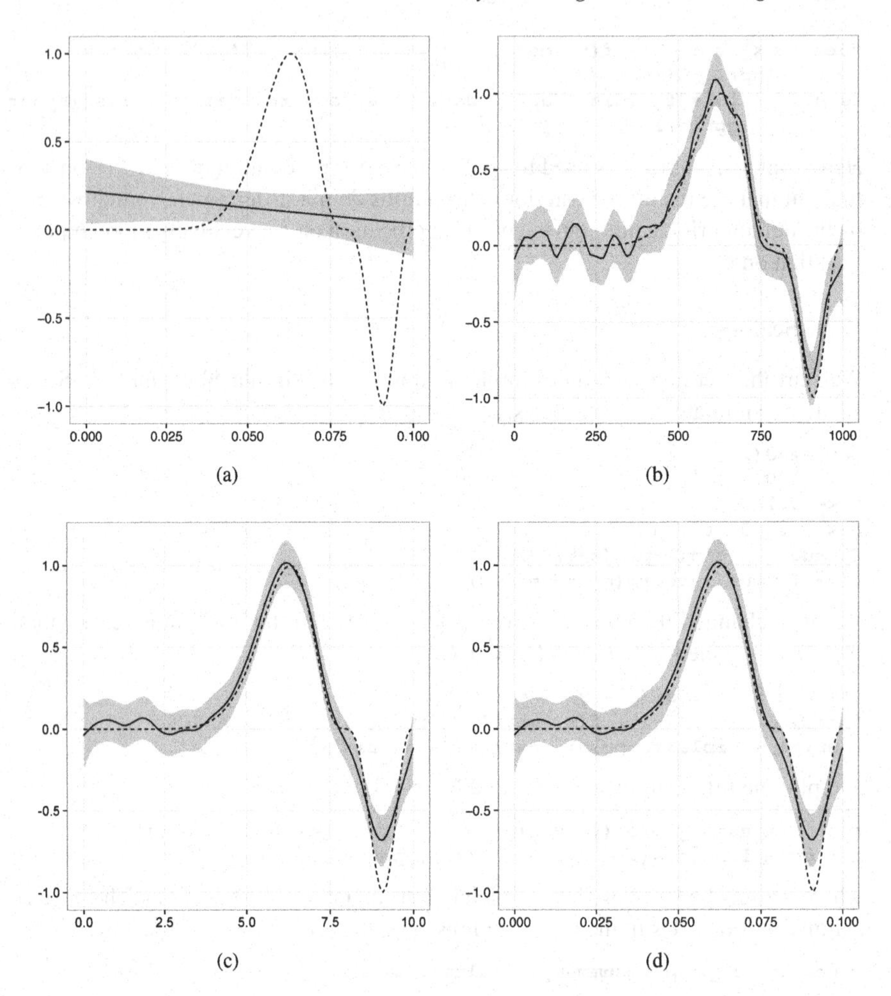

FIGURE 7.2
RW models: true mean function (dashed), posterior mean (solid) and 95% credible band (gray) when (a) $t = 0.1$ with unscaled prior; (b) $t = 1,000$ with unscaled prior; (c) $t = 10$ with unscaled prior; (d) $t = 0.1$ with scaled prior.

scale to $t = 1$, and obtain the same fit as before (see Figure 7.2(c)), but a different estimate of SD for x as we expected:

```
                                 mean     sd q0.025   q0.5 q0.975   mode
SD for the Gaussian observations 0.1900 0.0156 0.1616 0.1892 0.2228 0.1875
SD for x                         0.7609 0.2293 0.4159 0.7238 1.3109 0.6554
```

Obviously, it is not appropriate to use the default hyperprior on τ when the scale of x is too small ($t = 0.1$) or too big ($t = 1,000$). This is because the RW type models (not only RW2) penalize local deviation from a certain level, and the chosen hyperprior influences how large we allow this local deviation to be. The RW models of different scales, of course, require different ranges for their own deviations. The default prior is able to provide a good range when $t = 1$ or $t = 10$, but fails to do so when $t = 0.1$ or $t = 1,000$. Moreover, the hyperprior on τ does not have the same interpretation for different RW models because they reflect different neighborhood structures. It is therefore unreasonable to assign the same fixed hyperprior for those different models.

To solve the issue, Sørbye and Rue (2014) proposed to scale the hyperprior in such a way that its *generalized variance* (the geometric mean of the marginal variances) equals 1. As a result, the *scaled* hyperprior is invariant to the covariates of different scales and to the shape and size of the neighborhood structure for a specific model. It implies that the precision parameters in different models have a similar interpretation. This useful feature is built into INLA via the `scale.model` argument:

```
formula2 <- y ~ -1 + f(x, model = "rw2", constr = FALSE, scale.model =
    ↪ TRUE)
result2 <- inla(formula2, data = data.inla)
```

The posterior summary of SDs when $t = 0.1$ is given by:

```
round(bri.hyperpar.summary(result2), 4)
```

```
                                 mean     sd q0.025   q0.5 q0.975   mode
SD for the Gaussian observations 0.1900 0.0156 0.1616 0.1892 0.2228 0.1875
SD for x                         1.0006 0.3015 0.5469 0.9516 1.7240 0.8616
```

Compared to the result from the unscaled version, we obtain a much better estimate for the noise SD, and that estimate is the same as the one for $t = 10$ or $t = 1$. However, the estimate of SD for x does not reflect the original scale of x anymore because of the scaled hyperprior. We plot the corresponding fit in Figure 7.2(d), and we see it is exactly the same fit as for $t = 10$ or $t = 1$. We also tried the scaled model with $t = 1,000$ and obtained the same result as with $t = 0.1$.

The `scale.model` option is not only available for RW1 and RW2 models, but also for the RW2D and Besag models that will be introduced later. The downside to scaling is that the output has parameters in the new scale, but one might want results in the original scale. Unfortunately, INLA does not provide any information to do that. Also, there could be a situation that one won't want to scale a prior because the original scale is already appropriate. The problem might be better solved by using the default priors that scale with the data, but it may not be a good idea to use the data to say anything about the prior.

7.2.3 Random Walk Models for Non-Equally Spaced Locations

The RW models mentioned above only work for the data with equally-spaced locations. We would like to extend them to the applications where the locations are non-equally spaced. Wahba (1978) showed that the smoothing spline estimator is the solution to the following *stochastic differential equation* (SDE)

$$d^m f(x)/dx^m = \sigma_f dW(x)/dx, \quad m = 1,2, \tag{7.3}$$

where σ_f controls the scale of $f(x)$, and $W(x)$ is the standard Wiener process. Note that in (7.3) the left-hand side is the mth order derivative of f and the right-hand side is the Gaussian noise with mean zero and variance σ_f^2. The exact solution has a Bayesian representation to work with a Gaussian process prior. Unfortunately, that prior is computationally intensive because its covariance matrix is completely dense. To tackle this issue, we "weakly" solve the SDE in the way suggested in Rue and Held (2005) and Lindgren and Rue (2008a).

Let $x_1 < x_2 < \cdots < x_n$ be a sequence of observed locations, and $d_i = x_{i+1} - x_i$. Following Chapter 3 in Rue and Held (2005), the RW1 model can be viewed as the realization of a Wiener process at x_i from the SDE in (7.3) by letting

$$f(x_{i+1}) - f(x_i) \overset{iid}{\sim} N\left(0, d_i\sigma_f^2\right), \quad i = 1, \ldots, n-1.$$

Then, the prior distribution of $\boldsymbol{f} = (f(x_1), \ldots, f(x_n))'$ is (singular) multivariate normal with mean zeroes and precision matrix $\sigma_f^{-2}\boldsymbol{Q}_1$. The matrix $\boldsymbol{Q}_1$ is singular (rank $= n-1$) and has a banded structure given by

$$\boldsymbol{Q}_1[i,j] = \begin{cases} 1/d_{i-1} + 1/d_i & j = i \\ -1/d_i & j = i+1 \\ 0 & \text{otherwise} \end{cases},$$

for $1 < i < n$, with $\boldsymbol{Q}_1[1,1] = 1/d_1$ and $\boldsymbol{Q}_1[n,n] = 1/d_{n-1}$. As a result, the full conditional distribution of $f(x_i)$ is

$$N\left(\frac{d_i}{d_{i-1}+d_i} f(x_{i-1}) + \frac{d_{i-1}}{d_{i-1}+d_i} f(x_{i+1}), \frac{d_{i-1}d_i}{d_{i-1}+d_i}\sigma_f^2\right),$$

where the distances between locations affect the mean and variance of the distribution. The model can be interpreted as a discretely observed Wiener process that is adjusted for the non-equally spaced locations. We obtain the same result as in the equally spaced RW1 if $d_i = 1$ for all i.

Regarding the RW2 model, it is possible to find a solution to SDE (7.3) that has the conditional independence on an augmented space, but the computations take about 9/2 the time as for the RW2 model for equally spaced locations (see Rue and Held, 2005, Chapter 3.5 for details). To avoid the increased complexity, Lindgren and Rue (2008a) used a *finite element method* to derive a Gaussian model that approximates a continuous time integrated random walk. They first approximate $f(x)$ using a piecewise linear basis expansion, and then turn the SDE into a system of linear equations. It results in a RW2 model with a banded matrix $\boldsymbol{Q}_2$, whose non-zero elements of row i are

$$\boldsymbol{Q}_2[i,i-2] = \frac{2}{d_{i-2}d_{i-1}(d_{i-2}+d_{i-1})}, \quad \boldsymbol{Q}_2[i,i-1] = -\frac{2}{d_{i-1}^2}\left(\frac{1}{d_{i-2}} + \frac{1}{d_i}\right),$$

$$\boldsymbol{Q}_2[i,i] = \frac{2}{d_{i-1}^2(d_{i-2}+d_{i-1})} + \frac{2}{d_{i-1}d_i}\left(\frac{1}{d_{i-1}} + \frac{1}{d_i}\right) + \frac{2}{d_i^2(d_i+d_{i+1})},$$

with $\boldsymbol{Q}_2[i,i+1] \equiv \boldsymbol{Q}_2[i+1,i]$ and $\boldsymbol{Q}_2[i,i+2] \equiv \boldsymbol{Q}_2[i+2,i]$ due to the symmetry. For the elements in the upper left and lower right corner of $\boldsymbol{Q}_2$, we simply ignore the non-existing components, or, equivalently, let $d_{-1} = d_0 = d_n = d_{n+1} = \infty$. The matrix Q_2 is of rank $n-2$, with the null space spanned by $(1,\ldots,1)^T$ and $(x_1,\ldots,x_n)^T$. It coincides with the result obtained in Wahba (1978) for cubic smoothing splines. The consistency of this model has been shown in Lindgren and Rue (2008a) and Simpson et al. (2012).

Example: Munich Rental Guide

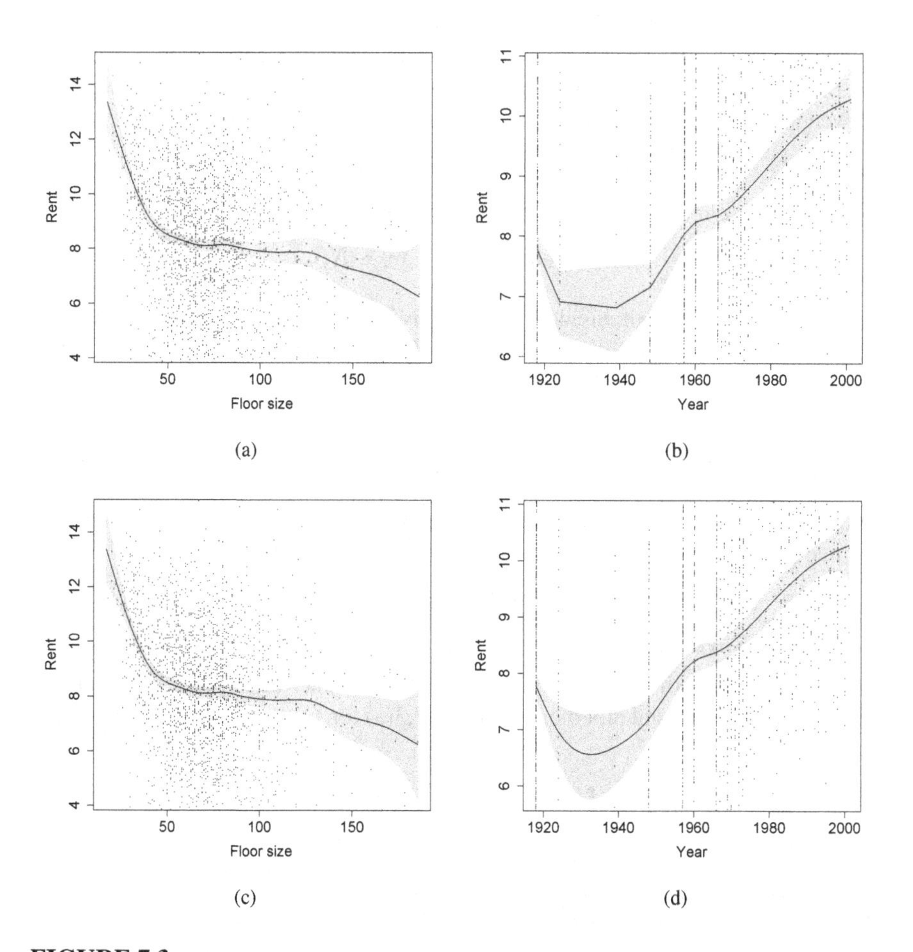

FIGURE 7.3
Munich rental guide using RW models: data points (dot), the posterior mean (solid line) and 95% credible interval (gray band) of the nonlinear effects of *floor size* (a and c) and *construction year* (b and d).

The German tenancy law puts restrictions on the increase of rents and forces landlords to keep the price in a range defined by apartments which are comparable in size, location and quality. To make it easier for tenants and owners to assess if the rent is appropriate for an apartment, the so-called rental guides are derived based on large samples of apartments. We here use the 2003 Munich rental dataset that has been analyzed in Rue and Held (2005). It is a data frame with 17 columns and 2035 rows. Each column represents a predictor and each row a sample of apartments. There are totally $n = 2,035$ flats from 380 districts in Munich. The response variable of interest is `rent` (rent per square meter in Euros) for a flat, and the predictors are `floor.size` (floor size from 17 to 185 square meters), `year` (construction year from 1918 to 2001), *spatial location* (380 districts) and various indicator variables such as an indicator for a flat with no central heating, no warm water, more refined kitchen equipment and so on. Type `?Munich` in R for details of these variables.

Based on the data plots (not shown), there appears to be a nonlinear relationship between `rent` and `floor.size`, and between `rent` and `year`. We therefore consider the following nonparametric regression models:

$$\begin{aligned} \texttt{rent}_i &= f_1(\texttt{floor.size}_i) + \varepsilon_{1i}, \quad \varepsilon_{1i} \sim N(0, \sigma^2_{\varepsilon_1}), \\ \texttt{rent}_i &= f_2(\texttt{year}_i) + \varepsilon_{2i}, \quad \varepsilon_{2i} \sim N(0, \sigma^2_{\varepsilon_2}), \end{aligned}$$

for $i = 1, \ldots, n$, although an additive model is obviously more appropriate and will be considered in Chapter 9. Note that both `floor.size` and `year` are non-equally spaced. The models are formulated using the RW2 priors described in Section 7.2.3 as follows:

```
formula1 <- rent ~ -1 + f(floor.size, model = 'rw2', constr = FALSE)
formula2 <- rent ~ -1 + f(year, model = 'rw2', constr = FALSE)
```

INLA is able to detect the presence of the non-equally spaced predictors, so we do not need to specify this fact. We then fit the models via INLA:

```
data(Munich, package = "brinla")
result1 <- inla(formula1, data = Munich)
result2 <- inla(formula2, data = Munich)
```

To visualize the estimated nonlinear effects, we extract their posterior means and 95% credible bands, and plot them in Figure 7.3(a) for `floor.size` and Figure 7.3(b) for `year`:

```
bri.band.plot(result1, name = 'floor.size', alpha = 0.05, xlab = '
    ↪ Floor size', ylab = 'Rent', type = 'random')
points(Munich$floor.size, Munich$rent, pch = 20, cex = 0.2)
bri.band.plot(result2, name = 'year', alpha = 0.05, xlab = 'Year',
    ↪ ylab = 'Rent', type = 'random')
points(Munich$year, Munich$rent, pch = 20, cex = 0.2)
```

In `bri.band.plot()` we use '`name`' to specify the variable to be plotted, '`alpha`' the significance level, and '`type`' the model component, which can be random effect (`random`), fitted values (`fitted`) and linear predictor (`linear`). As we can see, both `floor.size` and `year` show significant nonlinear patterns with `rent`. Due to the wars and other problems, there is very little data between 1920 and 1950. The fit becomes piecewise linear and the credible bands are wide in this period.

Regarding the hyperparameters in the two models, their posterior means, standard deviations (SDs), quantiles, and modes are calculated as:

```
round(bri.hyperpar.summary(result1), 4)
```

```
                                   mean     sd q0.025   q0.5 q0.975   mode
SD for the Gaussian observations 2.3386 0.0364 2.2681 2.3381 2.4111 2.3373
SD for floor.size                0.0156 0.0053 0.0078 0.0147 0.0284 0.0131
```

```
round(bri.hyperpar.summary(result2), 4)
```

```
                                   mean     sd q0.025   q0.5 q0.975   mode
SD for the Gaussian observations 2.3134 0.0360 2.2437 2.3129 2.3851 2.3121
SD for year                      0.0236 0.0093 0.0101 0.0221 0.0461 0.0192
```

When comparing the two models, the SD for the noise (`Gaussian observations`) is directly comparable since the response is the same. In contrast, the SD for the change in the response as the predictor varies depends on the scale of the predictor. Hence these two SDs should not be compared.

Prediction in INLA

Both `floor.size` and `year` are non-equally spaced. The former has a range from 17 to 185, and the latter from 1918 and 2001. There are gaps in the `year` observations. To fill these gaps, we make a vector of all possible integers in the range from 1918 to 2001. Although `floor.size` is more densely measured, we also demonstrate how predictions may be constructed in INLA. The points at which predictions will be made is specified using the '`values`' option in `f()`:

```
formula3 <- rent ~ -1 + f(floor.size, model = "rw2", values = seq(17,
    ↪ 185), constr = FALSE)
formula4 <- rent ~ -1 + f(year, model = "rw2", values = seq(1918,
    ↪ 2001), constr = FALSE)
```

Note that the vector used in `values` must contain every observed value in the data. We then fit the corresponding models:

```
result3 <- inla(formula3, data = Munich)
result4 <- inla(formula4, data = Munich)
```

and obtain the posterior results as before. We now have posterior estimates for every possible value in `floor.size` and `year`, even for those that have no observations. Figures 7.3(c) and 7.3(d) show the resulting fitted curves, together with their 95% credible bands. Compared to the previous results, the fitted curves are very similar in the regions of dense observation but much smoother in the sparser areas.

There is an alternative way to make predictions in INLA. Suppose we are interested in predicting the rents at year 1925, 1938 and 1945. Because making a prediction is the same as fitting a model with some missing data in INLA, we can simply set the response to be `NA` for the "locations" we want to predict at:

```
x.new <- c(1925, 1938, 1945)
xx <- c(Munich$year, x.new)
yy <- c(Munich$rent, rep(NA, length(x.new)))
```

and fit the model as before:

```
data.pred <- list(y = yy, x = xx)
formula5 <- y ~ -1 + f(x, model = 'rw2', constr = FALSE)
```

```
result5 <- inla(formula5, data = data.pred, control.predictor = list(
    compute = TRUE))
```

To obtain the posterior summary of the three predictions we must first find their indices in the INLA result:

```
ID <- result5$summary.random$x$ID
idx.new <- sapply(x.new, function(x) which(ID==x))
```

and then extract the corresponding output:

```
round(result5$summary.random$x[idx.new,], 4)
```

```
    ID   mean     sd 0.025quant 0.5quant 0.975quant   mode kld
3 1925 6.8328 0.3005     6.2184   6.8430     7.3863 6.8626   0
4 1938 6.7568 0.3603     6.0263   6.7652     7.4436 6.7855   0
6 1945 6.9825 0.2463     6.4854   6.9873     7.4516 6.9968   0
```

The marginal densities of them are stored in `result5$marginals.random$x`, from which other posterior quantities can be computed.

The predictive distributions above refer only to the mean rent for a given year (the linear predictor) so we are expressing only the uncertainty in this component through its credible interval. Suppose one wants to rent an apartment in a certain year, and needs a *prediction interval* such that the probability that the *monthly* rent (not mean rent) lies in that interval is, say 95%. This requires adding an error term ε to the linear predictor and knowledge of the uncertainty in ε. Unfortunately, INLA does not provide such prediction intervals directly, and we have to do it manually. One approach is to draw 100,000 samples from the posterior for the error precision $(1/\sigma_\varepsilon^2)$, convert this to an SD, and sample from the normal densities with these SDs. We then generate samples from posterior for the linear predictor computed earlier, and combine them with these randomly generated new ε's:

```
nsamp <- 10000
pred.marg <- NULL
for(i in 1:length(x.new)){
  error.prec <- inla.hyperpar.sample(nsamp, result5)[,1]
  new.eps <- rnorm(nsamp, mean = 0, sd = 1/sqrt(error.prec))
  pm.new <- result5$marginals.linear.predictor[[which(is.na(data.pred$
      y))[i]]]
  samp <- inla.rmarginal(nsamp, pm.new) + new.eps
  pred.marg <- cbind(samp, pred.marg)
}
```

The posterior quantities can be computed manually from these combined samples:

```
p.mean <- colMeans(pred.marg)
p.sd <- apply(pred.marg, 2, sd)
p.quant <- apply(pred.marg, 2, function(x) quantile(x, probs = c
    (0.025, 0.5, 0.975)))
data.frame(ID = x.new, mean = p.mean, sd = p.sd, '0.025quant' = p.
    quant[1,], '0.5quant' = p.quant[2,], '0.975quant' = p.quant
    [3,], check.names = FALSE)
```

```
    ID     mean       sd 0.025quant 0.5quant 0.975quant
1 1925 6.974585 2.331187   2.340086 6.974354   11.52660
2 1938 6.764161 2.353084   2.213512 6.746318   11.41703
3 1945 6.811205 2.332284   2.245040 6.821577   11.34970
```

Compared to the predictive distribution for the new linear predictor, the predictive distribution for the new response contains a much greater variation, reflected by its much bigger SD and a much wider 95% predictive interval.

7.3 Thin-Plate Splines

Consider a nonparametric regression model with two predictors

$$y_i = f(x_{1i}, x_{2i}) + \varepsilon_i, \quad i = 1, \ldots, n, \tag{7.4}$$

where f is an unknown but smooth function on the $\mathbf{R}^2$ domain, and ε_i is a noise term with zero mean. Without loss of generality, we scale x_1 and x_2 to be both in [0,1] ranges. An intuitive extension of cubic smoothing spline to the $\mathbf{R}^2$ space uses *thin-plate splines*. It is the solution to minimizing the following penalized least squares criterion

$$\sum_{i=1}^{n} \left(y_i - f(x_{1i}, x_{2i}) \right)^2 + \lambda P_2(f), \tag{7.5}$$

where $P_2(f)$ is the penalty function given by

$$\int\int_{\mathbf{R}^2} \left[\left(\frac{\partial^2 f}{\partial x_1^2} \right)^2 + 2 \left(\frac{\partial^2 f}{\partial x_1 \partial x_2} \right)^2 + \left(\frac{\partial^2 f}{\partial x_2^2} \right)^2 \right] dx_1 \, dx_2, \tag{7.6}$$

and the smoothing parameter λ controls the tradeoff between fidelity to the data from sum squared errors and function smoothness from the penalty. The value of the penalty function is not affected by changing the coordinates by rotation or translation in $\mathbf{R}^2$. It is always non-negative and equals zero if and only if $f(x_1, x_2)$ is a linear function of x_1 and x_2.

Thin-plate splines have a mechanical interpretation. Suppose that an infinite elastic flat plate interpolates a set of points $[\boldsymbol{x}_i, y_i], i = 1, \ldots, n$. Then the "bending energy" of the plate is proportional to the penalty (7.6), and the minimum energy solution is the thin-plate spline.

7.3.1 Thin-Plate Splines on Regular Lattices

We begin with the case where data are observed on regular lattices. Let $x_{11} < x_{12} < \cdots < x_{1n_1}$ and $x_{21} < x_{22} < \cdots < x_{2n_2}$ be two sequences of equally-spaced locations, which construct an $n_1 \times n_2$ lattice. As shown in Yue and Speckman (2010), a prior for Bayesian thin-plate splines in this scenario can be derived by, first, defining a few second-order difference operators on lattices:

$$
\begin{aligned}
\nabla^2_{(1,0)} f(x_{1i}, x_{2j}) &= f(x_{1i}, x_{2j}) - 2f(x_{1,i-1}, x_{2j}) + f(x_{1,i-2}, x_{2j}), \\
\nabla^2_{(0,1)} f(x_{1i}, x_{2j}) &= f(x_{1i}, x_{2j}) - 2f(x_{1i}, x_{2,j-1}) + f(x_{1i}, x_{2,j-2}), \\
\nabla^2_{(1,1)} f(x_{1i}, x_{2j}) &= f(x_{1i}, x_{2j}) - f(x_{1,i-1}, x_{2j}) - f(x_{1i}, x_{2,j-1}) + f(x_{1,i-1}, x_{2,j-1}),
\end{aligned}
$$

and then letting them independently follow a normal distribution with mean 0 and variance σ_f^2. The resulting prior distribution of $\boldsymbol{f} = (f(x_{11}, x_{21}), \ldots, f(x_{1n_1}, x_{2n_2}))'$ is again multivariate normal with mean zeroes and precision matrix $\sigma_f^{-2}\boldsymbol{Q}$. The matrix $\boldsymbol{Q}$ is the sum of three Kronecker products:

$$
\boldsymbol{Q} = \boldsymbol{I}_{n_2} \otimes \boldsymbol{B}^{(2)}_{n_1} + \boldsymbol{B}^{(2)}_{n_2} \otimes \boldsymbol{I}_{n_1} + 2\boldsymbol{B}^{(1)}_{n_1} \otimes \boldsymbol{B}^{(1)}_{n_2}, \tag{7.7}
$$

where $\boldsymbol{I}_n$ denotes $n \times n$ identity matrix, $\boldsymbol{B}^{(1)}_n = \boldsymbol{D}_1'\boldsymbol{D}_1$, $\boldsymbol{B}^{(2)}_n = \boldsymbol{D}_2'\boldsymbol{D}_2$, and $\boldsymbol{D}_1$ and $\boldsymbol{D}_2$ are as defined for RW1 and RW2 models, respectively, in Section 7.2.1. Letting $n = n_1 \times n_2$, $\boldsymbol{Q}$ is an $n \times n$ sparse and singular matrix of rank $n - 3$ with null space spanned by the two-dimensional second-order polynomials. Note that the dimension of $\boldsymbol{Q}$ can easily grow large, and therefore its sparsity is crucial for the efficient INLA computation. We term this prior a two-dimensional second-order RW model (RW2D) since it is an extension of the RW2 model to the two-dimensional space.

Using graphical notation, the full conditional distribution of $f(x_{1i}, x_{2j})$ in the interior of lattices is normal with mean

$$
\frac{1}{20}\left(8\begin{smallmatrix}\circ&\circ&\circ&\circ&\circ\\ \circ&\circ&\bullet&\circ&\circ\\ \circ&\bullet&\circ&\bullet&\circ\\ \circ&\circ&\bullet&\circ&\circ\\ \circ&\circ&\circ&\circ&\circ\end{smallmatrix} - 2\begin{smallmatrix}\circ&\circ&\circ&\circ&\circ\\ \circ&\bullet&\circ&\bullet&\circ\\ \circ&\circ&\circ&\circ&\circ\\ \circ&\bullet&\circ&\bullet&\circ\\ \circ&\circ&\circ&\circ&\circ\end{smallmatrix} - 1\begin{smallmatrix}\circ&\circ&\bullet&\circ&\circ\\ \circ&\circ&\circ&\circ&\circ\\ \bullet&\circ&\circ&\circ&\bullet\\ \circ&\circ&\circ&\circ&\circ\\ \circ&\circ&\bullet&\circ&\circ\end{smallmatrix}\right), \tag{7.8}
$$

and variance $\sigma_f^2/20$, where the locations denoted by "•" represent the neighbors that (x_{1i}, x_{2j}) depends on, and the number in front of each grid denotes the weight given to the corresponding "•" locations. It shows the conditional independence of RW2D model: each $f(x_{1i}, x_{2j})$ only conditionally depends on its 12 neighbors through different weights, and the closer the neighbor is, the bigger its (absolute) weight is.

Example: Simulated Data

The true mean function is a bimodal two-dimensional function defined by

$$
\begin{aligned}
f(x_1, x_2) &= \frac{0.75}{\pi\sigma_{x_1}\sigma_{x_2}} \exp\left[-\frac{1}{\sigma^2_{x_1}}(x_1 - 0.2)^2 - \frac{1}{\sigma^2_{x_2}}(x_2 - 0.3)^2\right] \\
&\quad + \frac{0.45}{\pi\sigma_{x_1}\sigma_{x_2}} \exp\left[-\frac{1}{\sigma^2_{x_1}}(x_1 - 0.7)^2 - \frac{1}{\sigma^2_{x_2}}(x_2 - 0.8)^2\right],
\end{aligned}
$$

where $\sigma_{x_1} = 0.3$ and $\sigma_{x_2} = 0.4$. We write this function in R as

```
test.fun <- function(x,z,sig.x,sig.z){.75/(pi*sig.x*sig.z)*exp(-(x -
    .2)^2/sig.x^2-(z - .3)^2/sig.z^2) + .45/(pi*sig.x*sig.z)*exp(-(
    x - .7)^2/sig.x^2-(z - .8)^2/sig.z^2)}
```

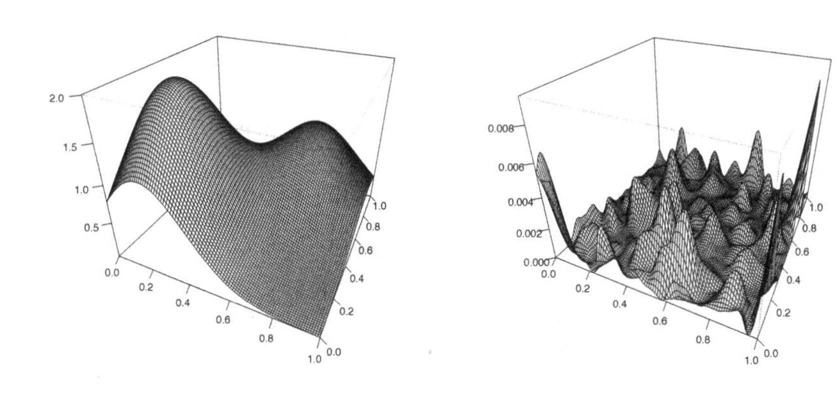

FIGURE 7.4
Thin-plate splines: true mean function (left) and squared errors between the true mean function and its thin-plate spline estimator (right).

We then simulate data from the following model

$$y = f(x_1, x_2) + \varepsilon, \quad \varepsilon \sim N(0, \sigma_\varepsilon^2),$$

given $\sigma_\varepsilon^2 = 0.09$. Both x_1 and x_2 are equally spaced within range $[0, 1]$, constructing a 100×100 lattice. We therefore generate function realizations on that lattice:

```
nrow <- 100
ncol <- 100
s.mat <- matrix(NA, nrow = nrow, ncol = ncol)
for(i in 1:nrow){
  for(j in 1:ncol){
    s.mat[i,j] <- test.fun(i/100, j/100, 0.3, 0.4)
  }
}
```

The data are simulated by adding Gaussian noises to the realizations:

```
set.seed(1)
noise.mat <- matrix(rnorm(nrow*ncol, sd = 0.3), nrow, ncol)
y.mat <- s.mat + noise.mat
```

All indices in INLA are one-dimensional so an appropriate mapping is required to get the data matrix into the ordering defined internally in INLA. It can be done by using `inla.matrix2vector()` function, which transforms a matrix to a vector with correct ordering:

```
y <- inla.matrix2vector(y.mat)
```

We then fit a Bayesian thin-plate spline model using the RW2D prior:

```
formula <- y ~ -1 + f(x, model="rw2d", nrow=nrow, ncol=ncol, constr=F)
data <- data.frame(y = y, x = 1:(nrow*ncol))
result <- inla(formula, data = data)
```

The posterior summary of function estimate is saved in `result$summary.random.x`. In Figure 7.4 we plot the true mean function (left panel):

```
persp(s.mat, theta = 25, phi = 30, expand = 0.8, xlab='', ylab='',
    ↪ zlab='', ticktype = 'detailed')
```

and the squared errors between the true mean function and its estimator (right panel):

```
fhat <- result$summary.random$x$mean
fhat.mat <- inla.vector2matrix(fhat, nrow, ncol)
persp((fhat.mat - s.mat)^2, theta = 25, phi = 30, expand = 0.8, xlab =
    ↪ '', ylab = '', zlab = '', ticktype = 'detailed')
```

Here we need `inla.vector2matrix()` to transform `fhat` vector to `fhat.mat` matrix in order to produce the image plot. We see that the estimated function is close to the true one in the light of the small scale of the squared errors.

The posterior summary of σ_ε and σ_f is given by:

```
round(bri.hyperpar.summary(result), 4)
```

```
                                  mean     sd q0.025   q0.5 q0.975   mode
SD for the Gaussian observations 0.3026 0.0021 0.2985 0.3026 0.3069 0.3026
SD for x                         0.0251 0.0019 0.0216 0.0250 0.0291 0.0248
```

We can see the estimate of σ_ε (0.3026) is quite close to the true value (0.3).

7.3.2 Thin-Plate Splines at Irregularly-Spaced Locations

The RW2D model is only appropriate for the data on regular lattices. We would like a more flexible model that is capable of applying thin-plate spline smoothing to the data at irregularly-spaced locations.

The thin-plate spline penalty function in (7.6) can be shown to be

$$\int\int \left[\left(\partial^2/\partial x_1^2 + \partial^2/\partial x_2^2\right) f\right]^2 dx_1 dx_2,$$

assuming that the integrable derivatives of f vanish at infinity (Wahba, 1990; Yue and Speckman, 2010). Such a penalty inspires Yue et al. (2014) to obtain a thin-plate spline estimator by solving the following stochastic partial differential equation (SPDE)

$$\left(\partial^2/\partial x_1^2 + \partial^2/\partial x_2^2\right) f(\boldsymbol{x}) = \sigma_f dW(\boldsymbol{x})/d\boldsymbol{x}, \tag{7.9}$$

where σ_f is the scale parameter, and $dW(\boldsymbol{x})/d\boldsymbol{x}$ is the spatial Gaussian white noise. Note that this SPDE can be viewed as a two-dimensional extension of SDE in (7.3) for smoothing spline. The SPDE is solved by a *finite element method* on a triangular mesh, and the resulting thin-plate spline (TPS) prior has a multivariate normal density with mean zero and precision matrix $\sigma_f^{-2}\boldsymbol{Q}$, a highly sparse matrix due to the local nature of basis functions used in the method. This TPS prior is a generalization of the RW2D prior, and as a matter of fact the two priors are essentially the same when the locations are on regular lattices. Here we only demonstrate how to implement their method using INLA with a toy example, and refer readers to Yue et al. (2014) and Lindgren et al. (2011) for details.

Example: SPDE Toy Data

Let's load the data and look at its structure:

```
data(SPDEtoy)
str(SPDEtoy)
```

```
'data.frame':        200 obs. of  3 variables:
 $ s1: num  0.0827 0.6123 0.162 0.7526 0.851 ...
 $ s2: num  0.0564 0.9168 0.357 0.2576 0.1541 ...
 $ y : num  11.52 5.28 6.9 13.18 14.6 ...
```

The data are a three column `data.frame` simulated from a Gaussian process. The first two columns are the coordinates and the third is the response variable simulated at these locations. Figure 7.5(a) shows the image plot of the data using the following commands:

```
library(fields)
quilt.plot(SPDEtoy$s1, SPDEtoy$s2, SPDEtoy$y)
```

We can see the locations are irregularly spaced. We observe y_i at location (x_{1i}, x_{2i}) for $i = 1, \ldots, n$ where $n = 200$ in this example.

To use the TPS prior we first need to build a triangular mesh by subdividing the $\mathbb{R}^2$ domain into a set of non-intersecting triangles, where any two triangles meet in at most a common edge or corner. This step, which is similar to choosing the integration points on a numeric integration algorithm, must be done carefully (see discussion in Lindgren and Rue, 2015):

```
coords <- as.matrix(SPDEtoy[,1:2])
mesh <- inla.mesh.2d(loc=coords, max.edge=c(0.15, 0.2), cutoff=0.02)
```

The `loc=coords` specifies that the observed locations `coords` are used as initial triangulation nodes. The `max.edge=c(0.15, 0.2)` specifies the maximum triangle edge length to be 0.15 for the inner domain and 0.2 for the outer extension. The specified lengths must be on the same scale unit as the coordinates. These two arguments are mandatory. For further control over the shape of the triangles, we use `cutoff=0.02` argument that defines the minimum distance allowed between points. It means that the points at a closer distance than 0.02 are replaced by a single vertex. As a result, it avoids building many small triangles around clustered input locations. Regarding the uses of other arguments, we refer readers to `help(inla.mesh.2d)` in R or Chapter 6 in Blangiardo and Cameletti (2015) for details. Let's take a look at the resulting mesh displayed in Figure 7.5(b):

```
plot(mesh, main='')
```

The mesh looks good because its triangles are somewhat regular and they are smaller where the observations are dense while larger where are more sparse. Note that there is no requirement that the measurement locations must be included as nodes in the mesh. The mesh can be designed from different principles, such as lattice points with no relation to the precise measurement locations.

Given the mesh constructed above, we are then able to define a TPS prior:

```
tps <- bri.tps.prior(mesh, theta.mean = 0, theta.prec = 0.001)
```

where the '`theta.mean = 0`' and '`theta.prec = 0.001`' specify a default normal

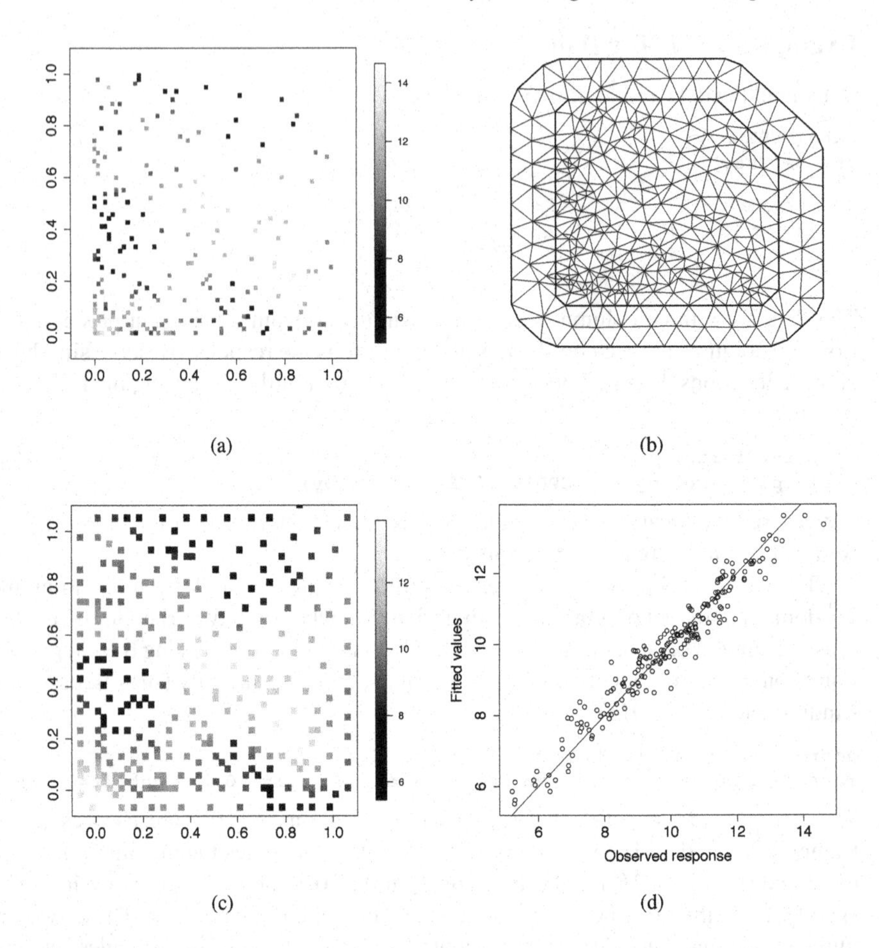

FIGURE 7.5
SPDE toy example: (a) simulated data; (b) triangular mesh; (c) posterior mean of function estimates at mesh locations; (d) plot of response against its fitted values.

prior for $\theta = \log(\sigma_f^{-2})$ with mean 0 and precision 0.001. It is also the default choice for prior. We then fit the model with INLA:

```
formula <- y ~ -1 + f(x, model = tps, diagonal = 1e-6)
data.inla <- list(y = SPDEtoy$y, x = mesh$idx$loc)
result <- inla(formula, data = data.inla, control.predictor = list(
    ↪ compute = TRUE))
```

Note that we use the `diagonal` option to add an extra constant to the diagonal of the precision matrix of the TPS model in order to guarantee the matrix is non-singular. In Figure 7.5(c) we present the thin-plate spline estimates (posterior means) at the locations of the mesh vertices:

```
fhat <- result$summary.random$x$mean
```

```
quilt.plot(mesh$loc[,1:2], fhat)
```

One should be aware that the mesh locations are usually different from the data locations. Therefore, we need to extract the posterior estimates at data locations from `result$summary.fitted`. In Figure 7.5(d) we plot the response observations against their estimates:

```
yhat <- result$summary.fitted$mean
plot(SPDEtoy$y, yhat, ylab='Fitted values', xlab='Observed response')
abline(0,1)
```

We see a nice match between the two. The posterior estimates of σ_ε^{-2} (error precision) and θ from the TPS prior are summarized below:

```
round(result$summary.hyperpar[,1:5], 3)
```

```
                                         mean    sd 0.025quant 0.5quant 0.975quant
Precision for the Gaussian observations 2.814 0.499      1.961    2.772      3.915
Theta1 for x                           -3.545 0.107     -3.757   -3.545     -3.335
```

Prediction. We would like to predict the response at unobserved locations (e.g., a fine grid) when we have spatial data collected at some locations. Suppose we are interested in the predictions at three target locations: $(0.1, 0.1), (0.5, 0.55), (0.7, 0.9)$, which are put in matrix form in R as:

```
loc.pre <- rbind(c(0.1, 0.1), c(0.5, 0.55), c(0.7, 0.9))
```

In Bayesian inference the prediction of a random function is usually done jointly with the parameter estimation process. This approach is made by the computation of the marginal posterior distribution of the random function at target locations. To do this in INLA, we first need to create `y.pre`, a vector of "missing" data at the target locations, and then combine it with the observed data into `y2`:

```
y.pre <- rep(NA, dim(loc.pre)[1])
y2 <- c(y.pre, SPDEtoy$y)
```

We also need to combine the target locations with the observed locations and build a mesh based on this combination of locations:

```
coords2 <- rbind(loc.pre, coords)
mesh2 <- inla.mesh.2d(coords2, max.edge = c(0.15, 0.2), cutoff = 0.02)
```

Then, we make a TPS prior with the new mesh:

```
tps2 <- bri.tps.prior(mesh2)
```

and fit this joint model as follows:

```
formula <- y ~ -1 + f(x, model = tps2)
data2.inla <- list(y = y2, x = mesh2$idx$loc)
result2 <- inla(formula, data = data2.inla, control.predictor = list(
    ↪ compute = TRUE))
```

To extract the posterior quantities regarding the three predictions, we must first know their indices in `y2`:

```
(idx.pre <- which(is.na(y2)))
```

```
[1] 1 2 3
```

indicating they are the first three elements in the vector. Then, their posterior summaries are given by:

```
round(result2$summary.fitted[idx.pre,], 3)
```

```
                        mean    sd 0.025quant 0.5quant 0.975quant   mode
fitted.Predictor.001 10.389 0.516      9.381   10.387     11.410 10.383
fitted.Predictor.002 12.733 0.835     11.101   12.730     14.384 12.723
fitted.Predictor.003  6.474 1.001      4.503    6.475      8.439  6.477
```

which are the first three rows of `result2$summary.fitted`. We may also extract the posterior samples of each prediction as follows:

```
pm.samp1 <- result2$marginals.fitted[[idx.pre[1]]]
pm.samp2 <- result2$marginals.fitted[[idx.pre[2]]]
pm.samp3 <- result2$marginals.fitted[[idx.pre[3]]]
```

Based on those samples, we are able to compute, say 95%, highest posterior density (HPD) intervals for the predictions:

```
inla.hpdmarginal(0.95, pm.samp1)
```

```
                   low     high
level:0.95 9.37652 11.40327
```

```
inla.hpdmarginal(0.95, pm.samp2)
```

```
                    low     high
level:0.95 11.09404 14.37404
```

```
inla.hpdmarginal(0.95, pm.samp3)
```

```
                    low     high
level:0.95 4.503045 8.436388
```

We can see that the HPD intervals are very similar to the credible intervals given by the quantiles. It is because the estimated marginal distributions are quite symmetric.

7.4 Besag Spatial Model

An important type of spatial data are so-called *areal* data, where the observations are related to geographic regions (e.g., the states of the US) with adjacency information. To smooth the data we need to construct a neighborhood structure. We say two regions are neighbors if they share a common border, but other ways to define neighbors are also possible. In the spirit of RW models, a Gaussian increment is defined between neighboring regions i and j as

$$f(\boldsymbol{x}_i) - f(\boldsymbol{x}_j) \sim N\left(0, \sigma_f^2 / w_{ij}\right),$$

where $\boldsymbol{x}_i$ and $\boldsymbol{x}_j$ represent the centroids of the regions, and w_{ij} are the positive and symmetric weights. We can let $w_{ij} = 1$ if we believe region i equally depends on its neighbors, or let w_{ij} be, for example, the inverse Euclidean distance between region centroids if we think the neighbors somehow contribute differently. Assuming the increments are independent, the resulting density of $\boldsymbol{f} = (f(\boldsymbol{x}_1), \ldots, f(\boldsymbol{x}_n))'$ is again multivariate normal with mean zeroes and precision matrix $\sigma_f^{-2}\boldsymbol{Q}$, where $\boldsymbol{Q}$ is the highly sparse matrix that has entries

$$\boldsymbol{Q}[i,j] = \begin{cases} w_{i+} & \text{if } i = j \\ -w_{ij} & \text{if } i \sim j \\ 0 & \text{otherwise} \end{cases},$$

where $w_{i+} = \sum_{j:j\sim i} w_{ij}$, the summation over neighbors of region i. Since the sum of each row is zero, $\boldsymbol{Q}$ is singular with rank $n-1$. We can show that the full conditional distribution of $f(\boldsymbol{x}_i)$ is normal

$$f(\boldsymbol{x}_i) \mid f(\boldsymbol{x}_{-i}), \tau \sim N\left(\frac{\sum_{j\sim i} w_{ij} f(\boldsymbol{x}_j)}{w_{i+}}, \frac{\sigma_f^2}{w_{i+}}\right),$$

where the conditional mean of $f(\boldsymbol{x}_i)$ depends on its neighboring nodes $f(\boldsymbol{x}_j)$ through weights w_{ij}, and its conditional variance depends on weight sum w_{i+}. We call this prior a *Besag* model because it is a special case of the intrinsic autoregressive models introduced by Besag and Kooperberg (1995).

Example: Munich Rental Guide

It is well known that "location" is an important factor with regard to apartment rent. We therefore study the potential spatial effect on `rent` using the model below:

$$\texttt{rent}_i = \beta_0 + f(\texttt{location}_i) + \varepsilon_i, \quad \varepsilon_i \sim N\left(0, \sigma_f^2\right), \tag{7.10}$$

for $i = 1, 2, \ldots, n$. The Besag model is an intuitive prior on the spatial effect since $\texttt{location}_i$ denotes i^{th} district in this case. Regarding their neighborhood structure we define any two districts as neighbors if they share the border. It can be defined via a graph file, which can be directly called in INLA:

```
data(Munich, package = "brinla")
g <- system.file("demodata/munich.graph", package = "INLA")
g.file <- inla.read.graph(g)
str(g.file)
```

```
List of 4
 $ n    : int 380
 $ nnbs: num [1:380] 5 6 4 6 3 2 4 5 5 1 ...
 $ nbs :List of 380
  ..$ : int [1:5] 92 136 137 138 298
  ..$ : int [1:6] 3 25 263 264 265 366
  ..$ : int [1:4] 2 263 366 369
  .. [list output truncated]
 $ cc   :List of 3
  ..$ id    : int [1:380] 1 1 1 1 1 1 1 1 1 1 ...
  ..$ n     : int 1
  ..$ nodes:List of 1
  .. ..$ : int [1:380] 1 2 3 4 5 6 7 8 9 10 ...
 - attr(*, "class")= chr "inla.graph"
```

Here `n` is the size of the graph, `nnbs` is the vector with the number of neighbors for each node, and `nbs` is a list-list with the neighbors for each node. The `cc` is the auto-generated list with connected component information, where `id` is the vector with the connected component id for each node, `n` is the number of connected components and `nodes` is a list-list of nodes belonging to each connected component. To understand the graph file above, there are `n=380` districts (nodes); the first district has `nnbs=5` neighbors, the second has `nnbs=6` neighbors and so on; the neighbors of the first district are those indexed by `nbs=92,136,137,138,298`.

It is also easy to make a graph file on your own. It can be defined in an ascii file, with the following format. The first entry is the number of nodes in the graph, `n`. The nodes in the graph are labelled `1,2,...,n`. The next entries specify for each node the number of neighbors, followed by the indices of those neighbors. Therefore, the first few lines of the graph file above should look like

```
380
1   5  92   136   137   138   298
2   6  3    25    263   264   265   366
3   4  3    25    263   264   265   366
...
```

Instead of storing it in a file, we can also specify the graph as a character string with one row after another in the file:

```
g <- inla.read.graph("380 1 5 92 136 137 138 298 2 6 3 25 263 264...")
```

More details on making the graph file can be found by typing `?inla.graph` in R.

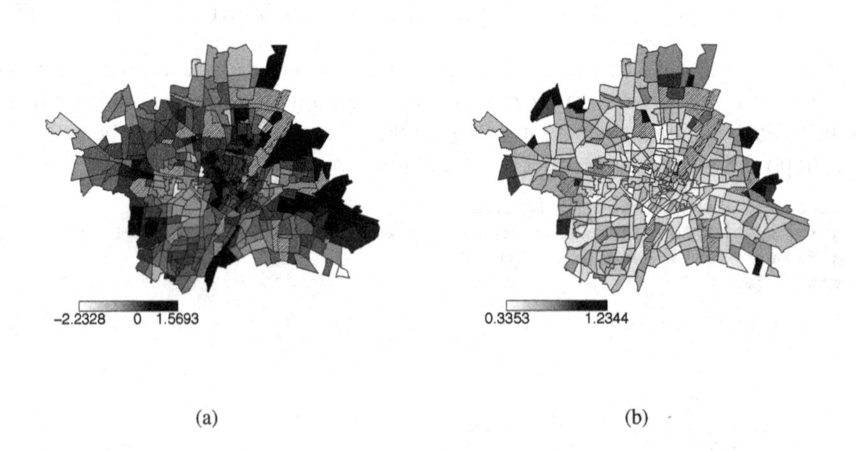

FIGURE 7.6
Munich rental guide using the Besag model: (a) posterior means of spatial effect; (b) posterior standard deviations of spatial effect.

Given the graph file we may easily fit model (7.10) using INLA:

```
formula <- rent ~ 1 + f(location, model = "besag", graph = g)
```

```
result <- inla(formula, data = Munich, control.predictor = list(
    compute = TRUE))
```

The posterior summary of β_0 (intercept) is given by

```
round(result$summary.fixed, 4)
```

```
              mean     sd 0.025quant 0.5quant 0.975quant   mode kld
(Intercept) 8.4973 0.0645     8.3712   8.4971     8.6245 8.4967   0
```

where we see the posterior mean $\hat{\beta}_0 = 8.497$ with standard deviation `sd = 0.0645` and 95% credible interval `(8.371,8.625)`. The posterior summary of the spatial effect at each district is saved in `result$summary.random$location`. Let us plot the posterior means (Figure 7.6(a)):

```
fhat <- result$summary.random$location$mean
map.munich(fhat)
```

and the posterior standard deviations (Figure 7.6(b)):

```
fhat.sd <- result$summary.random$location$sd
map.munich(fhat.sd)
```

We see that the rents are high in the center of Munich and some popular districts along the river Isar and near parks. In contrast, significantly negative effects are found for some districts on the borders of Munich. Regarding the uncertainty, the districts on the borders tend to have more variable rents than those at the center.

7.5 Penalized Regression Splines (P-Splines)

Since being introduced by Eilers and Marx (1996), the penalized regression splines (P-splines) approach has become popular for nonparametric regression due to its flexibility and efficient computation. In this approach, *B*-splines (De Boor, 1978) are combined with difference penalties on the estimated coefficients to give attractive properties. More specifically, we first approximate unknown function $f(x)$ by a *B*-spline of degree d with equally-spaced knots $x_{min} < t_1 < \cdots < t_r < x_{max}$, that is,

$$f(x) = \sum_{j=1}^{p} \beta_j B_j(x), \tag{7.11}$$

where B_j is the *B*-spline basis function and $p = d + r + 1$ is called the *degrees of freedom*. Note that B_j are constructed from polynomial pieces joined at the knots, and defined only locally in the sense that they are nonzero only on a domain spanned by $2 + d$ knots (see Eilers and Marx, 1996, Section 2). Such a function estimation is sensitive to the choice of the number and location of knots, and unfortunately it is hard to decide those issues in an automatic way. One solution is to somehow *penalize* β_j, making the selection of knots far less important than the choice of smoothing parameter. As suggested in the P-splines approach, an intuitive choice is to use $\nabla^m \beta_j$, the mth order difference operator on adjacent β_j for $j = m+1, \ldots, p$, as defined in

RW models in Section 7.2.1. This leads to a penalized least squares estimator that minimizes

$$\sum_{i=1}^{n}\left[y_i - \sum_{j=1}^{p}\beta_j B_j(x_i)\right]^2 + \lambda \sum_{j=m+1}^{p}(\nabla^m \beta_j)^2. \tag{7.12}$$

Clearly, the P-splines and smoothing splines are closely related based on the similarities between (7.2) and (7.12). However, they differ from each other in the following way. For smoothing splines, the observed unique x values are the knots and λ alone is used to control the smoothing. For P-splines, the knots of the B-splines used for the basis are typically much smaller in number than the sample size. Therefore, it can be computed more efficiently than smoothing splines. In practice one may use as many knots as needed to ensure the desired flexibility and let the penalty do the work to avoid overfitting.

The minimization problem in (7.12) has the following Bayesian representation

$$\boldsymbol{y} \mid \boldsymbol{\beta}, \sigma_\varepsilon^2 \sim N\left(\boldsymbol{B}\boldsymbol{\beta}, \sigma_\varepsilon^2 \boldsymbol{I}\right), \quad \boldsymbol{\beta} \mid \sigma_\beta^2 \sim N\left(\boldsymbol{0}, \sigma_\beta^2 \boldsymbol{Q}_m^-\right), \tag{7.13}$$

where $\boldsymbol{\beta} = (\beta_1, \ldots, \beta_p)'$, $\boldsymbol{B}$ is the $n \times p$ design matrix with entry $\boldsymbol{B}[i,j] = B_j(x_i)$, $\boldsymbol{Q}_m$ is the (singular) matrix as used in RW models, and the smoothing parameter $\lambda = \sigma_\varepsilon^2/\sigma_\beta^2$. It is actually a Bayesian linear regression model, with B-spline bases used as covariates and a RW prior taken on the regression coefficients.

Example: Simulated Data

We here use the P-spline models to estimate the nonparametric function from the simulated example used in Section 7.1. Again we simulate $n = 100$ data points from the model with Gaussian errors:

```
set.seed(1)
n <- 100
x <- seq(0, 1,, n)
f.true <- (sin(2*pi*x^3))^3
y <- f.true + rnorm(n, sd = 0.2)
```

We then generate cubic B-spline basis functions (intercept included) with $p = 25$ degrees of freedom:

```
library(splines)
p <- 25
B.tmp <- bs(x, df = p, intercept = TRUE)
```

and make the resulting design matrix $\boldsymbol{B}$ as in model (7.13):

```
attributes(B.tmp) <- NULL
Bmat <- as(matrix(B.tmp, n, p), 'sparseMatrix')
```

Note that we here convert `Bmat`, a sparse matrix of 100×25 dimension, to the particular sparse format for INLA to use. We then fit a P-spline model with RW1 penalty and ask INLA to compute the linear predictors using `compute = TRUE`:

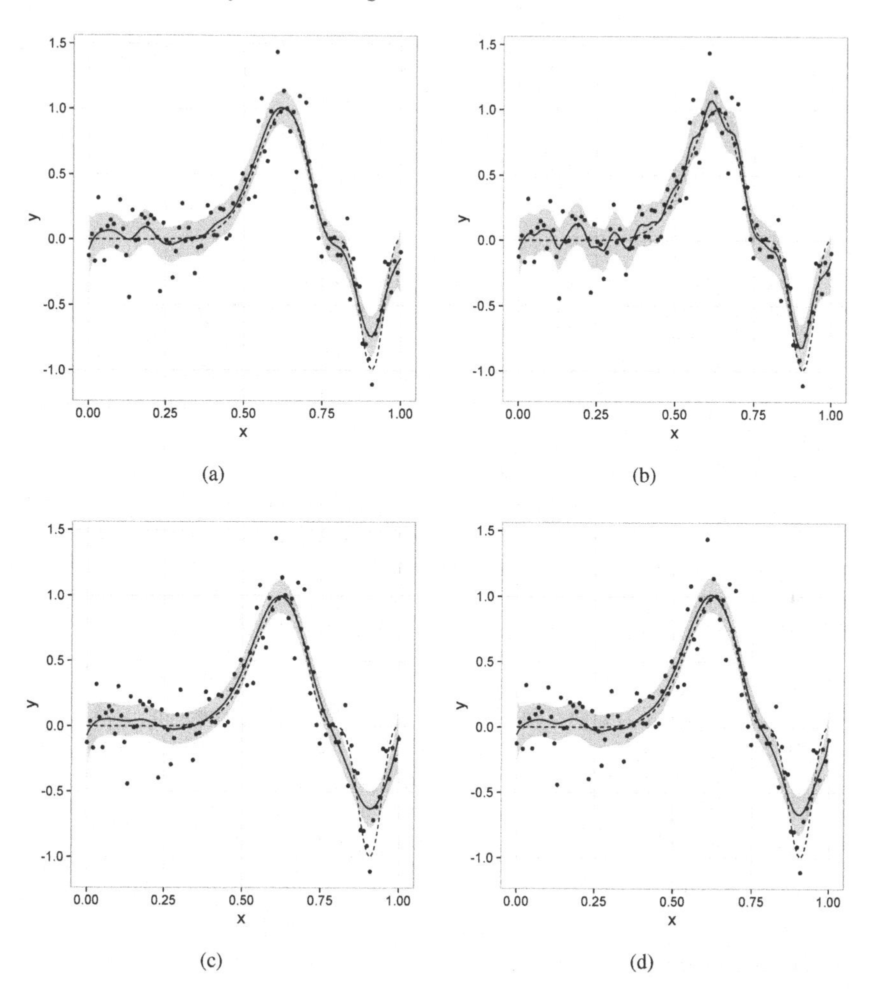

FIGURE 7.7
P-spline models: true function (dashed), posterior mean (solid), and 95% credible band (gray) of four scenarios: (a) RW1 penalty with 25 degrees of freedom (df); (b) RW1 penalty with 80 df; (c) RW2 penalty with 25 df; (d) RW2 penalty with 80 df.

```
data.inla <- list(y = y, x = 1:p)
formula <- y ~ -1 + f(x, model = 'rw1', constr = FALSE)
result <- inla(formula, data=data.inla, control.predictor = list(A =
    ↪ Bmat, compute = TRUE))
```

The posterior summary of β as in model (7.13) is saved in `summary.random`, while the summary of $B\beta$ is in the rows of `summary.linear.predictor` with name that begins with `Apredictor`. They are the first $n = 100$ rows, followed by $p = 25$ rows of `predictor` for β estimate:

```
round(head(result$summary.linear.predictor), 3)
```

```
                  mean    sd 0.025quant 0.5quant 0.975quant   mode kld
Apredictor.001 -0.081 0.147     -0.369   -0.080      0.207 -0.080   0
Apredictor.002 -0.019 0.096     -0.208   -0.019      0.170 -0.019   0
Apredictor.003  0.020 0.090     -0.157    0.020      0.197  0.020   0
Apredictor.004  0.043 0.083     -0.120    0.043      0.206  0.043   0
Apredictor.005  0.055 0.078     -0.097    0.055      0.208  0.055   0
Apredictor.006  0.062 0.076     -0.088    0.062      0.213  0.062   0
```

We plot the result in Figure 7.7(a):

```
p <- bri.band.ggplot(result, ind = 1:n, type = 'linear')
p + geom_point(aes(y = y, x = 1:n)) + geom_line(aes(y = f.true, x = 1:
  ↪ n), linetype = 2)
```

The `ind = 1:n` specifies the indices for the part of the function to be plotted. We can see that the fit is a little too rough for the flat part, but captures maximum and minimum pretty well.

To find out how sensitive the P-spline fit is to the choice of number of knots, we fit the same model but with 80 degrees of freedom, and plot the estimates in Figure 7.7(b). We see an clear overfit, although it captures the minimum better than the previous model does. This is because the RW1 penalty cannot provide sufficient shrinkage to the coefficients. For comparison purposes, we also fit a P-spline model with RW2 penalty using the following formula in INLA:

```
formula <- y ~ -1 + f(x, model = 'rw2', constr = FALSE)
```

Figures 7.7(c) and 7.7(d) present the results with degrees of freedom of 25 and 80, respectively. Compared to those using the RW1 penalty, the fits using the RW2 penalty are smoother: they perform better at the flat part but worse for extremes. The fits are also not that sensitive to the number of knots as those using RW1 penalty, although more knots seem to capture more local features of the function.

7.6 Adaptive Spline Smoothing

Despite their popularity, the smoothing spline and P-spline approaches are well known to perform mediocrely when estimating highly varying functions with peaks, jumps or frequent curvature transitions. This significant drawback stems from their usage of a single smoothing parameter, which applies a constant amount of smoothing across the function domain. Consequently, extensive research has been done to make those spline smoothing methods "adaptive," which means that different amounts of smoothing are applied to the function space as required by data. Although a variety of approaches exist, we here introduce the one proposed in Yue et al. (2014) that can be implemented by INLA.

The basic idea is to extend SDE (7.3) using a *smoothing function* that varies in space rather than a smoothing parameter. More specifically, we consider

$$d^2\lambda(x)f(x)/dx^2 = dW(x)/dx, \tag{7.14}$$

where the smoothing function $\lambda(x)$ can be seen as an instantaneous variance or local scaling. A small value of λ compresses the scale giving quick oscillations, while a large value stretches the function and decreases roughness. Note that we here only consider the cubic smoothing spline, that is $m = 2$ as in (7.3), because it is known to provide the best overall performance (e.g., Green and Silverman, 1994). As shown in Yue et al. (2014), the (weak) solution to (7.14) is multivariate normal with mean zeroes and precision matrix $\boldsymbol{Q}_\lambda$ that depends on the unknown function λ.

To implement a fully Bayesian inference, we need a prior taken on the smoothing function $\lambda(x)$. The prior is assumed to be continuous and differentiable, and must have a proper distribution to guarantee a proper posterior distribution (Yue et al., 2012). Since it is restricted to be positive, we model $\lambda(x)$ on its log scale: $\nu(x) = \log(\lambda(x))$, and then take a smooth prior on $\nu(x)$. Following Yue et al. (2014), we use B-spline basis expansion

$$\nu(x) = \gamma_k \sum_{k=1}^{q} B_k(x),$$

at knots $t'_1, t'_2, \ldots, t'_q$, and let the random weights $\boldsymbol{\gamma} = (\gamma_1, \ldots, \gamma_q)'$ follow a multivariate normal distribution with mean zero and precision matrix $\theta \boldsymbol{R}$, where θ is a fixed scale parameter. To ensure a proper posterior distribution, $\boldsymbol{R}$ must be a positive definite matrix. One simple choice is $\boldsymbol{R} = \boldsymbol{I}$, an identity matrix, which is equivalent to taking independent Gaussian priors on γ_k. Unfortunately, this choice makes the function estimation sensitive to the knots, although it has an easy computation. To relieve the issue, we may use a more sophisticated SPDE prior introduced in Lindgren et al. (2011), which sets $\boldsymbol{R}$ to be a sparse matrix (see details in Yue et al., 2014). With such a double-layer spline prior the model is able to apply data-driven adaptive smoothing when estimating the function.

Example: Simulated Data

For comparison purposes, we apply our adaptive smoothing model to the same simulated data used in the previous examples for smoothing splines and P-splines:

```
set.seed(1)
n <- 100
x <- seq(0, 1,, n)
f.true <- (sin(2*pi*x^3))^3
y <- f.true + rnorm(n, sd = 0.2)
```

We try four different scenarios with two types of $\boldsymbol{R}$ (independent and SPDE) and two numbers of knots (5 and 10). Below we present how the scenario of 5 knots and SPDE model is implemented in INLA. We first build the adaptive smoothing prior:

```
adapt <- bri.adapt.prior(x, nknot = 5, type = 'spde')
```

and then fit the model:

```
data.inla <- list(y = y, x = adapt$x.ind)
formula <- y ~ -1 + f(x, model = adapt)
result <- inla(formula, data = data.inla)
```

We extract the posterior mean as well as 95% credible band, and plot them in Figure

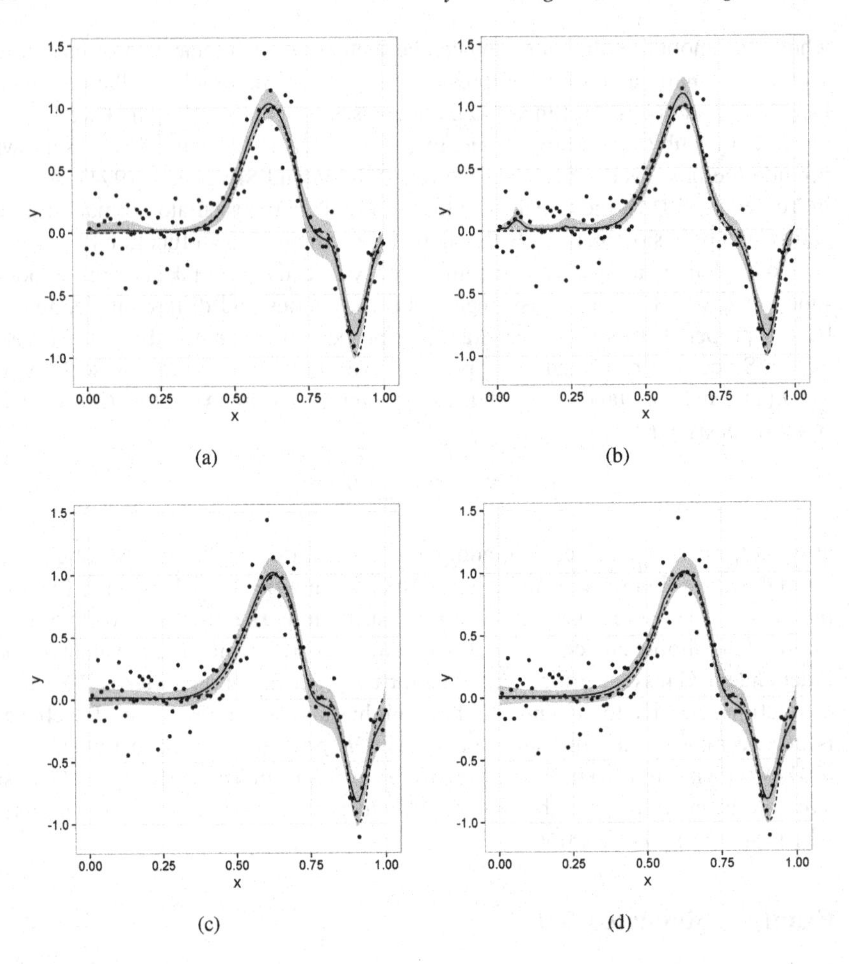

FIGURE 7.8
Adaptive smoothing models: true function (dashed), posterior mean (solid), and 95% credible band (gray) of four scenarios: (a) independent prior with 5 knots; (b) independent prior with 10 knots; (c) SPDE prior with 5 knots; (d) SPDE prior with 10 knots.

7.8. To apply the adaptive smoothing model to the other three scenarios we only need to change `nknot` and/or `type` in `bri.adapt.spline()` function. Note that `type = indpt` specifies the independent prior used in the model. The results are also presented in Figure 7.8. As we can see, the adaptive smoothing model yields an overall better performance than smoothing spline and P-spline models. Compared to the one with an independent prior, the adaptive model with SPDE prior yields a more accurate and less knot-sensitive estimation.

7.7 Generalized Nonparametric Regression Models

Generalized nonparametric regression (GNPR) models combine the idea of GLMs as seen in Chapter 4 with the nonparametric regression modeling seen earlier in this chapter. Suppose we have a response Y_i that follows a distribution from the exponential family. Letting $\mu_i = \mathrm{E}(Y_i)$, a GNPR model can be defined as

$$g(\mu_i) = f(x_i), \quad i = 1, \dots, n,$$

where f is the unknown but smooth function. If we impose a Gaussian prior on f, this falls within the latent Gaussian model framework required to use INLA. GNPR models are not easy to fit using a maximum likelihood approach while the Bayesian approach may find more acceptable solutions.

Example: Simulated Data

We begin with a few simulated examples. The true underlying function $f(x_i) = \sin(x_i)$ for $i = 1, \dots, n$, where $x_i \in [0, 6]$ are equally spaced. Two popular non-Gaussian response distributions are considered: binomial and Poisson distributions.

Binomial Response

We simulate data from the following model

$$y_i \sim \mathrm{Bin}(n_i, p_i), \quad g(p_i) = \sin(x_i),$$

for $i = 1, \dots, 200$, where $\mathrm{Bin}(n, p)$ denotes the binomial density with n trials and p success probability. Three commonly used link functions are considered: logit, probit and complementary log-log. The corresponding datasets are generated as follows:

```
set.seed(2)
n <- 200  #sample size
x <- seq(0, 6,, n)
eta <- sin(x)
Ntrials <- sample(c(1, 5, 10, 15), size = n, replace = TRUE)
prob1 <- exp(eta)/(1 + exp(eta))  ## logit link
prob2 <- pnorm(eta)  ## probit link
prob3 <- 1 - exp(-exp(eta))  ## complementary log-log link
y1 <- rbinom(n, size = Ntrials, prob = prob1)
y2 <- rbinom(n, size = Ntrials, prob = prob2)
y3 <- rbinom(n, size = Ntrials, prob = prob3)
data1 <- list(y = y1, x = x)
data2 <- list(y = y2, x = x)
data3 <- list(y = y3, x = x)
```

Using RW2 prior the three GNPMs are fitted by INLA:

```
formula <- y ~ -1 + f(x, model = "rw2", constr = FALSE)
result1 <- inla(formula, family = "binomial", data = data1, Ntrials =
   ↪ Ntrials, control.predictor = list(compute = TRUE))
```

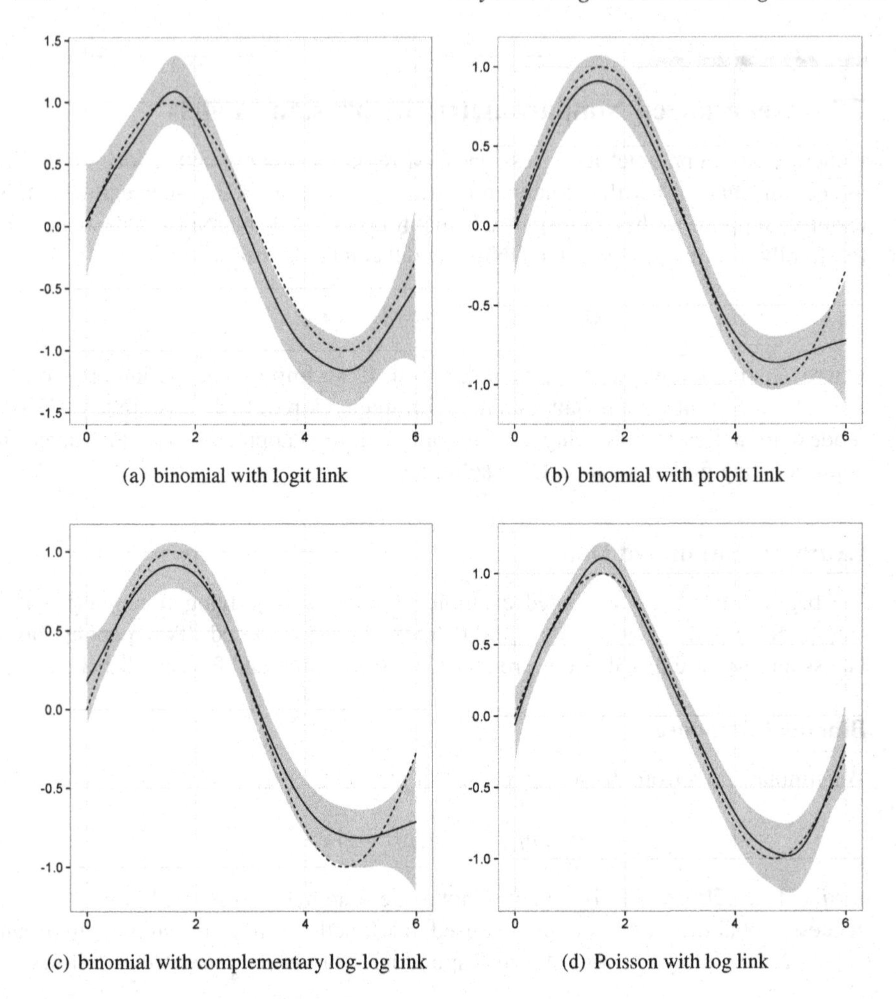

FIGURE 7.9
Simulation results of GNPR models: true function (dashed), posterior mean (solid), and 95% credible band (gray).

```
result2 <- inla(formula, family = "binomial", data = data2, Ntrials =
    ↪ Ntrials, control.predictor = list(compute = TRUE), control.
    ↪ family = list(link = 'probit'))
result3 <- inla(formula, family = "binomial", data = data3, Ntrials =
    ↪ Ntrials,  control.predictor = list(compute = TRUE), control.
    ↪ family = list(link = 'cloglog'))
```

Here the binomial likelihood is specified by `family = "binomial"`, and the link function is defined in `control.family = list(link = 'name')` given `name = logit, probit` or `cloglog`. Note that by default `logit` is used. We also use `control.predictor = list(compute = TRUE)` to compute the fitted values (estimated μ_i) that we may need later.

In Figure 7.9(a) we plot the true function (dashed), its posterior mean estimate (solid), and 95% credible band (gray) for using the logit link:

```
p1 <- bri.band.ggplot(result1, name = 'x', alpha = 0.05, type = '
    ↪ random')
p1 + geom_line(aes(y = eta), linetype = 2)
```

The results for using probit and complementary log-log link functions are presented in Figure 7.9(b) and 7.9(c), respectively. As we can see, INLA provides quite reasonable estimates with all three links, considering the sample size is relatively small. The posterior summary of estimated success probabilities $\hat{p}_i$ is given by, for instance, `result1$summary.fitted.values` for logit link.

Poisson Response

We consider the following model:

$$y_i \sim \text{Poisson}(\lambda_i), \quad \ln(\lambda_i) = \log(E_i) + \sin(x_i),$$

for $i = 1, \ldots, 200$, where $x_i \in [0, 6]$ and are equally spaced, and E_i are known constants. The data are simulated as follows:

```
set.seed(2)
n <- 200  #sample size
x <- seq(0, 6,, n)
E <- sample(1:10, n, replace = TRUE)
lambda <- E*exp(sin(x))
y4 <- rpois(n, lambda = lambda)
data4 <- list(y = y4, x = x)
```

We then fit this Poisson GNPR model in INLA:

```
formula <- y ~ -1 + f(x, model = "rw2", constr = FALSE)
result4 <- inla(formula, family = "poisson", data = data4, E = E,
    ↪ control.predictor = list(compute = TRUE))
```

Here the Poisson likelihood is specified by `family="poisson"`. In Figure 7.9(d) we show the result, and it looks good. Note that `result4$summary.fitted.values` gives the posterior summary of $\exp[\sin(x)]$, not mean λ, because the E_i are not all 1's. Since it is a simple linear transformation, we may easily obtain the posterior estimates (mean, SD, quantiles, etc.) of λ as follows:

```
lamb.hat <- E*result4$summary.fitted$mean
yhat.sd <- E*result4$summary.fitted$sd
lamb.lb <- E*result4$summary.fitted$'0.025quant'
lamb.ub <- E*result4$summary.fitted$'0.975quant'
```

Example: Tokyo Rainfall

The Tokyo rainfall data contain daily rainfall indicator counts for a period of two years. Each day during 1983 and 1984, it was recorded whether there was more than 1 mm rainfall in Tokyo. It is a data frame with 366 observations on three variables: `y` is the number of days with rain; `n` is the total number of days; `time` is the day of the year. Let us load the dataset:

```
data(Tokyo, package = 'INLA')
str(Tokyo)
```

```
'data.frame':      366 obs. of  3 variables:
 $ y   : int  0 0 1 1 0 1 1 0 0 0 ...
 $ n   : int  2 2 2 2 2 2 2 2 2 2 ...
 $ time: int  1 2 3 4 5 6 7 8 9 10 ...
```

Note that for `time` = 60, which corresponds to February 29, only one binary observation is available (`n=1`), while for all other calendar days there are two (`n=2`). We want to estimate the underlying probability of rainfall on a given calendar day.

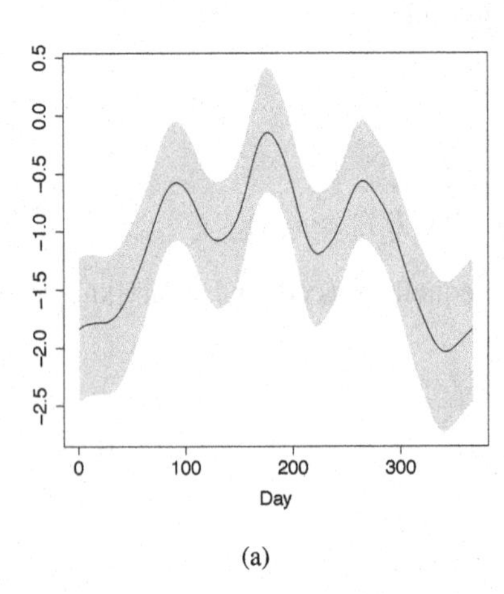

(a)

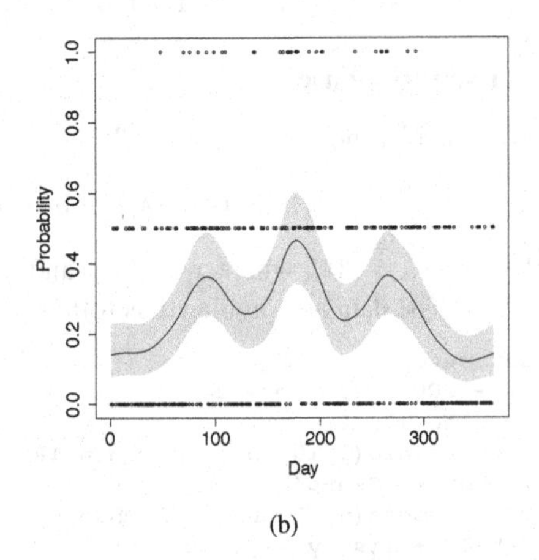

(b)

FIGURE 7.10
Tokyo rainfall data: (a) posterior mean of latent nonparametric function with 95% credible interval; (b) empirical and model-based binomial probability estimates, with 95% posterior predictive bounds. The empirical probability estimates are the proportion of observed rainfall days for each day of the year.

We assume that the rainfall follows a binomial distributions $\text{Bin}(n_i, p_i)$, and associate p_i with a *circular* RW2 model (see Rue and Held, 2005) using logit link. This explicitly connects the end and the beginning of the time series, because we expect smooth changes between the last week in December and the first week in January. We then build the model and fit it in INLA:

```
formula <- y ~ -1 + f(time, model = "rw2", cyclic = TRUE)
result <- inla(formula, family = "binomial", Ntrials = n, data = Tokyo
    ↪ , control.predictor = list(compute = TRUE))
```

Here `cyclic = TRUE` is used to select circular RW2 prior. We plot fitted curve (posterior mean) and 95% credible band of `time` effect (Figure 7.10(a)):

```
bri.band.plot(result, name = 'time', alpha = 0.05, type = 'random',
    ↪ xlab = 'Day', ylab = '')
```

A clear seasonal pattern can be seen. We also plot empirical and model-based binomial probability estimates, with 95% posterior predictive bounds (Figure 7.10(b)):

```
bri.band.plot(result, alpha = 0.05, type = 'fitted', ylim = c(0, 1),
    ↪ xlab = 'Day', ylab = 'Probability')
points(Tokyo$time, Tokyo$y/2, cex = 0.5)
```

Note that the empirical probability estimates are the proportion of observed rainfall days for each day of the year. It seems that the typical daily precipitation chance is no more than 50%, and is relatively high in April, June and August.

Prediction. It is interesting to predict the daily precipitation chances for the next few days based on what we have already observed. Suppose we want to see how likely it would rain on the first three days (January 1st, 2nd, and 3rd) in 1985 based on the data from the previous two years. Such predictions must be done as part of the fitting process in INLA. We therefore incorporate those three days as missing cases into the original data to make a new dataset `Tokyo.pred`:

```
time.new <- seq(length(Tokyo$time) + 1, length.out = 3)
time.pred <- c(Tokyo$time, time.new)
y.pred <- c(Tokyo$y, rep(NA, length(time.new)))
n.pred <- c(Tokyo$n, rep(1, length(time.new)))
Tokyo.pred <- list(y = y.pred, time = time.pred, n = n.pred)
```

We then use the same model as before and fit it to the new data:

```
result <- inla(formula, family = "binomial", Ntrials = n, data = Tokyo
    ↪ .pred, control.predictor = list(compute = TRUE, link = 1))
```

Here we use `link = 1` as a shortcut for

```
link <- rep(NA, length(y.pred))
link[which(is.na(y.pred))] <- 1
```

where we define "link" to be the *first* "family" specified in the model (which is binomial here) only for missing cases.

The posterior quantities for the predictions of `time` effect on those three days are summarized below:

```
ID <- result$summary.random$time$ID
idx.pred <- sapply(time.new, function(x) which(ID==x))
round(result$summary.random$time[idx.pred,], 4)
```

```
     ID    mean     sd 0.025quant 0.5quant 0.975quant    mode kld
367 367 -1.8384 0.3236    -2.4858  -1.8354    -1.2079 -1.8293   0
368 368 -1.8305 0.3236    -2.4772  -1.8277    -1.1995 -1.8222   0
369 369 -1.8231 0.3233    -2.4689  -1.8206    -1.1921 -1.8154   0
```

The corresponding daily precipitation chances are predicted as:

```
round(result$summary.fitted.values[which(is.na(y.pred)),], 4)
```

```
                        mean     sd 0.025quant 0.5quant 0.975quant   mode
fitted.predictor.367 0.1417 0.0392     0.0769   0.1376     0.2300 0.1300
fitted.predictor.368 0.1426 0.0394     0.0775   0.1385     0.2315 0.1308
fitted.predictor.369 0.1435 0.0396     0.0781   0.1394     0.2328 0.1316
```

We see it is getting more likely to rain as time goes by, although the chances are all small.

7.8 Excursion Set with Uncertainty

Suppose we want to find areas where the function studied exceeds a certain level or is different from some reference level, while accounting for estimation uncertainty. For example, with observations $\boldsymbol{y}$ from some function $f(\boldsymbol{x})$ we want to find a set D such that, with a given probability $1-\alpha$, $f(\boldsymbol{x}) > u$ for all $\boldsymbol{x} \in D$ for a given level u. It is easy, and quite common, to compute the marginal posterior probabilities $P(f(\boldsymbol{x}) > u \mid \boldsymbol{y})$ based on the conditional distribution for $f(\boldsymbol{x}) \mid \boldsymbol{y}$, and then specify D_m as the set where the probabilities exceed a threshold

$$D_m = \{\boldsymbol{x} : P(f(\boldsymbol{x}) > u \mid \boldsymbol{y}) \geq 1-\alpha\}. \tag{7.15}$$

However, the parameter α in this definition is the pointwise type I error rate, which does not give us information about the familywise error rate, and hence does not quantify the certainty of the level being exceeded at *all* points in the set simultaneously. From a frequentist point of view, this is the problem of multiple-hypothesis testing, and can be relieved by making an adjustment to α, e.g., type I error control thresholding (Adler, 1981).

Unfortunately, the hypothesis testing methods in frequentist settings do not transfer to a Bayesian hierarchical model framework. We therefore desire to formulate questions regarding excursions as properties of posterior distributions for functions. It can be done by using *joint* probability for exceeding or being different from the level in the entire set.

Fossil Data

Bralower et al. (1997) report data that reflect the global climate millions of years ago, through ratios of strontium isotopes found in fossil shells. We start by loading the data and summarizing them:

```
data(fossil, package = 'brinla')
str(fossil)
```

```
'data.frame':        106 obs. of  2 variables:
$ age: num  91.8 92.4 93.1 93.1 93.1 ...
$ sr : num  0.734 0.736 0.735 0.737 0.739 ...
```

The data frame has 106 observations on fossil shells, and two measurements on each shell: age of shell in millions of years (`age`) and its ratio of strontium isotopes (`sr`). The shells are dated by biostratigraphic methods, so the strontium ratio can be studied as a function of ages.

It is intuitive to use RW2 model for non-equally spaced ages. However, we observe that some ages are too close in terms of their range:

```
min(diff(sort(fossil$age)))/diff(range(fossil$age))
```

```
[1] 9.610842e-05
```

It will make the precision matrix of RW2 ill-conditioned (see Section 7.2.3), leading to bizarre behavior when estimating functions. We therefore use `inla.group()` to group `age` into 100 bins, which yields 54 unique values:

```
age.new <- inla.group(fossil$age, n = 100)
(length(unique(age.new)))
```

```
[1] 54
```

There are replicates for some of the new `age` values. Then, we take `sr` as the response variable and `age.new` as the predictor variable to fit a nonparametric regression model:

```
inla.data <- list(y = fossil$sr, x = age.new)
formula <- y ~ -1 + f(x, model = 'rw2', constr = FALSE, scale.model =
    ↪ TRUE)
result <- inla(formula, data = inla.data, control.compute = list(
    ↪ config = TRUE))
```

We use `control.compute = list(config = TRUE)` to store internal INLA approximations in the `result` for later computation. Also, we turn on the `scale.model` option in `f()` to make a scale adjustment to the hyperprior on RW2 model. In our experience this scaling procedure makes no difference in fitting ordinary nonparametric regression models. This example, however, is an exception, probably because of the replicates. We will find this option more useful in fitting additive models in Chapter 9 and provide more discussions there.

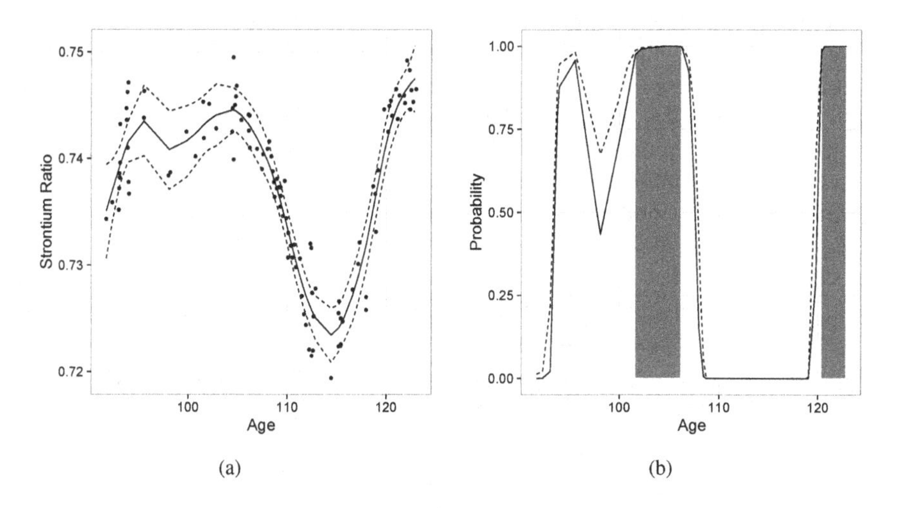

FIGURE 7.11
Fossil data using RW2: (a) fitted curve (solid) and 95% credible band (dashed) with data (dot); (b) excursion function (solid) and marginal probabilities (dashed) for mean *Strontium ratio* > 0.74 and excursion set (gray areas) at 5% level.

The fitted curve, 95% credible band and data points are plotted in Figure 7.11(a). We can see a clear nonlinear relationship between age and strontium ratio: it increases from 92 to 95 million years ago, and keeps this relatively high ratio until 105 million years ago, then has a substantial dip with a minimum near 115 million years ago, followed by an increase for fossils around 120 million years ago. The dip around 98 million years ago, however, is not significant because its credible interval is too wide. The features discovered by INLA coincide with those in Chaudhuri and Marron (1999).

Suppose we are interested in finding the ages where the ratios of strontium isotopes are greater than 0.74, accounting for estimation uncertainty. We compute the probability that each $f(\texttt{age}_i) > 0.74$ based on its marginal posterior distribution:

```
mar.x <- result$marginals.random$x  #marginal posterior
mar.prob <- 1 - sapply(mar.x, function(x) inla.pmarginal(0.74, x))
```

Then, we find a set of ages, each of which has at least 95% chance for its corresponding strontium ratio to be greater than 0.74:

```
result$summary.random$x$ID[mar.prob > 0.95]
```

```
 [1]  95.58932 101.53700 102.10000 102.81000 104.50698 104.78694 105.34000
 [8] 106.08018 106.19500 106.94000 120.30846 120.58692 120.99231 121.22769
[15] 121.83231 122.19961 122.46423 122.85615
```

Such a set of ages is defined as D_m in (7.15) with $\alpha = 0.05$. As mentioned, this α is only the pointwise type I error rate, which does not give us information about the familywise error rate, and hence does not quantify the certainty of the level being exceeded at all points in the set simultaneously.

The statistical problem here is to find a set D of ages such that $f(\texttt{age}_i) > 0.74$ with probability 0.95 or higher for *all* $\texttt{age}_i \in D$. There might be many such sets, so we might look for the largest of these. It is a so-called *excursion set*, a smaller set than D_m. Finding an excursion set in practice is often difficult because it requires the high dimensional integration that is computationally intensive, especially for the applications where the sample size is large. Bolin and Lindgren (2015) introduced an efficient computational method for latent Gaussian models. It is based on using a parametric family for the excursion sets in combination with a sequential importance sampling method for estimating joint probabilities. We here only show how to implement their method in R, and refer readers to their paper for details on methodology.

The excursion set of ages with probability at least 0.95 can be simply found using `excursions.brinla()` as follows:

```
res.exc <- excursions.brinla(result, name = 'x', u = 0.74, alpha =
    ↪ 0.05, type = '>', method = 'NI')
```

The `result` is the returned object from the INLA call, `name='x'` specifies the function in `result` for which to compute the joint probabilities, `u=0.74` shows the excursion level, `alpha=0.05` specifies the significance level, and `type='>'` gives the type of excursion set. Two other types, '>' and '≠', are also available. The `method` argument specifies the computational method, including "EB" = Empirical Bayes, "QC" = Quantile correction, "NI" = Numerical integration, 'NIQC' = Numerical integration with quantile correction, and "iNIQC" = Improved integration with quantile correction. The EB method is the simplest and may be sufficient in many situations. The QC method is based on correcting the limits of the integral, and is as easy to implement as the EB method and should perform better in most scenarios. The NI method is more computationally demanding but should also be the most exact method for problems with Gaussian likelihoods. For non-Gaussian likelihoods, the NI with QC and improved NI with QC methods can be used for improved results. The improved NI with QC is slightly more computationally demanding but should also perform better in practice for models with non-Gaussian likelihoods. In our case, the sample size is small and we therefore chose NI to obtain the most accurate possible result.

The resulting excursion set can be extracted from:

```
res.exc$E
```

```
 [1]  95.58932 101.53700 102.10000 102.81000 104.50698 104.78694 105.34000
 [8] 106.08018 106.19500 120.30846 120.58692 120.99231 121.22769 121.83231
[15] 122.19961 122.46423 122.85615
```

It shows the values of `age` that are jointly greater than 0.74 with at least 0.95 probability. Note that this set is smaller than the one derived from marginal posterior probabilities, as it is supposed to be. Such an excursion set, however, is not sufficient because it does not provide any information about the ages that are not contained in the sets. We'd better have something similar to the p-values, i.e., the marginal probabilities of exceeding the level, but which can be interpreted simultaneously. Bolin and Lindgren (2015) therefore defined a so-called *excursion function* to serve this purpose. The function takes values between 0 and 1, which are saved in `res.exc$F`. The first few values are given by:

```
round(head(res.exc$F), 4)
```

```
[1] 0.0001 0.0003 0.0206 0.7921 0.8773 0.9581
```

The excursion function is not equal to the marginal probability function. The former is always smaller than the latter. In addition, we may use `res.exc$G` to return a propositional variable where `TRUE` means the corresponding location is in the excursion set and `FALSE` otherwise. The marginal probabilities, excursion function and excursion set at 5% level are plotted in Figure 7.11(b):

```
bri.excursions.ggplot(res.exc)
```

Note that we only see two `age` intervals (101.53, 106.19) and (120.30, 122.85) marked by gray bars, where mean `sr` is greater than 0.74 with at least 0.95 joint probability. There is actually one additional age 95.58 identified in the excursion set, but we fail to display this single point in the figure. Also, we see the excursion function (closely) covered by the marginal probability curve, as we expected.

Tokyo Rainfall Data

Suppose we are interested in finding the days whose precipitation chances are more than 0.3 in a year with uncertainty accounted for. We fit the model as in Section 7.7:

```
data(Tokyo, package = 'INLA')
formula <- y ~ -1 + f(time, model = "rw2", cyclic = TRUE)
result <- inla(formula, family = "binomial", Ntrials = n, data = Tokyo
    ↪ , control.predictor = list(compute=TRUE), control.compute =
    ↪ list(config = TRUE))
```

Note that the estimated precipitation chances are fitted values in the INLA result. We therefore compute for each day the marginal probability that the precipitation chance on that day is greater than 0.3 as follows:

```
u.fitted <- 0.3
mar.fitted <- result$marginals.fitted.values
mar.prob<- 1-sapply(mar.fitted,function(x) inla.pmarginal(u.fitted,x))
```

Let's find out the days that have the marginal probabilities being at least 0.95:

```
Tokyo$time[mar.prob >= 0.95]
```

```
 [1] 163 164 165 166 167 168 169 170 171 172 173 174 175 176 177 178 179 180 181
[20] 182 183 184 185 186 187 188 189 190 191 192 193 194
```

They are 32 consecutive days between May and June.

The excursion set of days we need is also associated with the fitted values. However, the excursion function is still computed based on the random effect rather than the fitted values. So we need to first transform the threshold 0.3 to logit(0.3):

```
u.pred <- log(u.fitted/(1 - u.fitted))
```

Then we use this value in the following computation:

```
res.exc <- excursions.brinla(result, name = 'time', u = u.pred, type =
    ↪ '>', method = 'NIQC', alpha = 0.05)
```

Here `method = 'NIQC'` is used for non-Gaussian data. The resulting excursion set is given by:

```
res.exc$E
```

```
 [1] 165 166 167 168 169 170 171 172 173 174 175 176 177 178 179 180 181
[18] 182 183 184 185 186 187 188 189 190 191 192
```

The probability that all the days have more than 0.3 precipitation is at least 0.95. These are 28 consecutive days between May and June, three days fewer than those identified by the marginal probabilities.

8

Gaussian Process Regression

As the title suggests, Gaussian process regression (GPR) places priors on the functions describing the relationship between the predictor(s) and the response which are Gaussian processes. We can implement these methods in INLA but we need to make some compromises to achieve reasonable computational efficiency. In this chapter, we introduce the methodology and demonstrate some extensions.

8.1 Introduction

Suppose we have a regression model:

$$y_i = f(\mathbf{x}_i) + \varepsilon_i \qquad i = 1, \ldots, n.$$

We specify a Gaussian process on f with mean $m(\mathbf{x})$ and covariance $k(\mathbf{x}, \mathbf{x}')$, where $\mathbf{x}'$ is another point in the predictor space, so that

$$f(\mathbf{x}) \sim \mathcal{GP}(m(\mathbf{x}), k(\mathbf{x}, \mathbf{x}')),$$

which has the property that any finite number of $f(\mathbf{x}_i)$'s have a joint Gaussian distribution with mean and covariance defined by $m()$ and $k()$. An introduction to Gaussian processes may be found in Rasmussen and Williams (2006).

At first glance, this problem seems readily solved by INLA. The values $f(\mathbf{x}_i)$ have a Gaussian distribution and would satisfy the requirements of a latent Gaussian model (LGM), but there are two problems. Firstly, we are likely to be interested in what happens to f at more than just the points of observation $\mathbf{x}_i$. Secondly, we require more than just an LGM for INLA to be effective. For many choices of $k()$, the precision matrix on $f(\mathbf{x})$ will be dense. For anything but small data problems, there will be no shortcut computation. This means we must restrict our choice of $k()$ to allow fast computation. Fortunately, we still have some good choices that would result in a sparse precision matrix permitting a rapid computation.

A solution to these problems arises from an unexpected source. Consider the stochastic partial differential equation (SPDE) defined by:

$$(\kappa^2 - \Delta)^{\alpha/2}(\tau f(\mathbf{x})) = \mathcal{W}(\mathbf{x}), \quad \mathbf{x} \in \mathbb{R}^d, \tag{8.1}$$

where κ is a scale parameter, Δ is the Laplacian, α is a parameter that affects the smoothness, and τ relates to the variance of f. $\mathcal{W}(\mathbf{x})$ is a Gaussian white noise process and d is the dimension of the space. f is the stationary solution to this SPDE. Surprisingly, Whittle (1954) showed that the stationary solution has a Matérn covariance:

$$\mathrm{cov}(f(\mathbf{0}), f(\mathbf{x})) = \frac{\sigma^2}{2^{\nu-1}\Gamma(\nu)}(\kappa\|\mathbf{x}\|)^{\nu} K_{\nu}(\kappa\|\mathbf{x}\|),$$

where K_ν is a Matérn family kernel (also called a modified Bessel function of the second kind), $\nu = \alpha - d/2$ and σ^2 is the marginal variance defined by

$$\sigma^2 = \frac{\Gamma(\nu)}{\Gamma(\alpha)(4\pi)^{d/2}\kappa^{2\nu}\tau^2}.$$

Standard choices of α are one, which corresponds to the exponential kernel (which looks like a double exponential distribution) and two, which is the default choice in INLA. The squared exponential or Gaussian kernel corresponds to $\alpha = \infty$ but this choice results in a dense precision matrix because the Gaussian kernel has unbounded support. This is not practical to compute using INLA. In contrast, any choice in $0 \leq \alpha \leq 2$ is available in INLA as these result in kernels with a compact support.

Finding the stationary solution requires an approximation. We restrict solutions to the form:

$$f(\mathbf{x}) = \sum_{k=1}^{b} \beta_k B_k(\mathbf{x}).$$

The B-spline basis functions, $B_k(\mathbf{x})$, have a compact support. We choose the coefficients, β_k, to approximate the solution to the SPDE. We need to choose the number of basis functions b to be sufficiently large so as to obtain a good approximation, but not so large as to unnecessarily burden the computation. The mechanics of the computation require the use of a finite element method whose details we do not explain here. See Lindgren and Rue (2015) for more details.

This method requires that we specify three priors: on τ, κ and the precision of ε (the measurement error). The parameter α controls the shape of the kernel — we just set this. τ and κ are difficult to visualize so we prefer to use more intuitive parameters. Let σ be the standard deviation of f and let ρ be the "range." This is the distance at which the correlation in the function has fallen to about 0.13. The choice of this particular value may seem odd but results in the clean formulation shown below. Informally, one might think about the change in x necessary for the correlation to fall to a small but not negligible value, i.e., 0.13. We can link these parameters together with these relations:

$$\begin{aligned}\log\tau &= \frac{1}{2}\log\left(\frac{\Gamma(\nu)}{\Gamma(\alpha)(4\pi)^{d/2}}\right) - \log\sigma - \nu\log\kappa,\\ \log\kappa &= \frac{\log(8\nu)}{2} - \log\rho.\end{aligned}$$

We can see that it is convenient to specify the priors in terms of $\log\sigma$ and $\log\rho$. It is helpful to specify a guess at these two parameters and allow the prior to vary around these guesses. We can achieve this by:

$$\begin{aligned}\log\sigma &= \log\sigma_0 + \theta_1,\\ \log\rho &= \log\rho_0 + \theta_2. \end{aligned} \tag{8.2}$$

We now have the two initial guesses σ_0 and ρ_0 and we can assign a mean zero joint Gaussian prior on (θ_1, θ_2). We will see how to do this in our example to follow.

The data for our example come from Bralower et al. (1997) who reported the ratio of strontium isotopes found in fossil shells in the mid-Cretaceous period from about 90 to 125 million years ago. This data was analyzed using the SiZer method in Chaudhuri and Marron (1999). They rescaled the strontium ratio which we replicate here. We load and plot the data as seen in Figure 8.1.

```
data(fossil, package="brinla")
fossil$sr <- (fossil$sr-0.7)*100
library(ggplot2)
pf <- ggplot(fossil, aes(age, sr)) + geom_point() + xlab("Age") +
      ylab("Strontium Ratio")
pf+geom_smooth(method="gam", formula = y ~ s(x, bs = "cs"))
```

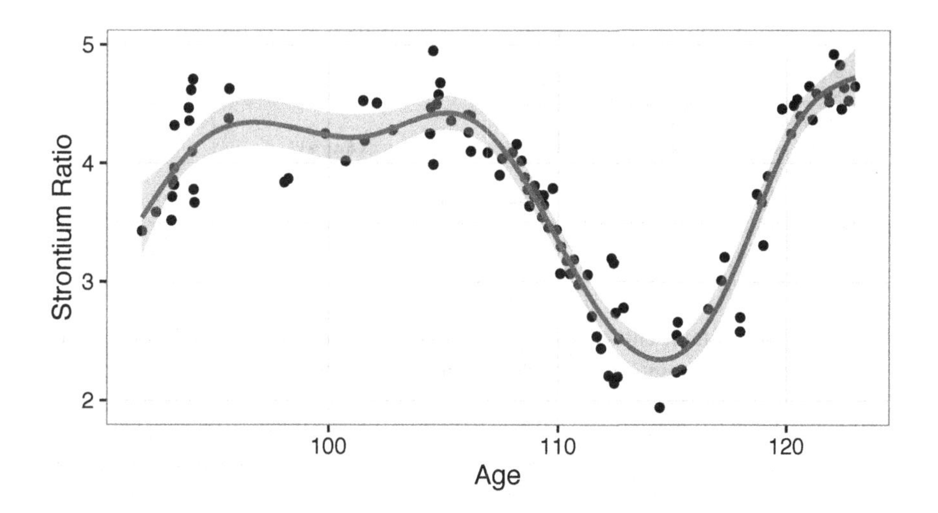

FIGURE 8.1
The age of a shell in millions of years and its (standardized) ratio of strontium isotopes. The grayed area indicates 95% pointwise confidence bands.

We have used a standard *B*-spline smoother which also generates 95% confidence intervals. We can see that the shape of the curve is not much in doubt for the older half of the curve but there is substantially more uncertainty for the more recent period. This dataset is also modeled using a random walk regression in Chapter 7.

Fitting a GPR model in INLA requires several steps. Once these are understood, much of the process can be automated but it is worth understanding these in case we wish to modify the procedure according to the circumstances.

We will need the INLA package throughout this chapter and will also use our brinla package. From now on, we will assume these have already been loaded. If you forget, you will get a "function not found" error message.

```
library(INLA); library(brinla)
```

The first step is to set up the *B*-spline basis. We have chosen $b = 25$ basis functions:

```
nbasis <- 25
mesh <- inla.mesh.1d(seq(min(fossil$age), max(fossil$age),
        length.out = nbasis),degree =2)
```

We have chosen a support covering the range of the data. The degree of two for the *B*-splines means we will get a smoother fit than the choices of zero or one that are also available. If we choose too few basis functions, we will lack the flexibility to fit the function. If we choose more than necessary, the computation will be inefficient. In some regression spline applications, b is used to control the amount of smoothing. Here ρ fills that role mostly, while b just needs to be large enough to allow sufficient flexibility in the fit. If you make b too large, you will waste compute time but the fit will not change much.

The next step is to create the SPDE object. This is where the priors are set:

```
alpha <- 2
nu <- alpha - 1/2
sigma0 <- sd(fossil$sr)
rho0 <- 0.25*(max(fossil$age) - min(fossil$age))
kappa0 <- sqrt(8 * nu)/rho0
tau0 <- 1 / (4 * kappa0^3 * sigma0^2)^0.5
spde <- inla.spde2.matern(mesh, alpha=alpha,
                          B.tau = cbind(log(tau0), 1, 0),
                          B.kappa = cbind(log(kappa0), 0, 1),
                          theta.prior.prec = 1e-4)
```

We make the default choice of $\alpha = 2$ which is the smoothest kernel available to us. ν is simply a function of α and the dimension (which is one here). We set σ_0 to the SD of the response. It would be preferable to use only prior knowledge of the response which with expert opinion would be available to us. Failing that we use a small amount of information from the data to make this choice. If we have good prior knowledge, we should override this choice. We set ρ_0 to be a quarter of the range of the predictor. Again, prior knowledge would be better employed. We can also be assured that the result is not very sensitive to these choices as you can readily check. Although we set the priors in terms of σ and ρ, we need to convert these to τ and κ.

We now need to specify a model on $\log\tau$ taking the form:

$$\log\tau = \gamma_0 + \gamma_1\theta_1 + \gamma_2\theta_2.$$

The coefficients for $(\gamma_0, \gamma_1, \gamma_2)$ are set by `B.tau` as `(log(tau0), 1, 0)` here (which is the default choice). A similar model is specified for $\log\kappa$ using `B.kappa`. More complex choices are possible here as we shall see later. Finally, we need to set a precison for the Gaussian priors on the θ's using `theta.prior.prec`. We have chosen a very small value here expressing our uncertainty about the priors on τ and κ. In some instances, we might have stronger information in which case a larger choice would be more appropriate.

Further setup steps are required. We must specify the points at which we require posterior distributions. Here we choose the points of observation although in some cases one might want a denser grid or to include particular values of interest. The `A` matrix allows the computation at these grid points based on the spline basis we set up earlier.

```
A <-  inla.spde.make.A(mesh, loc=fossil$age)
```

In two dimensional problems, there are tedious problems keeping track of the spatial relationships. This is solved by setting up an index. In one dimension, the problem is not so difficult but we must still set the index (which we must give a name - `sinc` in this case):

```
index <- inla.spde.make.index("sinc", n.spde = spde$n.spde)
```

A further housekeeping step whose purpose is to stack matrices as vectors is:

```
st.est <- inla.stack(data=list(y=fossil$sr), A=list(A),
          effects=list(index),  tag="est")
```

For now, we are satisfied to estimate the curve at the points of observation only so the `effects` are just the index points which we give the tag `est` for later reference. We are now ready to fit the model:

```
formula <- y ~ -1 + f(sinc, model=spde)
data <- inla.stack.data(st.est)
result <- inla(formula, data=data,  family="gaussian",
       control.predictor= list(A=inla.stack.A(st.est), compute=TRUE))
```

The formula is of `spde` type and we must use the name for the index we set up earlier. An intercept term would be redundant so we exclude this with a `-1`. Further stacking is necessary using `inla.stack.data`.

We can extract a plot of the posterior mean together with 95% credible bands as seen in Figure 8.2:

```
ii <- inla.stack.index(st.est, "est")$data
plot(sr ~ age, fossil)
tdx <- fossil$age
lines(tdx, result$summary.fitted.values$mean[ii])
lines(tdx, result$summary.fitted.values$"0.025quant"[ii], lty = 2)
lines(tdx, result$summary.fitted.values$"0.975quant"[ii], lty = 2)
```

We see some roughness in the fit where the data are sparse. We also see that the credible bands are wider in these sparse regions. Compare this to the first panel of Figure 7.11 for the random walk fit. The function `bri.gpr()` in the `brinla` package gathers the computation into a single call. We can reproduce Figure 8.2 with:

```
fg <- bri.gpr(fossil$age, fossil$sr)
plot(sr ~ age, fossil, pch=20)
lines(fg$xout, fg$mean)
```

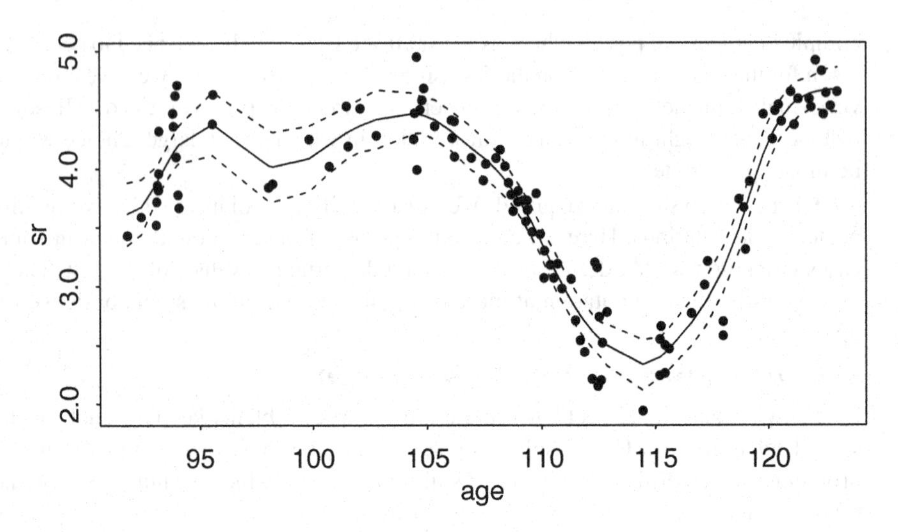

FIGURE 8.2
Posterior mean shown as solid line with 95% credible bands shown as dashed lines.

```
lines(fg$xout, fg$ucb,lty=2)
lines(fg$xout, fg$lcb,lty=2)
```

Even so, one may need to modify the calculation so it is worth understanding the steps involved in constructing this.

8.2 Penalized Complexity Priors

In Simpson et al. (2017) a principled approach to the setting of priors is presented. The details of how this applies to GPR models is found in Fuglstad et al. (2015). These priors are specified by setting four constants based in the relations:

$$\begin{aligned} P(\rho < \rho_0) &= p_\rho, \\ P(\sigma > \sigma_0) &= p_\sigma. \end{aligned}$$

To set these, we need some knowledge of the scale of the response and the predictor. Recall that ρ is the distance such that $\text{corr}(f(x), f(x+\rho)) \approx 0.13$. We pick p_ρ as a small probability and so ρ_0 represents a small range — we do not expect that ρ lies below this but we do not exclude the possibility. Thus ρ_0 reflects our prior opinion about the smoothness of the function. The main purpose is to provide some weak information about the correlation but there is still much flexibility for the posterior on ρ to be quite a bit different from this. Our choice of ρ_0 is not destiny. The choice of σ_0 is similarly constructed based on our prior knowledge of the variability in f. Again we just need to get the general scale correct.

Implementation of this PC prior requires that the SPDE is set up slightly differently. Everything else is much the same as before. We choose $p_\rho = p_\sigma = 0.05$. We set $\rho_0 = 5$ because this is a relatively small distance on the predictor scale. We'd be surprised if the correlation dropped so quickly after only 5 million years. We set $\sigma_0 = 2$ because this is quite large and so we would be surprised if the SD of f exceeded this. The observed response varies from about 2 to 5, so this choice is quite conservative. We reuse most of the prior setup steps leaving just the following commands which need to be run:

```
spde <- inla.spde2.pcmatern(mesh,alpha=alpha,prior.range=c(5,0.05),
        prior.sigma=c(2,0.05))
formula <- y ~ -1 + f(sinc, model=spde)
resultpc <- inla(formula, data=data,  family="gaussian",
 control.predictor= list(A=inla.stack.A(st.est), compute=TRUE))
```

The plot can be constructed exactly as previously. We do not show it here since it is extremely similar to Figure 8.2. The function `bri.gpr()` in the `brinla` package has an option `pcprior` which performs the whole computation like this:

```
pcmod <-  bri.gpr(fossil$age, fossil$sr, pcprior=c(5,2))
```

8.3 Credible Bands for Smoothness

In Chaudhuri and Marron (1999), the apparent dip seen in Figure 8.2 around 97 million years was deemed insignificant. The credible bands shown in the plot are not much help in answering this question as a large variety of curves can be drawn that lie between the two bands — some quite rough and some rather smooth. The existence of the dip depends crucially on how much we smooth the data. In most smoothing methods, this choice of smoothness is either chosen by the user or by some automated method which produces a point estimate of the best choice. With a GPR model, we can do better than this.

First let us extract the posterior distributions of the hyperparameters. We work with the error SD:

```
errorsd <- bri.hyper.sd(result$marginals.hyperpar[[1]])
```

Many remaining quantities of interest can be extracted with

```
mres <- inla.spde.result(result,"sinc",spde)
```

from which we may obtain the posterior for σ (we need to square root the variance):

```
mv <- mres$marginals.variance.nominal[[1]]
sigmad <- as.data.frame(inla.tmarginal(function(x) sqrt(x), mv))
```

and that for ρ:

```
rhod <- mres$marginals.range.nominal[[1]]
```

All these can be plotted as seen in Figure 8.3:

```
plot(y ~ x, errorsd, type="l", xlab="sr", ylab="density")
plot(y ~ x,sigmad, type="l",xlab="sr",ylab="density")
plot(rhod,type="l",xlab="age",ylab="density")
```

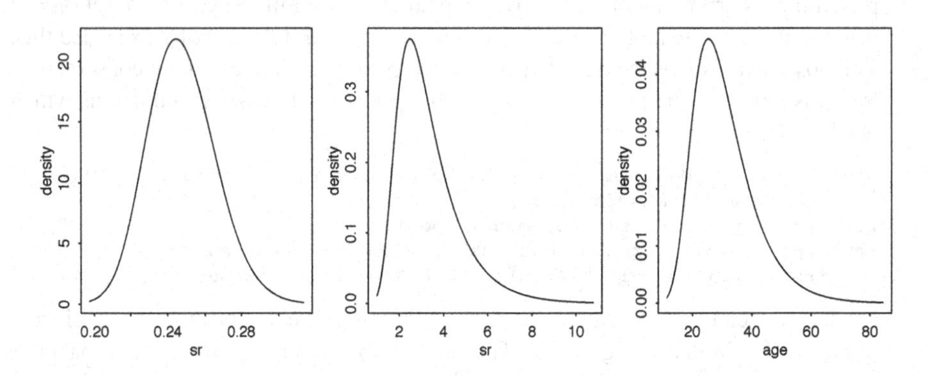

FIGURE 8.3
Posterior densities of the hyperparameters from the fossil data fit. The first plot shows the error SD, the second the function SD and the third the range.

We see there is not much uncertainty in the error SD but there is considerable uncertainty about the range ρ. We can easily derive a credible interval for κ (and hence ρ) as:

```
exp(mres$summary.log.kappa[c(4,6)])
```

```
         0.025quant 0.975quant
kappa.1     0.06098    0.21686
```

but it is difficult to interpret this or divine the impact on the shape of the fitted curve.

Let's fix σ, the function SD at its median posterior value and calculate the posterior mean curves corresponding to the two endpoints of the credible interval for ρ:

```
kappa0 <- exp(mres$summary.log.kappa['0.025quant'])[,]
sigma02 <- exp(mres$summary.log.variance.nominal['0.5quant'])[,]
tau0 <- 1 / (4 * kappa0^3 * sigma02)^0.5
spde <- inla.spde2.matern(mesh, alpha=alpha, constr = FALSE,
                          B.tau = cbind(log(tau0)),
                          B.kappa = cbind(log(kappa0)))
formula <- y ~ -1 + f(sinc, model=spde)
resulta <- inla(formula, data=data,  family ="gaussian",
          control.predictor = list(A=inla.stack.A(st.est),
          compute=TRUE))
```

The point specification of `B.tau` and `B.kappa` mean these hyperparameters are not allowed to vary and are fixed at the choices we have made. Notice that τ_0 depends on κ_0 and σ_0 so we must compute this value. We repeat the same calculation but for the upper end of the κ credible interval:

```
kappa0 <- exp(mres$summary.log.kappa['0.975quant'])[,]
sigma02 <- exp(mres$summary.log.variance.nominal['0.5quant'])[,]
```

```
tau0 <- 1 / (4 * kappa0^3 * sigma02)^0.5
spde <- inla.spde2.matern(mesh, alpha=alpha, constr = FALSE,
                          B.tau = cbind(log(tau0)),
                          B.kappa = cbind(log(kappa0)))
formula <- y ~ -1 + f(sinc, model=spde)
resultb <- inla(formula, data=data, family="gaussian",
         control.predictor= list(A=inla.stack.A(st.est),
         compute=TRUE))
```

We plot the two posterior means corresponding to the ends of the κ credible interval as seen in Figure 8.4.

```
ii <- inla.stack.index(st.est, "est")$data
plot(sr ~ age, fossil, pch=20)
tdx <- fossil$age
lines(tdx, resulta$summary.fitted.values$mean[ii],lty=2)
lines(tdx, resultb$summary.fitted.values$mean[ii],lty=1)
```

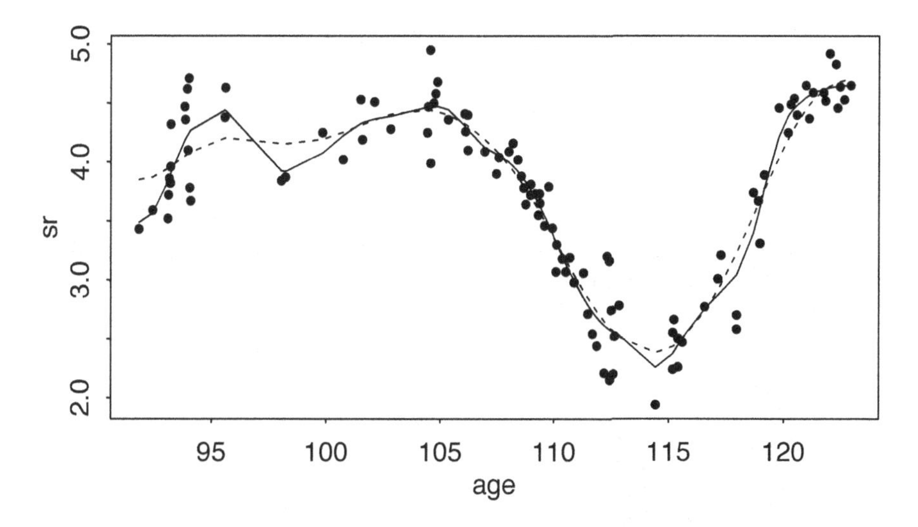

FIGURE 8.4
95% Credible bands for smoothness. The solid line is the posterior mean curve corresponding to the 97.5% percentile of ρ, while the dashed line is for the 2.5% percentile.

We see that uncertainty over the smoothness has little impact on the shape of the curve on the right half of the plot but makes a substantial difference on the first part. In particular, we can see that the smoothness has a just perceptable dip at 97 million years whereas the dip is very clear in the rougher end of the fit. Since both bands contain a dip here, we conclude there really is a dip in the curve at this age. The function `bri.smoothband()` in the `brinla` package gathers these computations into a single call. We can reconstruct Figure 8.4 with:

```
fg <- bri.smoothband(fossil$age, fossil$sr)
plot(sr ~ age, fossil, pch=20)
lines(fg$xout, fg$rcb,lty=1)
```

```
lines(fg$xout, fg$scb,lty=2)
```

The credible bands for smoothness tell us nothing about the position of the curve at any particular points — for this we need the standard bands shown in Figure 8.2. But if we are more interested in the uncertainty about the shape of the curve, these new bands are more helpful. For more discussion see Faraway (2016a).

8.4 Non-Stationary Fields

We may wish to model functions that exhibit variations in amplitude and smoothness over the domain. We can allow for this by generalizing the equations (8.2) and to:

$$\begin{aligned}\log(\sigma(\mathbf{x})) &= \gamma_0^{\sigma}(\mathbf{x}) + \sum_{k=1}^{p} \gamma_k^{\sigma}(\mathbf{x})\theta_k, \\ \log(\rho(\mathbf{x})) &= \gamma_0^{\rho}(\mathbf{x}) + \sum_{k=1}^{p} \gamma_k^{\rho}(\mathbf{x})\theta_k,\end{aligned}$$

where $\gamma_k^{\sigma}()$ and $\gamma_k^{\rho}()$ are sets of basis functions. The coefficients θ are held in common between the two sets but could be defined so that both vary freely if required. Let's illustrate how this can be applied.

To show that the method works effectively, we need a function with clearly varying smoothness. We use the same test function as used in Chapter 7:

```
set.seed(1)
n <- 100
x <- seq(0, 1, length=n)
f.true <- (sin(2*pi*(x)^3))^3
y <- f.true + rnorm(n, sd = 0.2)
td <- data.frame(y = y, x = x, f.true)
```

The data are simulated from the following model:

$$y_i = \sin^3(2\pi x_i^3) + \varepsilon_i, \quad \varepsilon_i \sim N(0, \sigma_\varepsilon^2),$$

where $x_i \in [0,1]$ are equally spaced and $\sigma_\varepsilon^2 = 0.04$. We simulate $n = 100$ data points.

For comparison purposes let's construct the fit with a stationary field. The process is the same as followed in the introduction although we need to change the scaling for this data:

```
nbasis <- 25
mesh <- inla.mesh.1d(seq(0,1,length.out = nbasis),degree =2)
alpha <- 2
nu <- alpha - 1/2
sigma0 <- sd(y)
rho0 <- 0.1
kappa0 <- sqrt(8 * nu)/rho0
tau0 <- 1 / (4 * kappa0^3 * sigma0^2)^0.5
```

```
spde <- inla.spde2.matern(mesh, alpha=alpha,
                          B.tau = cbind(log(tau0), 1, 0),
                          B.kappa = cbind(log(kappa0), 0, 1),
                          theta.prior.prec = 1e-4)
A <-  inla.spde.make.A(mesh, loc=td$x)
index <- inla.spde.make.index("sinc", n.spde = spde$n.spde)
st.est <- inla.stack(data=list(y=td$y), A=list(A),
          effects=list(index),  tag="est")
formula <- y ~ -1 + f(sinc, model=spde)
data <- inla.stack.data(st.est)
result <- inla(formula, data=data,  family="gaussian",
    control.predictor= list(A=inla.stack.A(st.est), compute=TRUE))
```

We display this in the first plot of Figure 8.5:

```
ii <- inla.stack.index(st.est, "est")$data
plot(y ~ x, td, col=gray(0.75))
tdx <- td$x
lines(tdx, result$summary.fitted.values$mean[ii])
lines(tdx,f.true,lty=2)
```

We see that the fitted function is too rough on the left but too smooth on the right. This function is difficult to fit because it has varying smoothness. All stationary smoothers struggle with this test function for this reason. We now introduce non-stationarity:

```
basis.T <-as.matrix(inla.mesh.basis(mesh, type="b.spline",
  n=5, degree=2))
basis.K <-as.matrix(inla.mesh.basis(mesh, type="b.spline",
  n=5, degree=2))
spde <- inla.spde2.matern(mesh, alpha=alpha,
             B.tau = cbind(basis.T[-1,],0),
        B.kappa = cbind(0,basis.K[-1,]/2),
                  theta.prior.prec = 1e-4)
formula <- y ~ -1 + f(sinc, model=spde)
result <- inla(formula, data=data,  family="gaussian",
  control.predictor= list(A=inla.stack.A(st.est), compute=TRUE))
```

We use *B*-spline bases for both τ and κ. We do not need much flexibility so only five basis functions are used. Using more than five would just increase the cost of computation here. The division by two in the specification of `B.kappa` is there because `inla.spde2.matern` works with $\log(\kappa)$, whereas `inla.spde.create` works with $\log(\kappa^2)$. We have used zeroes in the specification of the bases so that each part has its own set of θ's. We plot the posterior mean as seen in the second plot of Figure 8.5.

```
plot(y ~ x, td, col=gray(0.75))
lines(tdx, result$summary.fitted.values$mean[ii])
lines(tdx,f.true,lty=2)
```

We see that the posterior mean follows the true function very well. In particular, it is able to follow the initial flat part of the test function while still capturing the two optima and point of inflexion seen later in the function. Compare this to Figure 7.8 which shows the random walk smoothing method applied to this same problem. The function `bri.nonstat()` executes these computations more conveniently although any modifications will require running through the steps above. We can reproduce the second panel of Figure 8.5 with:

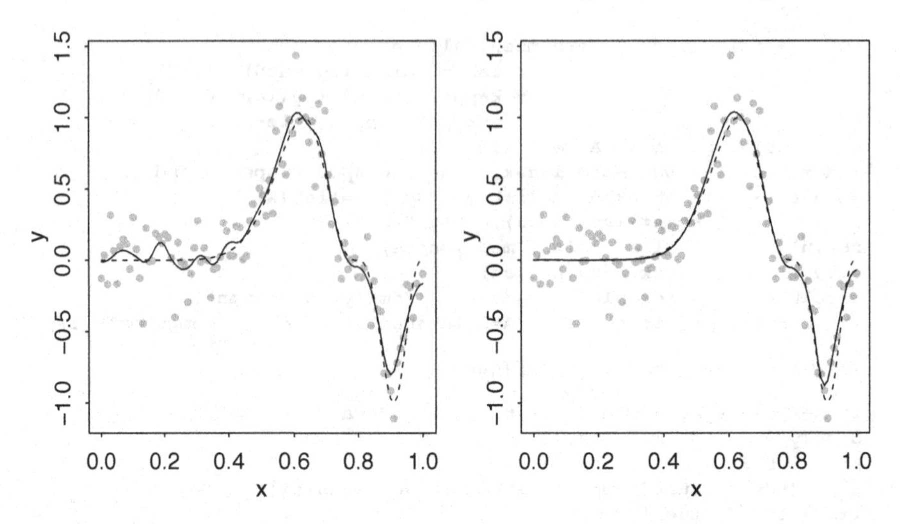

FIGURE 8.5
Stationary and non-stationary fits. The posterior mean is shown as a solid line and the true function as a dashed line.

```
fg <- bri.nonstat(td$x, td$y)
plot(y ~ x, td, col=gray(0.75))
lines(f.true ~ x, td, lty=2)
lines(fg$xout, fg$mean)
```

8.5 Interpolation with Uncertainty

Suppose we observe pairs $(\mathbf{x}_i, y_i)$ from the relationship $y = f(\mathbf{x})$ and there is no (or at least very little) measurement error. At first glance, this may not seem like a statistics problem at all but rather an exercise in function approximation. The usual methods of nonparametric regression do not apply. Nevertheless, although we know the value of y exactly at the points of observation, we would be uncertain about f between these points. We should like to quantify this uncertainty. This is an example of what applied mathematicians call *uncertainty quantification*. In the *computer experiments* problem, observations $(\mathbf{x}, f(\mathbf{x}))$ require time-consuming computer simulations and so one obtains limited information about f. We want to estimate f from a limited sample. We want an expression of our uncertainty about f.

We can solve this problem using GPR by forcing the error in the observations towards zero. We can also allow a small amount of measurement error if appropriate. To illustrate how the method works, we generate a small artificial dataset:

```
x <- c(0,0.1,0.2,0.3,0.7,1.0)
y <- c(0,0.5,0,-0.5,1,0)
```

```
td <- data.frame(x,y)
```

The first part of the solution follows the same path as previous examples. We use a large number of basis functions. This will be computationally expensive but the method will adapt to our extravagance. If you need this to run faster, you might economize here.

The choice of the prior on f is critical to the outcome. The choice α determines the shape of the kernel — two is the smoothest available to us. We set up the mesh with degree two splines which is again the smoothest choice available. We use penalized complexity priors because this makes it easier to specify our prior uncertainty. We have specified a small chance that the range is quite small indicating a preference for a smoother fit. We have allowed a small chance that the variance of the function is much larger than the data would suggest. Finally, we set up the `A` matrix to compute the fit at the points of observation:

```
nbasis <- 100
alpha <- 2
mesh <- inla.mesh.1d(seq(0,1,length.out = nbasis),degree = 2)
spde <- inla.spde2.pcmatern(mesh, alpha=alpha,
         prior.range=c(0.05,0.1), prior.sigma=c(5,0.05))
A <-  inla.spde.make.A(mesh, loc=td$x)
```

The purpose of this analysis is to investigate f between the points of observation so we need estimates in these ranges. We specify a fine grid of 101 points and define an `A` to convert between the mesh and these points. This method is useful also in other examples where we want estimates at more than the points of observation. We need to keep track of two sets of estimates now — one for the points of observation and one for the grid. We use the stacking functions to keep track of this:

```
ngrid <- 101
Ap <-  inla.spde.make.A(mesh, loc=seq(0,1,length.out = ngrid))
index <- inla.spde.make.index("sinc", n.spde = spde$n.spde)
st.est <- inla.stack(data=list(y=td$y), A=list(A),
          effects=list(index),  tag="est")
st.pred <- inla.stack(data=list(y=NA), A=list(Ap),
          effects=list(index),  tag="pred")
formula <- y ~ -1 + f(sinc, model=spde)
sestpred <- inla.stack(st.est,st.pred)
```

Now we are ready to fit the function. We specify that the hyperparameter associated with the observation error has a very large fixed precision. This forces the observation error down towards zero.

```
result <- inla(formula, data=inla.stack.data(sestpred),
 family="gaussian",
 control.predictor= list(A=inla.stack.A(sestpred), compute=TRUE),
 control.family(hyper=list(prec = list(fixed = TRUE, initial = 1e8))))
```

We now plot the resulting fit. We use the predicted response on the grid. We display the posterior mean along with the 95% credible bands as seen in the first panel of Figure 8.6. We are certain of the function at the points of observation but there is considerable uncertainty between these points.

```
ii <- inla.stack.index(sestpred, tag='pred')$data
```

```
plot(y ~ x, td,pch=20,ylim=c(-2,2))
tdx <- seq(0,1,length.out = ngrid)
lines(tdx, result$summary.linear.pred$mean[ii])
lines(tdx, result$summary.linear.pred$"0.025quant"[ii], lty = 2)
lines(tdx, result$summary.linear.pred$"0.975quant"[ii], lty = 2)
```

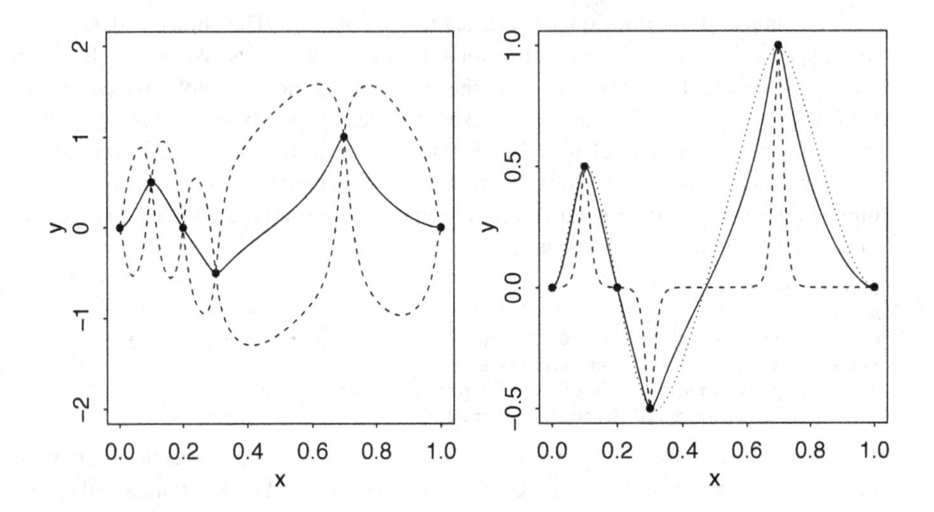

FIGURE 8.6
On the left, the solid line shows the posterior mean and the dotted lines show the 95% credible interval for the function. On the right, the solid line again shows the posterior mean while the dashed line shows the roughest end of the 95% credible interval on the range. The dotted line shows the smoothest end of that interval.

Should we need it, we can extract the posterior distribution at any point on the grid. For example, we can get this distribution at $x = 0.5$ by:

```
jj <- which(tdx == 0.5)
margpred <- result$marginals.linear[ii]
plot(margpred[[jj]],type="l", ylab="Density")
```

We have not shown the plot as it is the usual bell-shaped posterior we would expect here.

With a small dataset, the results are likely to be more sensitive to the prior. Readers may find it helpful to experiment with the choices we have made above. The choice of α makes a large difference while moderate changes to the prior on the range and SD have relatively less impact.

We can also construct bands that express our uncertainty regarding the smoothness of the fitted function. We can obtain summary information about the posterior distribution on the range:

```
mres <- inla.spde.result(result,"sinc",spde)
exp(mres$summary.log.range.nominal[c(2,4,5,6,7)])
```

```
                  mean 0.025quant 0.5quant 0.975quant    mode
range.nominal.1 0.1327   0.033269  0.14456     0.3462 0.18881
```

We need to exponeniate since the original summary is on the log scale. It only makes sense to exponentiate some of the summary statistics as we have selected above. We see the posterior median for the range is 0.145 but the 95% credible interval is quite wide. We can obtain similar information about the SD of the function:

```
sqrt(exp(mres$summary.log.variance.nominal[c(2,4,5,6,7)]))
```

```
                     mean 0.025quant 0.5quant 0.975quant    mode
variance.nominal.1 0.52937    0.30277  0.53804      0.857 0.56354
```

As before, we need to exponentiate and hence the selection of statistics to report. We have also taken the square root to get the results on the SD scale (row title is misleadling). We see that this 95% credible interval is relatively narrower since our six observations do give us some good information about the variance.

It is difficult to visualize the consequences of this uncertainty so, as in the previous section, we find it helpful to plot the estimated functions corresponding to the ends of the credibility interval. We fix the hyperparameters for range and SD, first using the lower end of the range credibility interval. We use the median for the SD. This will give us the roughest fit within the credibility interval for range:

```
spde <- inla.spde2.pcmatern(mesh,alpha=alpha,
  prior.range=c(0.033269,NA),prior.sigma=c(0.53804,NA))
resultl <- inla(formula, data=inla.stack.data(sestpred),
 family="gaussian",
 control.predictor= list(A=inla.stack.A(sestpred), compute=TRUE),
 control.family(hyper=list(prec = list(fixed = TRUE, initial = 1e8))))
```

We repeat the calculation but now for the upper end of the interval:

```
spde <- inla.spde2.pcmatern(mesh,alpha=alpha,
  prior.range=c(0.3462,NA),prior.sigma=c(0.53804,NA))
resulth <- inla(formula, data=inla.stack.data(sestpred),
 family="gaussian",
 control.predictor= list(A=inla.stack.A(sestpred), compute=TRUE),
 control.family(hyper=list(prec = list(fixed = TRUE, initial = 1e8))))
```

We now plot the result as seen in Figure 8.6.

```
plot(y ~ x, td,pch=20)
tdx <- seq(0,1,length.out = ngrid)
lines(tdx, result$summary.linear.pred$mean[ii])
lines(tdx, resultl$summary.linear.pred$mean[ii],lty=2)
lines(tdx, resulth$summary.linear.pred$mean[ii],lty=3)
```

We see that the roughest fit means that the correlation between points which are well-separated must drop towards zero. This is achieved by a constant fit as seen in the figure. The fitted function represented by the dotted line moves up and down to catch the observations but quickly returns to the constant. The smoothest fit represented by the dashed line is clearly smoother than the posterior mean fit. The plot gives us a different view of the uncertainty from the previous plot. In the first plot, we claim a 95% probability that the true function lies between the two bands (in a pointwise sense). In the second plot, the claim is regarding the smoothness of the true function rather than its position.

8.6 Survival Response

In the previous examples, the response had a Gaussian distribution. We can generalize to other distributions by incorporating the GPR into the linear predictor. Consider the accelerated failure time models described in Section 6.3 with a Weibull response. In equation (6.5) a parametric linear relationship between the mean response and the covariates is described. We can replace this with:

$$\log \lambda = f(x),$$

where λ is the rate parameter of the Weibull distribution and $f(x)$ is the GPR term. We can add parametric or additional GPR terms to accommodate other predictors as needed.

Consider the larynx cancer example introduced in Section 6.3. We model the survival time using only the age covariate. The construction of the SPDE solution is very similar to previous examples. We use penalized complexity priors as described in Section 8.2. These require some notion of the scaling in the data with the prior on the range depending on the range of ages seen and the prior on `sigma` depending on the survival times. We make conservative choices in both cases. We also need to set up the survival response in `inla.surv` which requires both the time and the censoring variable (where 1=completed observation and 0=censored).

```
data(larynx, package="brinla")
nbasis <- 25
alpha <- 2
xspat <- larynx$age
mesh <- inla.mesh.1d(seq(min(xspat), max(xspat), length.out = nbasis),
        degree = 2)
spde <- inla.spde2.pcmatern(mesh,alpha=alpha,prior.range=c(20,0.1),
        prior.sigma=c(10,0.05))
A <-  inla.spde.make.A(mesh, loc=xspat)
index <- inla.spde.make.index("sinc", n.spde = spde$n.spde)
st.est <- inla.stack(data=list(time=larynx$time,censor=larynx$delta),
          A=list(A),  effects=list(index),  tag="est")
formula <- inla.surv(time,censor) ~  0 +f(sinc, model=spde)
data <- inla.stack.data(st.est)
result <- inla(formula, data=data,  family="weibull.surv",
   control.predictor= list(A=inla.stack.A(st.est), compute=TRUE))
```

Required quantities such as the hazard and survival function can be computed via the linear predictor. Here we demonstrate the computation of the mean survival time along with 95% credible intervals. The index tells us which cases are needed from the linear predictor object to construct the estimate. The shape parameter α is a hyperparameter which is needed to compute the survival time in terms of λ. The resulting estimate along with the credibility bands is shown in the first panel of Figure 8.7. We can see that the linear term for age used in the parametric model is a reasonable choice as there is no strong evidence of nonlinearity.

```
ii <- inla.stack.index(st.est, "est")$data
lcdf <- data.frame(result$summary.linear.predictor[ii,],larynx)
```

```
alpha <- result$summary.hyperpar[1,1]
lambda <- exp(lcdf$mean)
lcdf$exptime <- lambda^(-1/alpha)*gamma(1/alpha + 1)
lambda <- exp(lcdf$X0.025quant)
lcdf$lcb <- lambda^(-1/alpha)*gamma(1/alpha + 1)
lambda <- exp(lcdf$X0.975quant)
lcdf$ucb <- lambda^(-1/alpha)*gamma(1/alpha + 1)
p <- ggplot(data=lcdf,aes(x=age,y=time)) + geom_point()
p + geom_line(aes(x=age,y=exptime)) +
    geom_line(aes(x=age,y=ucb),linetype=2) +
    geom_line(aes(x=age,y=lcb),linetype=2)
```

We can also compute and plot the hazard with 95% credible bands. Since we have no intercept in this model, the baseline hazard is one. The plot is shown in the second panel of Figure 8.7. We see how the hazard increases with age.

```
lambda <- exp(lcdf$mean)
lcdf$hazard <- alpha * lcdf$age^(alpha-1) * lambda
lambda <- exp(lcdf$X0.025quant)
lcdf$hazlo <- alpha * lcdf$age^(alpha-1) * lambda
lambda <- exp(lcdf$X0.975quant)
lcdf$hazhi <- alpha * lcdf$age^(alpha-1) * lambda
ggplot(data=lcdf,aes(x=age,y=hazard)) + geom_line() +
      geom_line(aes(x=age,y=hazlo),lty=2) +
      geom_line(aes(x=age,y=hazhi),lty=2)
```

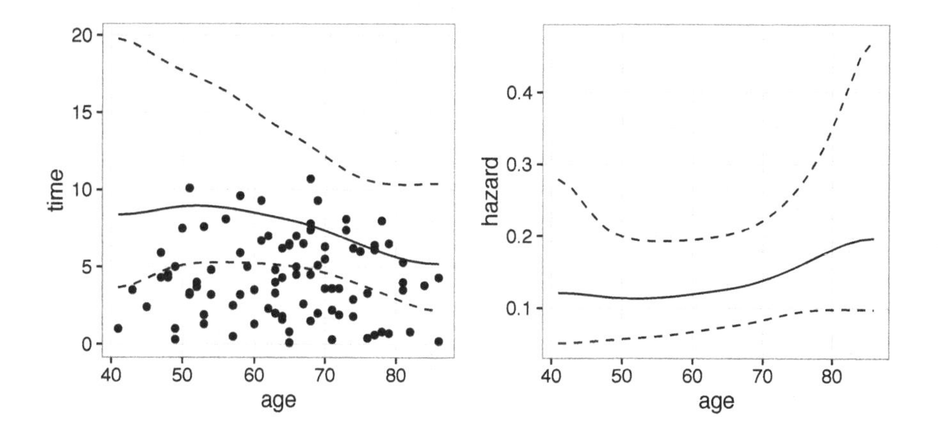

FIGURE 8.7
Mean survival time with age is shown on the left and hazard rate varying with age is shown on the right. Both are shown with 95% credible intervals.

We could add additional terms to the linear predictor to model more covariates.

9

Additive and Generalized Additive Models

The smoothing methods introduced in Chapter 7 can be extended further into higher dimensions, but they become rather computationally intensive as the dimension exceeds two. In multiple dimensions, the variance of nonparametric estimators becomes a real problem. Nonparametric methods often suffer from variance that scales exponentially with the number of predictors, p; This is known as the *curse of dimensionality*. Because of this, it is often desirable to introduce some sort of structure into the model. Introducing structure will certainly introduce bias if the structure does not accurately describe reality; however, it can result in a dramatic reduction in variance. By far the most common approach is to introduce an *additive* structure, resulting in so-called additive and generalized additive models. In this chapter, we show how to make Bayesian inference on these models using INLA.

9.1 Additive Models

Given a response variable y and predictor variables $x_1, \ldots, x_p$, a linear regression model takes the form:

$$y_i = \beta_0 + \sum_{j=1}^{p} \beta_j x_{ij} + \varepsilon_i,$$

for $i = 1, \ldots, n$, where the error ε_i has mean 0 and variance σ_ε^2. To make this model more flexible, we may include transformations and combinations of the predictor variables. Given the wide choice of possible transformations, it can often be difficult to find a good model. Alternatively, we may try a systematic approach of fitting a family of transformations, such as polynomials of the predictor variables. But, the number of terms will become very large if we particularly include interactions.

Instead, we may try a nonparametric regression approach by fitting the following model

$$y_i = f(x_{i1}, \ldots, x_{ip}) + \varepsilon_i.$$

Although it avoids the parametric assumptions about the function f, fitting such a model is simply impractical for p bigger than two, due to the requirement of a large sample size. A good compromise between these two extremes is the *additive model*

(Friedman and Stuetzle, 1981):

$$y_i = \beta_0 + \sum_{j=1}^{p} f_j(x_{ij}) + \varepsilon_i, \tag{9.1}$$

where β_0 is the intercept. Here f_j's may be the functions with a parametric form (e.g., a polynomial), or may be specified simply as "smooth functions" to be estimated by nonparametric means. The additive models are more flexible than the linear models, but still interpretable since f_j can be plotted to give a sense of the marginal relationship between the predictor and the response. It is also easy to accommodate categorical variables in the model using the usual regression approach. However, when strong interactions exist, the additive models will perform poorly. The terms like $f_{ij}(x_i x_j)$ or $f_{ij}(x_i, x_j)$ can be added to the model in that situation. We can also have an interaction between a factor and a continuous predictor by fitting a different function for each level of that factor.

In the frequentist paradigm, the additive models can be taken as a form of nonparametric regression, and be fitted using the *backfitting* algorithm (Buja et al., 1989; Hastie and Tibshirani, 1990). The backfitting algorithm allows a variety of smoothing methods (e.g., smoothing splines) to estimate the functions, and can be implemented in R via the `gam` package. Alternatively, we may first represent the smooth arbitrary functions with a family of spline basis functions, and then estimate the function coefficients based on the penalized likelihood method (Wood, 2006). This penalized smoothing approach allows us to efficiently estimate the degree of smoothness of the model components using generalized cross validation (Wood, 2008), and can be implemented in R via the `mgcv` package. To compare the two methods, the backfitting approach allows for more choice in the smoothers we may use, while the penalized smoothing approach has an automatic choice in the degree of smoothness as well as wider functionality.

Bayesian additive models

We assume that in model (9.1) the random error $\varepsilon_i \sim N(0, \sigma_\varepsilon^2)$ and take a prior on each function $\boldsymbol{f}_j$, i.e., $p(\boldsymbol{f}_j | \boldsymbol{\tau}_j)$, that depends on the unknown hyperparameter(s) $\boldsymbol{\tau}_j$. Then, the joint posterior distribution of this Bayesian additive model is given by

$$\begin{aligned} p(\boldsymbol{f}_1, \ldots, \boldsymbol{f}_p, \boldsymbol{\theta} \mid \boldsymbol{y}) &= \mathcal{L}(\boldsymbol{f}_1, \ldots, \boldsymbol{f}_p, \sigma_\varepsilon^2, \beta_0; \boldsymbol{y}) \times \\ & \quad p(\beta_0) p(\sigma_\varepsilon^2) \prod_{j=1}^{p} p(\boldsymbol{f}_j \mid \boldsymbol{\tau}_j) p(\boldsymbol{\tau}_j), \end{aligned}$$

where $\boldsymbol{\theta}$ denotes all the unknown parameters, $\mathcal{L}$ is the Gaussian likelihood function, and $p(\beta_0)$, $p(\sigma_\varepsilon^2)$ and $p(\boldsymbol{\tau}_j)$ are the priors.

The *Bayesian backfitting* algorithm (Hastie and Tibshirani, 2000) combined with MCMC simulations is often used to sample the marginal posterior distributions. It allows for a wide choice of the smoothing priors for f_j and automatically choosing the amount of smoothing through the data and the priors. However, it is well known that the Markov chains in this case tend to converge slowly and have poor mixing property. It is because the samples of $f_j(x)$ depend on those of the rest of $f_k(x)$ $(k \neq j)$. In addition, it is difficult for general users to implement the MCMC algorithms when the additive models in need are complicated.

INLA, therefore, is a good alternative for fitting the additive models. When taking on each $\boldsymbol{f}_j$ the Gaussian prior introduced in Chapters 7 and 8, the additive models fall in the class of latent Gaussian models and thus can be fitted by INLA. INLA does not suffer from the drawbacks in the MCMC method due to its approximation nature. It is also fairly efficient in computation because all the Gaussian priors used in the model have sparse precision matrices and can be combined together as another big but sparse Gaussian prior to facilitate the approximations.

Simulated data

Let's load the `mgcv` package and simulate data from the so-called "Gu and Wahba 4 univariate term example":

```
library(INLA); library(brinla)
library(mgcv)
set.seed(2)
dat <- gamSim(1, n = 400, dist = "normal", scale = 2)
str(dat)
```

```
'data.frame':        400 obs. of  10 variables:
 $ y : num  7.13 2.97 3.98 10.43 14.57 ...
 $ x0: num  0.185 0.702 0.573 0.168 0.944 ...
 $ x1: num  0.6171 0.5691 0.154 0.0348 0.998 ...
 $ x2: num  0.41524 0.53144 0.00325 0.2521 0.15523 ...
 $ x3: num  0.132 0.365 0.455 0.537 0.185 ...
 $ f : num  8.39 7.52 3.31 10.86 14.63 ...
 $ f0: num  1.097 1.609 1.947 1.008 0.351 ...
 $ f1: num  3.44 3.12 1.36 1.07 7.36 ...
 $ f2: num  3.853084 2.786858 0.000331 8.782269 6.923314 ...
 $ f3: num  0 0 0 0 0 0 0 0 0 0 ...
```

The data frame contains 400 observations of 10 variables, where `y` is the response variable, `xi` (`i` = 0,1,2,3) are the predictor variables, `fi` (`i` = 0,1,2,3) are the true functions, and `f` is the true linear predictor. The noise is Gaussian with mean 0 and SD 2. We then fit the following additive model to the data

$$y = \beta_0 + f_0(x_0) + f_1(x_1) + f_2(x_2) + f_3(x_3) + \varepsilon, \quad \varepsilon \sim N(0, \sigma_\varepsilon^2),$$

where we take the diffuse normal prior on β_0, the diffuse inverse gamma prior on σ_ε^2, and the RW2 prior on each f_i function. Those priors are explained in detail in Chapter 7. The model can be formulated in INLA as

```
formula <- y ~ f(x0, model = 'rw2', scale.model = TRUE) + f(x1, model
    ↪ = 'rw2', scale.model = TRUE) + f(x2, model = 'rw2', scale.model
    ↪ = TRUE) + f(x3, model = 'rw2', scale.model = TRUE)
```

where we scale the RW2 priors to account for the potential different scales required by different functions. We also set the sum-to-zero constraint (the default choice) to each RW2 prior in order to make it identifiable from the intercept. Note that some x

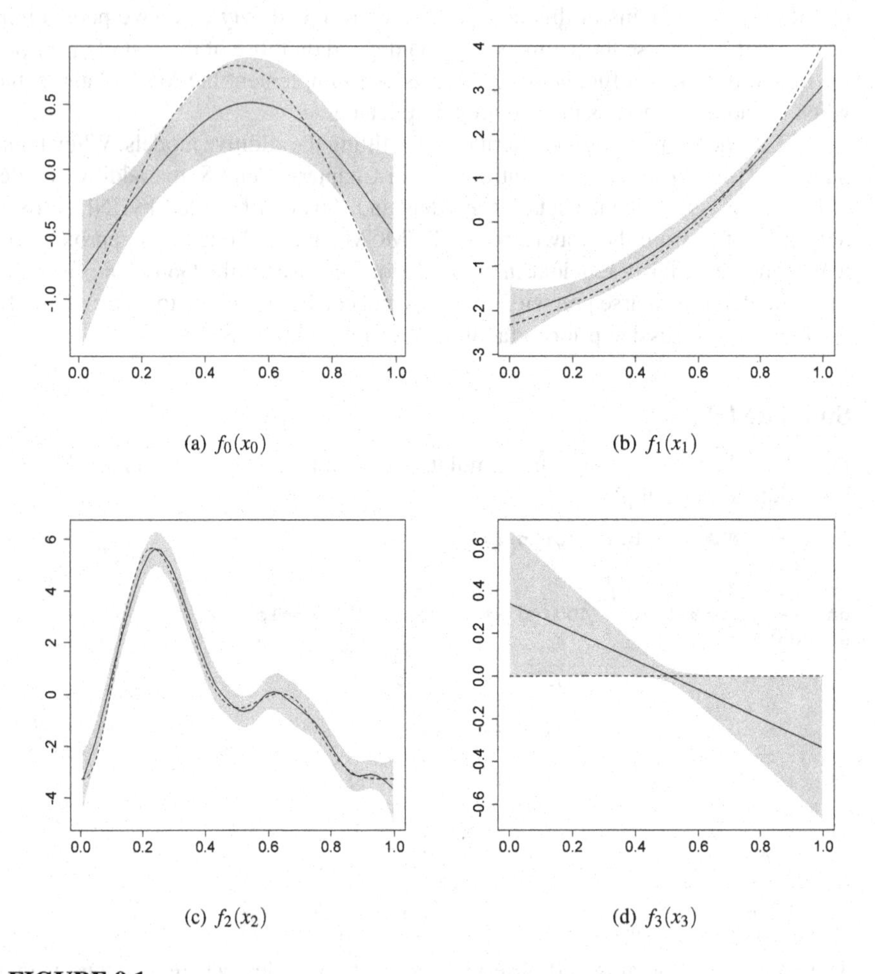

FIGURE 9.1
Simulation results of additive models using RW2 priors: true function (dashed), posterior mean (solid), and 95% credible band (gray).

values are too close to be used for building the RW2 priors. We therefore group them into a few bins first:

```
n.group <- 50
x0.new <- inla.group(dat$x0, n = n.group, method = 'quantile')
x1.new <- inla.group(dat$x1, n = n.group, method = 'quantile')
x2.new <- inla.group(dat$x2, n = n.group, method = 'quantile')
x3.new <- inla.group(dat$x3, n = n.group, method = 'quantile')
```

and then use those bins to fit the model:

```
dat.inla <- list(y = dat$y, x0 = x0.new, x1 = x1.new, x2 = x2.new, x3
    ↪ = x3.new)
result <- inla(formula, data = dat.inla)
```

The posterior summary of β_0 is given by:

```
round(result$summary.fixed, 4)
```

```
              mean     sd 0.025quant 0.5quant 0.975quant   mode kld
(Intercept) 7.7696 0.0995     7.5741   7.7696      7.965 7.7697   0
```

We plot the fitted curve (posterior mean) and 95% credible interval for each function in Figure 9.1. The true function is also plotted for comparison purposes. The R code for plotting Figure 9.1(a) is shown below, and the other plots can be made similarly:

```
bri.band.plot(result, name = 'x0', type = 'random', xlab='', ylab='')
lines(sort(dat$x0), (dat$f0 - mean(dat$f0))[order(dat$x0)], lty = 2)
```

Note that we center each true function to satisfy the sum-to-zero constraint. We can see that the functions are estimated well, considering most of the fits are within the credible bands. They are comparable with those obtained using the `mgcv` package (not shown). The posterior summary of the SDs is given below:

```
round(bri.hyperpar.summary(result), 4)
```

```
                                 mean     sd q0.025   q0.5 q0.975   mode
SD for the Gaussian observations 1.9796 0.0713 1.8443 1.9775 2.1246 1.9733
SD for x0                        0.2164 0.1072 0.0754 0.1933 0.4893 0.1541
SD for x1                        0.1796 0.0986 0.0530 0.1578 0.4318 0.1191
SD for x2                        3.7413 0.9482 2.2800 3.5964 5.9867 3.3188
SD for x3                        0.0105 0.0063 0.0039 0.0087 0.0276 0.0063
```

The estimated SD of the Gaussian noises is close to the true value $\sigma_\varepsilon = 2$. The estimated SD in the RW2 prior for each f function reflects its degrees of smoothness: the smoother the estimated function is, the higher its estimated SD is.

Munich rental guide

We have fitted a few nonparametric regression models for this dataset in Chapter 7. We considered the predictors like *floor size*, *construction year* and *spatial location*, and saw how they individually impacted the *rent*. However, it would be better to build an additive model including all of those predictors as well as the categorical predictors provided in the dataset. We thus consider

$$
\begin{aligned}
\texttt{rent} &= f_1(\texttt{location}) + f_2(\texttt{year}) + f_3(\texttt{floor.size}) \\
&\quad + (\texttt{Gute.Wohnlage})\beta_1 + (\texttt{Beste.Wohnlage})\beta_2 + (\texttt{Keine.Wwv})\beta_3 \\
&\quad + (\texttt{Keine.Zh})\beta_4 + (\texttt{Kein.Badkach})\beta_5 + (\texttt{Besond.Bad})\beta_6 \\
&\quad + (\texttt{Gehobene.Kueche})\beta_7 + (\texttt{zim1})\beta_8 + (\texttt{zim2})\beta_9 \\
&\quad + (\texttt{zim3})\beta_{10} + (\texttt{zim4})\beta_{11} + (\texttt{zim5})\beta_{12} + (\texttt{zim6})\beta_{13} \\
&\quad + \varepsilon, \quad \varepsilon \sim N(0, \sigma_\varepsilon^2),
\end{aligned} \tag{9.2}
$$

where β's are the regression coefficients of the linear effects, f_1 is the unknown function of the spatial effect in `location`, and f_2 and f_3 are the functions of nonlinear effects in `year` and `floor.size`, respectively. Note that no intercept is used in the model because of the dummy predictors. We take a Besag prior on f_1, and RW2 priors on f_2 and f_3. From the nonparametric regression examples, we see that the two nonlinear effects and the spatial effect have quite different estimated precisions. We therefore must scale their priors in order to make assigning the same gamma prior for their precisions a reasonable approach. As a result, we express this model and fit it in INLA as follows:

```
data(Munich, package = "brinla")
g <- system.file("demodata/munich.graph", package = "INLA")
formula <- rent ~ f(location, model = "besag", graph = g, scale.model
    ↪ = TRUE) + f(year, model = "rw2", values = seq(1918, 2001),
    ↪ scale.model = TRUE) + f(floor.size, model = "rw2", values = seq
    ↪ (17, 185), scale.model = TRUE) + Gute.Wohnlage + Beste.Wohnlage
    ↪  + Keine.Wwv + Keine.Zh + Kein.Badkach  + Besond.Bad + Gehobene
    ↪ .Kueche + zim1 + zim2 + zim3 + zim4 + zim5 + zim6 - 1
result <- inla(formula, data = Munich, control.predictor = list(
    ↪ compute = TRUE))
```

Note that the sum-to-zero constraint is added to each of the Besag and RW2 models (the default choice in INLA) in order to make them identifiable from the dummy predictors.

Let's look at the estimated regression coefficients of the categorical variables:

```
round(result$summary.fixed, 3)
```

```
                  mean    sd 0.025quant 0.5quant 0.975quant   mode kld
Gute.Wohnlage    0.621 0.109      0.405    0.621      0.834  0.622   0
Beste.Wohnlage   1.773 0.317      1.151    1.773      2.394  1.773   0
Keine.Wwv       -1.942 0.278     -2.488   -1.942     -1.397 -1.942   0
Keine.Zh        -1.373 0.191     -1.747   -1.373     -0.999 -1.373   0
Kein.Badkach    -0.552 0.115     -0.777   -0.552     -0.327 -0.552   0
Besond.Bad       0.493 0.160      0.179    0.493      0.807  0.493   0
Gehobene.Kueche  1.136 0.175      0.793    1.136      1.478  1.136   0
zim1             7.917 0.290      7.347    7.917      8.487  7.917   0
zim2             8.198 0.223      7.760    8.198      8.635  8.199   0
zim3             8.019 0.201      7.624    8.020      8.413  8.020   0
zim4             7.590 0.207      7.182    7.591      7.997  7.591   0
zim5             7.722 0.319      7.094    7.722      8.348  7.722   0
zim6             7.504 0.565      6.395    7.504      8.611  7.504   0
```

We see all the variables are statistically useful, because the 95% credible intervals of their β's do not cover zero.

There are three random effects in the model: `year`, `floor.size` and `location`. Their results are all saved in `result1$summary.random`. Now let's look at them one by one. We plot the fitted curves and 95% credible bands for `floor.size` and `year` in Figure 9.2(a) and 9.2(b), respectively:

```
bri.band.ggplot(result, name = 'floor.size', type = 'random')
bri.band.ggplot(result, name = 'year', type = 'random')
```

It appears that the effect of `floor.size` on `rent` decreases as it increases to 50, followed by a flat pattern around 0 to the end. The effect of `year` has a U-shape pattern between 1920 and 1960, and then keeps increasing to the end. In Figure 9.2(c) and 9.2(d) we present the posterior mean and SD maps of the spatial effect based on the R commands:

```
map.munich(result$summary.random$location$mean)
map.munich(result$summary.random$location$sd)
```

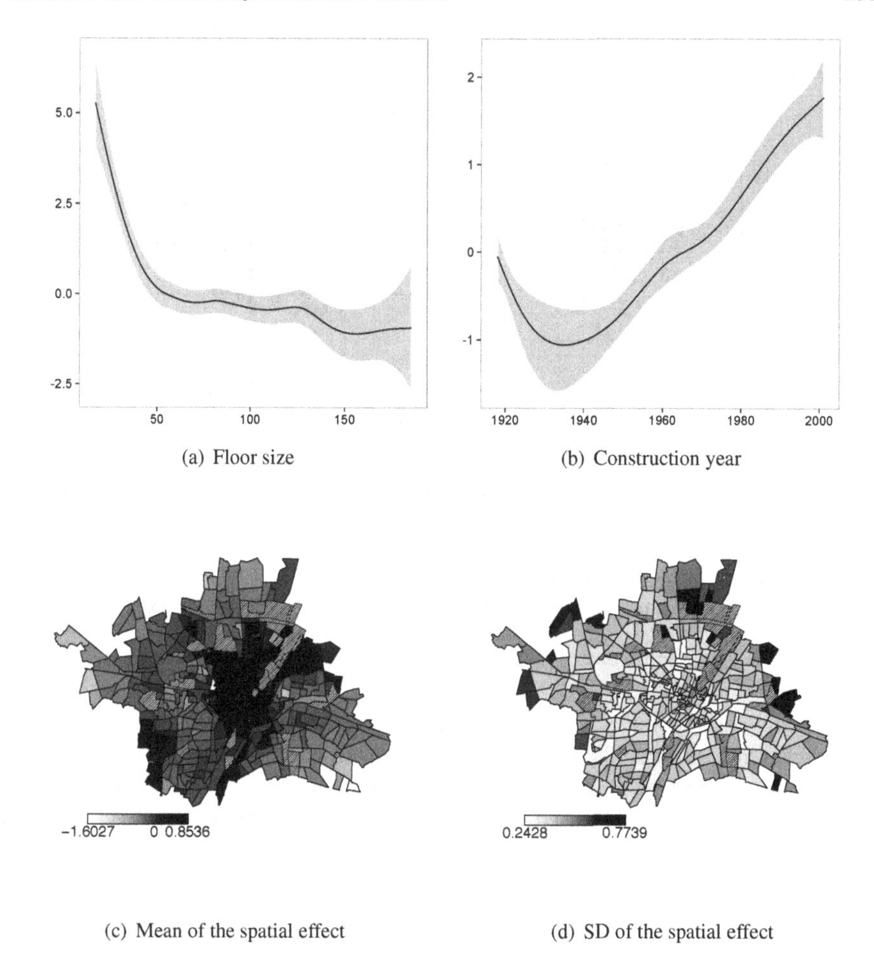

(a) Floor size

(b) Construction year

(c) Mean of the spatial effect

(d) SD of the spatial effect

FIGURE 9.2
Munich rental guide: (a) estimated nonlinear effect of floor size and its 95% credible band; (b) estimated nonlinear effect of year of construction and its 95% credible band; (c) posterior means of the spatial effect; (d) posterior SDs of the spatial effect.

Note that the `map.munich` is the function created for plotting the maps for this Munich rental guide dataset only. It will not work for a different dataset. We can see that the apartment rents are higher on average in the central regions of Munich than those in the suburban areas. The rent variability, however, shows a reverse pattern: the SDs are lower in the central regions than those in the suburban areas.

For diagnostics purposes, we show the residual plot in the left panel of Figure 9.3 and the plot of observed response against estimated mean response in the right panel using:

```
yhat <- result$summary.fitted.values$mean
```

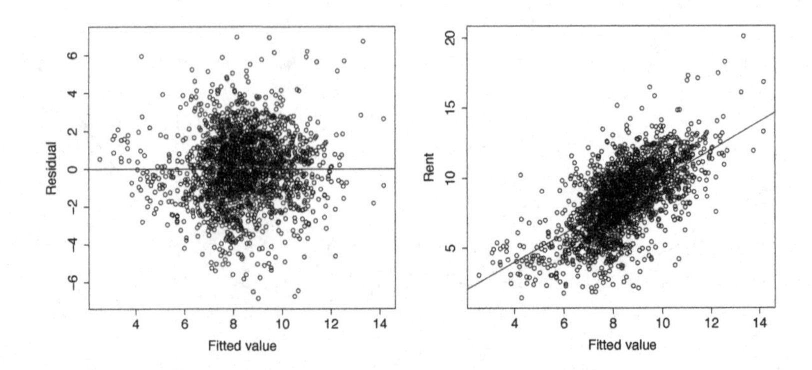

FIGURE 9.3
Munich rental guide: residual vs. fitted value (left); rent vs. fitted value (right).

```
residual <- Munich$rent - yhat
plot(yhat, residual, ylab = 'Residual', xlab = 'Fitted value')
abline(0,0)
plot(yhat, Munich$rent, ylab = 'Rent', xlab = 'Fitted value')
abline(0,1)
```

The residual plot shows no systematic pattern, which is the evidence of a good fit. However, it seems that the model slightly overestimates low rents but underestimates high rents. We believe it is because the distribution of rents is skewed to the right, and the normality assumption used in the model is not quite appropriate. But, generally speaking, the model provides reasonable results.

9.2 Generalized Additive Models

A generalized additive model (GAM) is a generalized linear model (GLM) (see Chapter 4), in which the linear predictor depends on the unknown smooth functions of some predictor variables. Assuming that response variable y_i, for $i = 1, \ldots, n$, follows a distribution from the exponential family (e.g., binomial or Poisson distribution) with mean $E(y_i) = \mu_i$, a GAM is given by

$$g(\mu_i) = \eta_i, \quad \eta_i = \beta_0 + \sum_{j=1}^{p} f_j(x_{ij}), \tag{9.3}$$

where g is the link function, connecting μ_i to the linear predictor η_i. GAMs were originally developed by Hastie and Tibshirani (1990), in order to blend the properties of the GLMs with the additive models introduced in the previous section. The GAM family is quite broad, and most current models belong to that family.

A GAM, like the additive model, can be fitted using the mgcv and the gam packages, but with different approaches. The mgcv takes a likelihood approach and the amount of smoothing applied to each function is decided by some information criterion, e.g., generalized cross validation (Wood, 2008). The gam package uses a backfitting approach based on the iterative reweighted least squares (IRWLS) fitting algorithm as used in GLM. It is straightforward to extend the INLA method of fitting the additive models to the GAMs. Actually, only a Laplace approximation to the non-Gaussian likelihood needs to be added to the algorithm. In the following sections, we will use a few real-data examples to demonstrate how to fit GAMs using INLA.

9.2.1 Binary Response

Bell et al. (1994) studied the result of multiple-level thoracic and lumbar laminectomy, a corrective spinal surgery commonly performed on children. The data in the study consist of retrospective measurements on 83 patients. The specific outcome of interest is the presence (1) or absence (0) of Kyphosis, defined as a forward flexion of the spine of at least 40 degrees from vertical. The predictor variables are age in months at the time of the operation (Age), the starting of vertebrae levels involved in the operation (StartVert), and the number of levels involved (NumVert). Let us load the dataset:

```
data(kyphosis, package = 'brinla')
str(kyphosis)
```

```
'data.frame':        83 obs. of  4 variables:
 $ Age      : int  71 158 128 2 1 1 61 37 113 59 ...
 $ StartVert: int  5 14 5 1 15 16 17 16 16 12 ...
 $ NumVert  : int  3 3 4 5 4 2 2 3 2 6 ...
 $ Kyphosis : int  0 0 1 0 0 0 0 0 0 1 ...
```

In the data frame, we see 4 integer variables and each variable has 83 observations.

Because they are binary, we assume the response observations follow Bernoulli distributions and use the logit link:

$$\texttt{Kyphosis\_i} \sim \text{Bin}(1,\ p_i), \quad \log\left(\frac{p_i}{1-p_i}\right) = \eta_i$$

where p_i is the probability of the presence of Kyphosis for i^{th} patient, and η_i is the linear predictor related to p_i. Because we are not certain about what kind of relationship each predictor has with the response variable, we first model all three predictors with nonparametric functions as follows:

$$\eta_i = \beta_0 + f_1(\texttt{Age}_i) + f_2(\texttt{StartVert}_i) + f_3(\texttt{NumVert}_i). \tag{9.4}$$

The default normal prior provided by INLA is taken on β_0, and an RW2 prior on each f function. We then formulate the model and fit it using INLA:

```
formula1 <- Kyphosis ~ 1 + f(Age, model = 'rw2') + f(StartVert, model
    ↪ = 'rw2') + f(NumVert, model = 'rw2')
result1 <- inla(formula1, family='binomial', data = kyphosis, control.
    ↪ predictor = list(compute = TRUE), control.compute = list(waic =
    ↪  TRUE))
```

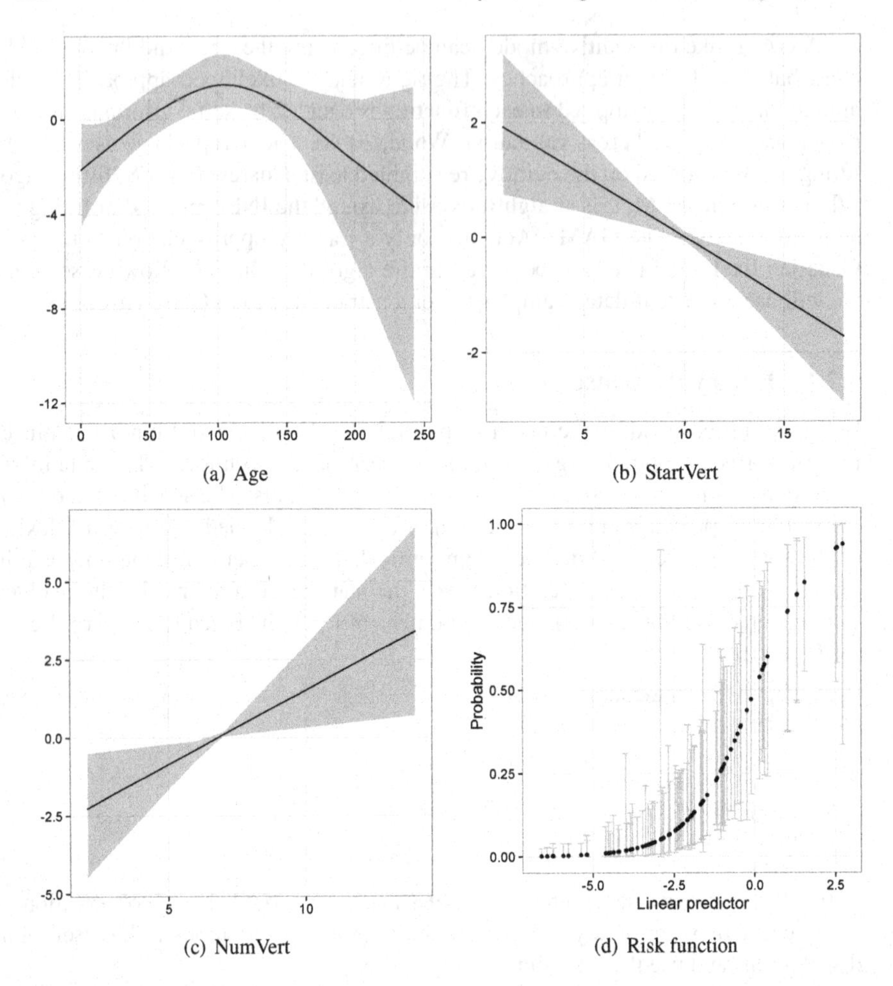

(a) Age

(b) StartVert

(c) NumVert

(d) Risk function

FIGURE 9.4
Risk factors for kyphosis: posterior mean (black line) and 95% credible band (gray band) for (a) Age, (b) StartVert and (c) NumVert; (d) estimated risk function (dot) and 95% credible band (gray vertical line).

By default the logit link is used for the binomial likelihood, and the sum-to-zero constraint is added to each RW2 prior to make it identifiable from the intercept. We also ask INLA to compute the WAIC score for model comparison purposes. Let's look at the three nonlinear effects by plotting their fitted curves and 95% credible bands (Figures 9.4(a), 9.4(b) and 9.4(c)):

```
bri.band.ggplot(result1, name = 'Age', type = 'random')
bri.band.ggplot(result1, name = 'StartVert', type = 'random')
bri.band.ggplot(result1, name = 'NumVert', type = 'random')
```

It seems that the risk of kyphosis increases with `Age` until it reaches maximum around 120 months, and then falls down as the `Age` continues to increase. After accounting for the uncertainty represented by the credible band, the effect of `Age` is negative before 30 months, but is positive between 60 and 160 months, because the band does not cover 0 in those intervals. We also note that the width of the credible band increases dramatically after 200 months due to the lack of observations. The `StartVert` and `NumVert`, however, simply show linear patterns associated with the risk.

To understand how the linear predictor η affects the risk of kyphosis, we extract the posterior mean of η:

```
eta <- result1$summary.linear.predictor$mean
```

and the posterior mean and the 95% credible interval of p_i (probability of presence of kyphosis for i^{th} patient):

```
phat <- result1$summary.fitted.values$mean
phat.lb <- result1$summary.fitted.values$'0.025quant'
phat.ub <- result1$summary.fitted.values$'0.975quant'
```

and then plot them:

```
data.plot <- data.frame(eta, phat, phat.lb, phat.ub)
ggplot(data.plot, aes(y = phat, x = eta)) + geom_errorbar(aes(ymin =
    ↪ phat.lb, ymax = phat.ub), width = 0.2, col = 'gray') + geom_
    ↪ point() + theme_bw(base_size = 20) + labs(x = 'Linear predictor
    ↪ ', y = 'Probability')
```

Figure 9.4(d) shows the plot and we can see the risk (dot) increases slowly ($\hat{p} < 0.2$) until $\hat{\eta} = -2$, and then goes up sharply to 1. The credible interval for each risk level becomes wider as $\hat{\eta}$ increases due to the sparse data observed at the high levels.

One of the best things about GAMs is that we can use them to suggest simpler parametric models, which are better for interpretation, stability, prediction and ease of use. In this case, the GAM seems to suggest the following quadratic model

$$\eta_i = \beta_0 + \texttt{StartVert}_i\beta_1 + \texttt{NumVert}_i\beta_2 + \texttt{Age}_i\beta_3 + \texttt{Age}_i^2\beta_4. \tag{9.5}$$

Such a model can be fitted using INLA as follows:

```
kyphosis$AgeSq <- (kyphosis$Age)^2
formula2 <- Kyphosis ~ 1 + StartVert + NumVert + Age + AgeSq
result2 <- inla(formula2, family='binomial', data = kyphosis, control.
    ↪ predictor = list(compute = TRUE), control.compute = list(waic =
    ↪  TRUE))
```

The posterior summary of the linear effects is given by:

```
round(result2$summary.fixed, 4)
```

```
               mean     sd 0.025quant 0.5quant 0.975quant    mode kld
(Intercept) -5.1887 2.0379    -9.5830  -5.0425    -1.5791 -4.7333   0
StartVert   -0.2214 0.0708    -0.3680  -0.2187    -0.0896 -0.2134   0
NumVert      0.5132 0.2191     0.1294   0.4958     0.9899  0.4587   0
Age          0.0899 0.0334     0.0320   0.0870     0.1632  0.0808   0
AgeSq       -0.0004 0.0002    -0.0008  -0.0004    -0.0001 -0.0004   0
```

The point estimates of β_1 and β_2 are -0.2214 and 0.5132, respectively. Their 95% credible intervals are (-0.3680, -0.0896) and (0.1294, 0.9899), none of which contains 0. We therefore draw the conclusion that given `Age` the increase in `StartVert` will decrease the risk, but the increase in `NumVert` will increase the risk. Both linear and quadratic coefficients for Age are nonzero because their credible intervals do not cover 0 either. It indicates that there is a quadratic relationship between Age and the risk of kyphosis.

To compare quadratic model (9.5) to GAM (9.4), we check their WAIC scores:

```
c(result1$waic$waic, result2$waic$waic)
```

```
[1] 65.91754 65.46884
```

It shows that the quadratic model slightly outperforms the GAM. It seems that the simplicity advantages of the quadratic model outweigh the small difference in the fit given by the GAM. However, without the GAM we might not think of the quadratic or know that it is the right choice. The GAM analysis was therefore definitely useful.

9.2.2 Count Response

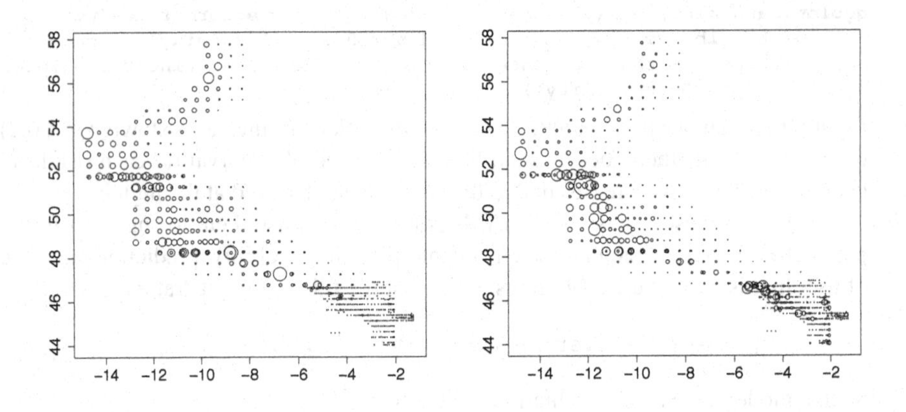

FIGURE 9.5
Mackerel egg survey: egg count (left) and egg density (right) at each sampled station. The size of each circle shows the relative magnitude of each observation.

Most commercially exploitable fish stocks in the world have been over exploited. To effectively manage stocks we must have sound fish stock assessment, but counting the number of catchable fish of any given species is almost impossible. One solution is to assess the number of eggs produced by a stock, and then try to figure out the number (or the mass) of adult fish required to produce this number. Such egg data are obtained by sending out scientific cruise ships to sample eggs at each station of some predefined sampling grid over the area occupied by a stock.

We here consider the data about the distribution of mackerel eggs and were collected as part of the 1992 mackerel survey. At each of a number of stations, mackerel eggs were sampled by hauling a fine net up from deep below the sea surface to the sea surface. The egg count data are obtained from the resulting samples, and these have been converted to (Stage I) eggs produced per meter squared per day - the egg density data. Other possibly useful predictor variables have been recorded, along with identification information. The `gamair` R package needs to be installed in order to use the dataset.

Let's load the dataset and look at its structure:

```
data(mack, package = 'gamair')
str(mack, vec.len = 2)
```

```
'data.frame':        634 obs. of  16 variables:
 $ egg.count   : num  0 0 0 1 4 ...
 $ egg.dens    : num  0 0 ...
 $ b.depth     : num  4342 4334 ...
 $ lat         : num  44.6 44.6 ...
 $ lon         : num  -4.65 -4.48 -4.3 -2.87 -2.07 ...
 $ time        : num  8.23 9.68 ...
 $ salinity    : num  35.7 35.7 ...
 $ flow        : num  417 405 377 420 354 ...
 $ s.depth     : num  104 98 101 98 101 ...
 $ temp.surf   : num  15 15.4 15.9 16.6 16.7 ...
 $ temp.20m    : num  15 15.4 15.9 16.6 16.7 ...
 $ net.area    : num  0.242 0.242 0.242 0.242 0.242 ...
 $ country     : Factor w/ 4 levels "EN","fr","IR",..: 4 4 4 4 4 ...
 $ vessel      : Factor w/ 4 levels "CIRO","COSA",..: 2 2 2 2 2 ...
 $ vessel.haul: num  22 23 24 93 178 ...
 $ c.dist      : num  0.84 0.859 ...
```

The data frame has 16 columns and 634 rows. Each column represents a predictor and each row corresponds to one sample of eggs. The egg count and egg density from each sample are recorded in `egg.count` and `egg.dens`, respectively. The location of each station is defined in `lon` (longitude) and `lat` (latitude). We plot the locations where eggs were sampled, and the relative magnitudes (represented by circle size) of egg counts and egg densities recorded at that location:

```
loc.obs <- cbind(mack$lon, mack$lat)
plot(loc.obs, cex = 0.2+mack$egg.count/50, cex.axis = 1.5)
plot(loc.obs, cex = 0.2+mack$egg.dens/150, cex.axis = 1.5)
```

Figure 9.5 shows the results. The information about the other predictors can be found by typing `?mack` in R.

We would like to predict the number of mackerel eggs produced by a stock. As we can see in Figure 9.5, there seems to be a spatial effect on the distribution of eggs. Besides `lon` and `lat`, we also consider the predictors of egg abundance, such as the saltiness of the water (`salinity`), the water temperature at a depth of 20 meters (`temp.20m`), and the distance from the 200-meter seabed contour (`c.dist`). The `c.dist` predictor reflects the biologists' belief that the fish like to spawn near the edge of the continental shelf, conventionally considered to end at a seabed depth of 200 meters. Following Wood (2006) we assume that each `egg.count` follows a Poisson distribution and model its mean λ_i as follows:

$$\begin{aligned} \texttt{egg.count\_i} &\sim \text{Poisson}(\lambda_i), \quad \log(\lambda_i) = \eta_i \\ \eta_i &= \log(\texttt{net.area}_i) + \beta_1 \texttt{salinity}_i + \beta_2 \texttt{c.dist}_i \\ &\quad + f_1(\texttt{temp.20m}_i) + f_2(\texttt{lon}_i, \texttt{lat}_i), \end{aligned} \tag{9.6}$$

where `net.area` (area of each net) is used as an offset, β_1 and β_2 are the coefficients of the linear effects of `salinity` and `c.dist`, respectively, f_1 is the nonlinear effect of `temp.20m`, and f_2 is the spatial effect of `(lon,lat)`.

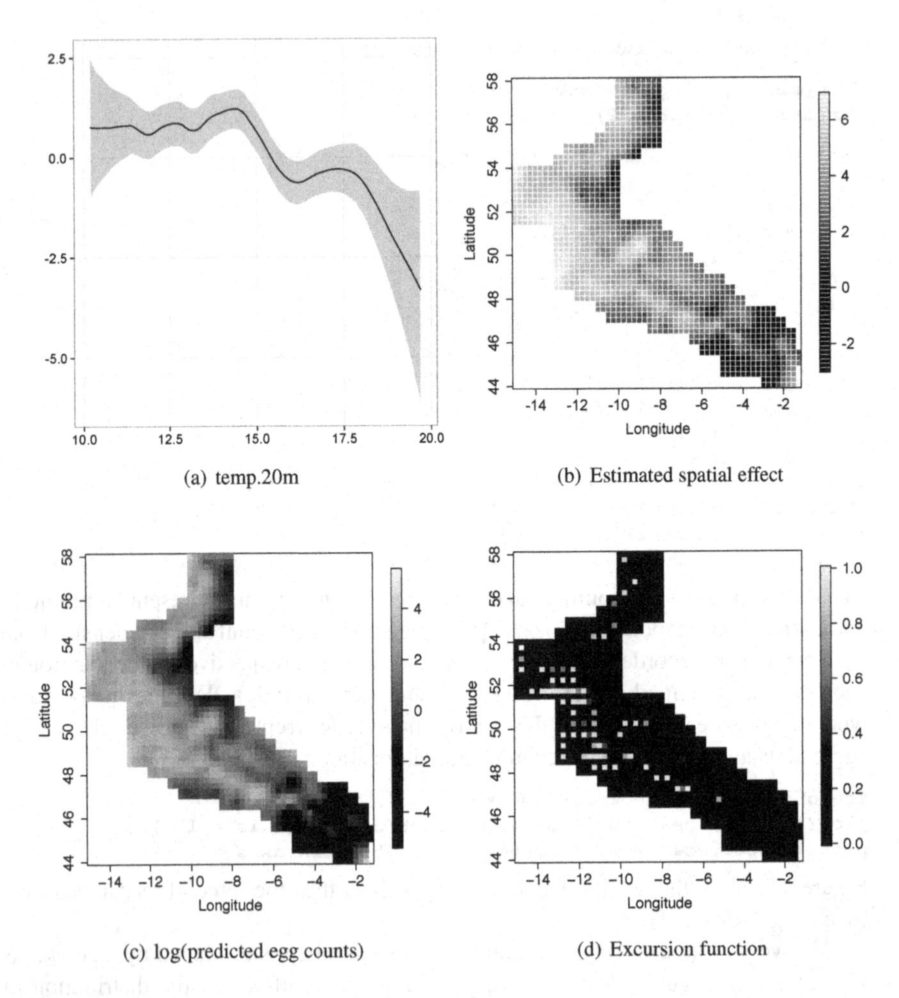

FIGURE 9.6
Mackerel egg survey: (a) posterior mean and 95% credible band of the *temp.20m* effect; (b) posterior mean of the spatial effect; (c) predicted mean egg counts on log scale; (d) excursion function showing the probability that there are at least 10 eggs for a given location.

We take INLA default priors on the linear effects, an RW2 prior on the nonlinear effect, and a thin-plate spline (TPS) prior on the spatial effect. The triangular mesh needed for the TPS prior is constructed as follows:

```
mesh <- inla.mesh.2d(loc.obs, cutoff = 0.05, max.edge = c(.5,1))
```

Here we set `cutoff=0.05` to replace by a single vertex the points that are less than

0.05 far apart, and allow the largest triangle edge lengths to be 0.5 and 1 to obtain a smooth triangulation. We then build the TPS prior and fit model (9.6) using INLA:

```
tps <- bri.tps.prior(mesh)
node <- mesh$idx$loc
formula <- egg.count ~ -1 + salinity + c.dist + f(temp.20m, model = '
    ↪ rw2') + f(node, model = tps, diagonal = 1e-6)
result <- inla(formula, family = 'poisson', data = mack, offset = log(
    ↪ net.area))
```

The posterior summary of the linear effects is given by:

```
round(result$summary.fixed, 4)
```

```
              mean     sd 0.025quant 0.5quant 0.975quant    mode kld
salinity  0.0058 0.0025     0.0008   0.0058     0.0107  0.0058   0
c.dist   -0.9915 0.5634    -2.1084  -0.9884     0.1068 -0.9821   0
```

We can see `salinity` is a significant effect with mean .0058 and 95% credible interval (.0008, .0107), while `c.dist` is insignificant because its credible interval (-.983, .106) covers 0. For the nonlinear effect of `temp.20m`, we plot its fitted curve and 95% credible band:

```
bri.band.ggplot(result, name = 'temp.20m', type = 'random')
```

Figure 9.6(a) shows the result and we see that the egg density seems to slowly increase with the temperature until it reaches around 14.5 degrees, then drops sharply until 16 degrees, followed by another increase until 18 degrees and a big drop to the end.

Now let's take a look at the spatial effect. To make an image plot, we need a new dataset `mackp`, which provides a regular spatial grid as well as some other predictor variables within the area covered by the survey. We load the data and project posterior means of the spatial effect onto that grid:

```
data(mackp, package = 'gamair')
proj <- inla.mesh.projector(mesh, loc = cbind(mackp$lon, mackp$lat))
spa.mean <- inla.mesh.project(proj, result$summary.random$node$mean)
```

and plot them using `quilt.plot` from the `fields` package:

```
library(fields)
quilt.plot(mackp$lon, mackp$lat, spa.mean, nx = length(unique(mackp$
    ↪ lon)), ny = length(unique(mackp$lat)))
```

Figure 9.6(b) shows the result and we see that how the egg abundance depends on the spatial locations.

Prediction. One main purpose of this study is to assess the total stock of eggs. Therefore, a simple map of predicted egg densities is useful. It needs us to make predictions at unobserved locations using the fitted model. INLA can do it jointly with the estimation process as follows. We define

```
n.pre <- dim(mackp)[1]
y.pre <- c(mack$egg.count, rep(NA, n.pre))
```

to be the number of predictions, and the vector of "missing" response variables, respectively. We then combine the variables from `mack` with those from `mackp`:

```
z1.pre <- c(mack$salinity, mackp$salinity)
```

```
z2.pre <- c(mack$c.dist, mackp$c.dist)
x.pre <- inla.group(c(mack$temp.20m, mackp$temp.20m), n = 100)
E.pre <- c(mack$net.area, rep(0.25^2, n.pre))
```

Note that we group `x.pre` into `n = 100` bins to remove too close values. We also need to combine the locations from both datasets to build a new mesh, and construct a TPS prior based on that mesh:

```
loc.pre <- rbind(cbind(mack$lon,mack$lat), cbind(mackp$lon,mackp$lat))
mesh2 <- inla.mesh.2d(loc.pre, cutoff = 0.05, max.edge = c(.5, 1))
node2 <- mesh2$idx$loc
tps2 <- bri.tps.prior(mesh2)
```

As a result, a new GAM is formulated in INLA as follows:

```
mack.pre <- list(egg.count = y.pre, salinity = z1.pre, c.dist = z2.pre
    ↪ , temp.20m = x.pre, node = node2)
formula <- egg.count ~ -1 + salinity + c.dist + f(temp.20m, model = '
    ↪ rw2') + f(node, model = tps2, diagonal = 1e-6)
```

Before we fit the model we need to specify the link function that will be used for the predictions:

```
link <- rep(NA, length(y.pre))
link[which(is.na(y.pre))] <- 1
```

Here '`1`' is the reference to the first '`family`' specified in `inla` function. It will thus be the log link used in the Poisson family. We are now ready to fit this GAM:

```
result2 <- inla(formula, family = 'poisson', data = mack.pre, offset =
    ↪  log(E.pre), control.predictor = list(link = link, compute =
    ↪ TRUE), control.compute = list(config = TRUE))
```

Here we explicitly ask INLA to compute the marginals of the linear predictor and the fitted values (`compute = TRUE`), and store internal approximations for later use (`config = TRUE`). Note that this computation is a little demanding, and thus the result may not appear quickly.

The posterior summary regarding the predictions can be extracted as follows:

```
idx.pre <- which(is.na(y.pre))
res.pre <- result2$summary.fitted.values[idx.pre, ]
```

where `idx.pre` is the vector of indices of the prediction locations. We plot the posterior mean on log scale (`log(res.pre$mean)`) in Figure 9.6(c), where we can see the egg density is predicted to be high on the western boundary of the survey area, while relatively low in the southeast corner. It has a pattern similar to that of the spatial effect (see Figure 9.6(b)), which means the egg abundance depends more on the locations than the other predictors.

Excursion set. Suppose we are interested in finding the squares in the grid (0.25 degree lon-lat squares) where there are 10 or more eggs. We may use the excursion method described in Section 7.8 to find an excursion set of squares that has that many eggs with a joint probability of at least 0.95:

```
res.exc <- excursions.brinla(result2, name = 'Predictor', ind = idx.
    ↪ pre, u = log(10), type = '>', alpha = 0.05, method = 'NIQC')
```

Note that the threshold u must be specified on log scale, i.e., u = log(10), due to the log link used in the model. We plot the excursion function:

```
quilt.plot(mackp$lon, mackp$lat, res.exc$F, nx = length(unique(mackp$
   ↪ lon)), ny = length(unique(mackp$lat)))
```

Figure 9.6(d) shows the result and we see that the lighter the square is, the more likely it has 10 or more eggs. The indices of the squares in the resulting excursion set are given by:

```
res.exc$E
```

```
 [1]  262  264  328  330  382  384  386  392  446  451  452  453  459  512  519  570
[17]  572  574  576  636  638  676  720  722  724  772  774  775  776  777  778  780
[33]  782  818  820  822  854  858  938  988 1092 1112
```

9.3 Generalized Additive Mixed Models

A generalized additive mixed model (GAMM) manages to combine the idea of GAM seen earlier in this chapter with the mixed modeling ideas as seen in Chapter 5. The response y can be non-Gaussian, having a distribution from the exponential family, and the error structure can allow for grouping and hierarchical arrangements in the data. It is straightforward to extend GAM (9.3) to a GAMM as follows:

$$g(\mu_i) = \beta_0 + \sum_j f_j(x_{ij}) + \sum_k g_k(u_{ij}), \tag{9.7}$$

where the random effects $g_k(u_{ij})$ can be constructed in various ways that introduce different patterns of correlation in the response as appropriate for the particular application. The common correlation structures have been described in Chapter 5. Provided we assign Gaussian priors on its components, GAMM (9.7) falls in the latent Gaussian model framework required by INLA.

Sole eggs in the Bristol Channel

Assessing fish stocks is difficult because it is not easy to survey adult fish. Fisheries biologists therefore try to count fish eggs, and work back to the number of adult fish required to produce the estimated egg population. The data concerned in this section are the density measurements of sole eggs per square meter of sea surface in each of 4 identifiable egg developmental stages, at each of a number of sampling stations in the Bristol Channel on the west coast of England. The samples were taken from 5 research cruises over the spawning season.

Let's load the dataset and look at its structure:

```
data(sole, package = 'gamair')
str(sole)
```

```
'data.frame':	1575 obs. of  7 variables:
```

```
 $ la    : num  50.1 50.1 50.1 50.2 50.2 ...
 $ lo    : num  -5.87 -6.15 -6.39 -6.15 -5.9 ...
 $ t     : num  49.5 49.5 49.5 49.5 49.5 49.5 49.5 49.5 49.5 49.5 ...
 $ eggs : num  0 0 0 0 0 0 0 0.147 0.524 0 ...
 $ stage: int  1 1 1 1 1 1 1 1 1 1 ...
 $ a.0   : num  0 0 0 0 0 0 0 0 0 0 ...
 $ a.1   : num  2.4 2.4 2.4 2.4 2.4 2.4 2.4 2.4 2.4 2.4 ...
```

In this dataset, `la` and `lo` are the latitude and longitude, respectively, of sampling station, `t` is the time of midpoint of the cruise on which this sample was taken, `eggs` is the egg density per square meter of sea surface, `stage` is one of 4 stages the sample relates, `a.0` is the age of the youngest possible egg in this sample, and `a.1` is of the oldest.

Following Wood (2006), we calculate the width (`off`) and average age (`a`) of the corresponding egg class:

```
solr <- sole
solr$off <- log(sole$a.1 - sole$a.0)
solr$a <- (sole$a.1 + sole$a.0)/2
```

and use the former as an offset term and the latter as a predictor in the model. It has been shown that there exist interactions between the coordinates (`lo`, `la`) and time `t`. We therefore need to include some polynomial terms in the model and must translate and scale a few predictors for numerical stability:

```
solr$t <- solr$t - mean(sole$t)
solr$t <- solr$t/var(sole$t)^0.5
solr$la <- solr$la - mean(sole$la)
solr$lo <- solr$lo - mean(sole$lo)
```

We also need to consider a “sampling station” effect, because at each station the counts for the four different egg stages are all taken from the same net sample, and therefore the data for different stages at a station cannot be treated as independent. As a result, we make a `station` variable and take it as random effect in the model:

```
solr$station <- as.numeric(factor(with(solr, paste(-la, -lo, -t, sep="
    ↪ "))))
```

The response variable `eggs` is not the raw count, but rather the density per m^2 sea surface. We therefore multiply `eggs` by 1000 to make integers, and adjust the offset accordingly:

```
solr$eggs <- solr$eggs*1000
solr$off <- solr$off + log(1000)
```

Noting that there are over 70% zeros in the response variable:

```
length(which(solr$eggs == 0))/length(solr$eggs)
```

```
[1] 0.7612698
```

The overdispersion is probably present. So, we propose the following GAMM based on the zero-inflated negative binomial (ZINB) likelihood

$$\begin{aligned} \texttt{eggs}_i &\sim \text{ZINB}(\rho, n, p_i), \quad \mu_i = n(1-p_i)/p_i \\ \log(\mu_i) &= \log(\texttt{off}_i) + \beta_0 + \texttt{a}_i + \texttt{la}_i * \texttt{t}_i + \texttt{la}_i * \texttt{t}_i^2 + \texttt{lo}_i * \texttt{t}_i + \texttt{lo}_i * \texttt{t}_i^2 + \\ &\quad \texttt{a}_i f_1(\texttt{t}_i) + f_2(\texttt{lo}_i, \texttt{la}_i) + \texttt{station}_{j(i)}, \end{aligned} \tag{9.8}$$

where ρ is the zero-probability parameter, n is the overdispersion parameter, p_i is the probability of "success" in i^{th} trial, and μ_i is the mean of ith egg density.

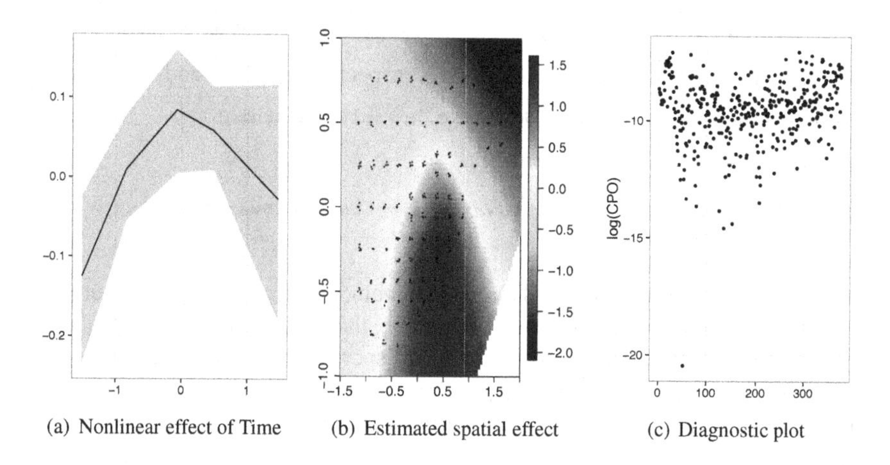

(a) Nonlinear effect of Time (b) Estimated spatial effect (c) Diagnostic plot

FIGURE 9.7
Sole eggs survey: (a) fitted curve and 95% credible band of *time* effect; (b) map of posterior means of spatial effect (black dots are the sampling stations); (c) an index plot of the CPO statistics on log scale.

Regarding priors, we assign an iid prior for `station`, an RW2 prior for f_1, a TPS prior for f_2:

```
loc <- cbind(solr$lo, solr$la)
mesh <- inla.mesh.2d(loc, max.edge = c(0.1, 0.2))
node <- mesh$idx$loc
tps <- bri.tps.prior(mesh, constr = TRUE)
```

The INLA default priors are used for remaining unknown parameters. We then express this ZINB GAMM in the INLA formula:

```
formula1 <- eggs ~ a + I(la*t) + I(la*t^2) + I(lo*t) + I(lo*t^2) + f(t
    ↪ , a, model = 'rw2', scale.model = TRUE) + f(node, model = tps,
    ↪ diagonal = 1e-6) + f(station, model = 'iid')
```

To see if it is necessary to account for the `station` effect we also consider the following GAM (without the random effect):

```
formula0 <- eggs ~ a + I(la*t) + I(la*t^2) + I(lo*t) + I(lo*t^2) + f(t
    ↪ , a, model = 'rw2', scale.model = TRUE) + f(node, model = tps,
    ↪ diagonal = 1e-6)
```

Now let's fit these two ZINB models:

```
result0 <- inla(formula0, family = 'zeroinflatednbinomial0', offset =
    ↪ off, data = solr, control.predictor = list(compute = TRUE),
    ↪ control.compute = list(dic = TRUE, waic = TRUE, cpo = TRUE))
result1 <- inla(formula1, family = 'zeroinflatednbinomial0', offset =
    ↪ off, data = solr, control.predictor = list(compute = TRUE),
    ↪ control.compute = list(dic = TRUE, waic = TRUE, cpo = TRUE))
```

Here we ask INLA to compute DIC and WAIC scores for model comparison purposes. To see if the overdispersion exists, we also fit a zero-inflated Poisson GAMM:

```
result2 <- inla(formula1, family = 'zeroinflatedpoisson0', offset =
    ↪ off, data = solr, control.predictor = list(compute = TRUE),
    ↪ control.compute = list(dic = TRUE, waic = TRUE, cpo = TRUE))
```

We compare it to the first two models with respect to their DIC and WAIC scores:

```
c(result0$dic$dic, result1$dic$dic, result2$dic$dic)
```

```
[1]   7878.029   7814.104 182594.471
```

```
c(result0$waic$waic, result1$waic$waic, result2$waic$waic)
```

```
[1]   7878.994   7813.335 218587.232
```

We can clearly see that the ZINB model that includes the `station` effect outperforms the other two models in both criteria, indicating that we need to consider the `station` effect in the model and account for the overdispersion existing in the data.

We here only present the analysis from `result1`, because it gives us the best performance. The posterior summary of the fixed effects is given by:

```
result <- result1
round(result$summary.fixed, 4)
```

```
               mean      sd 0.025quant 0.5quant 0.975quant    mode kld
(Intercept) -0.5701 19.0117   -37.8964  -0.5706    36.7251 -0.5701   0
a           -0.1970  0.0335    -0.2610  -0.1975    -0.1323 -0.2012   0
I(la * t)    0.6681  0.2863     0.1063   0.6678     1.2307  0.6672   0
I(la * t^2)  1.4153  0.2928     0.8417   1.4147     1.9906  1.4127   0
I(lo * t)   -0.5884  0.2512    -1.0821  -0.5883    -0.0953 -0.5883   0
I(lo * t^2) -1.1310  0.2622    -1.6456  -1.1307    -0.6173 -1.1291   0
```

We see `a` has a negative linear effect on `eggs`, `la` and `t` have positive interaction, and `lo` and `t` interact with each other in a negative way. In Figure 9.7(a) we plot the nonlinear effect of `t`:

```
bri.band.ggplot(result, name = 't', type = 'random')
```

We see a concave pattern having maximum around mean time. We also plot the estimated spatial effect:

```
proj <- inla.mesh.projector(mesh)
spa.mean <- inla.mesh.project(proj, result$summary.random$node$mean)
library(fields)
image.plot(proj$x, proj$y, spa.mean, xlim=c(-1.5, 2), ylim=c(-1, 1))
points(loc, pch = 19, cex = 0.2)
```

In Figure 9.7(b), we see the relatively high egg densities come from the stations in the southeast of the survey region. The posterior distributions of the hyperparameters are summarized using:

```
tmp <- bri.hyperpar.summary(result)
row.names(tmp) <- c("Overdispersion", "Zero-probability", 'SD for t',
    ↪ 'Theta1 for node', 'SD for station')
round(tmp, 4)
```

```
                     mean     sd  q0.025    q0.5  q0.975    mode
Overdispersion     1.9130 0.1623  1.6123  1.9057  2.2480  1.8949
Zero-probability   0.7610 0.0107  0.7395  0.7610  0.7816  0.7613
SD for t           0.0410 0.0235  0.0110  0.0358  0.1011  0.0260
Theta1 for node   -1.1959 0.2890 -1.7800 -1.1909 -0.6482 -1.1685
SD for station     0.5755 0.0701  0.4488  0.5715  0.7239  0.5642
```

We see the zero-probability estimate $\hat{\rho} = 0.7610$, which is quite close to its empirical estimate 0.7612.

Regarding diagnostics, we can check individual observations using conditional predictive ordinate (CPO) statistics (see Section 1.4). This is $P(y_i \mid y_{-i})$, a "leave-one-out" predictive measure of fit, and we should pay attention to the low values of this statistic. Although INLA provides CPO statistics in `result$cpo$cpo` directly, we should always check the quality of these values using `result$cpo$failure`, where non-zero values indicate some degree of suspicion (see details in Section 2.5). It turns out that there are 68 suspicious CPO measures that violate INLA assumptions in our case:

```
length(which(result$cpo$failure > 0))
```

```
[1] 68
```

Therefore, we need to recalculate each of those measures with explicit leave-one-out procedure. It can be efficiently implemented by

```
improved.result <- inla.cpo(result)
```

In Figure 9.7(c) we plot the improved CPO statistics on log scale to make them more distinguished:

```
idx <- which(solr$eggs != 0)
dat.plot <- data.frame(x = 1:length(idx), y = log(improved.result$cpo$
    ↪ cpo[idx]))
ggplot(dat.plot, aes(x = x, y = y)) + geom_point()
```

We can see there is one particular point with very low probability, and it is

```
round(solr[which.min(improved.result$cpo$cpo),], 4)
```

```
         la      lo       t  eggs stage a.0 a.1    off   a station
664 -0.4486 -0.1237 -0.8377 17048     1   0 2.4 7.7832 1.2     326
```

Let's consider this case among all the other observations in that station:

```
solr[solr$station==326, c('eggs', 'stage', 'a', 't')]
```

```
      eggs stage   a          t
664  17048     1 1.2 -0.8376579
6611   319     2 2.9 -0.8376579
6621   319     3 4.8 -0.8376579
6631   159     4 7.3 -0.8376579
```

We see it has many more eggs than the other three surveys. This is certainly unusual.

10

Errors-in-Variables Regression

Data measured with errors occur frequently in many scientific fields. Standard regression models assume that the independent variables have been measured exactly, in other words, observed without error. Those models account only for errors in the response variable. However, the presence of measurement errors in independent variables causes biased and inconsistent parameter estimates and leads to erroneous conclusions to various degrees using standard statistical analysis. Errors-in-variables regression models refer to regression models that account for measurement errors in the predictors.

10.1 Introduction

One could think of several examples in which measurement error can be a concern:

- *In medicine:* The NHANES-I epidemiological study is a cohort study consisting of thousands of women who were investigated about their nutrition habits and then evaluated for evidence of cancer. The primary predictor of interest in the study is the "long-term" saturated fat intake which was known to be imprecisely measured. Indeed, NHANES-I was one of the first studies where the measurement error model approach was used (Carroll et al., 2006). Even more comprehensive studies, NHANES-II and NHANES-III, have been published later.

- *In bioinformatics:* Gene microarray techniques are very popular in genetics research. A microarray consists of an arrayed series of thousands of microscopic spots of DNA molecules (genes). A gene present in the RNA sample finds its DNA counterpart on the microarray and binds to it. The spots then become fluorescent, and a microarray scanner is used to "read" the intensities in the microarray. The whole process to obtain the fluorescent intensities in a microarray study is subject to measurement error. Correction of measurement error is a critical step for microarray data analysis.

- *In chemistry:* A Massachusetts acid rain monitoring project was first described by Godfrey et al. (1985), where water samples were collected from about 1800 water bodies and chemical analyses were accomplished by 73 laboratories. Measuring values in chemistry typically involves error, therefore external calibration/validation data were collected based on blind samples sent to the lab with "known" values. In the statistical analysis of the study, one faces the problem of measurement errors in the predictors. The essential perceptiveness underlying the solution of the measurement error problem is to recover the parameter of the latent variables using extraneous information.

- *In astronomy:* Most astronomical data come with information on their measurement errors. Morrison et al. (2000) studied galaxy formation with a large survey of stars in the Milky Way. The investigators were interested in the velocities of stars, which represent the "fossil record" of their early lives. The observed velocities involved heteroscedastic measurement errors. To verify the galaxy formation theories, one is to estimate the density function from contaminated data that are effective in unveiling the numbers of bumps or components.

- *In econometrics:* Stochastic volatility model has been fairly successful in modeling financial time series. As a basis for analyzing the risk of financial investments, it is an important technique used in finance to model asset price volatility over time. It can be shown that a stochastic volatility model can be rewritten as a regression model with errors-in-variables (Comte, 2004). Therefore, the techniques in measurement error models are able to be used for solving finance time series problems.

The consequences of ignoring measurement error include, for example, masking the important features of the data which makes graphical model analysis confusing; losing the power to detect relationships among variables; and bringing forth bias in function/parameter estimation (Carroll et al., 2006). In past decades, many statistical approaches have been proposed to model and correct for measurement error. These include, for example, method-of-moments corrections (Fuller, 1987), regression calibration (Carroll and Stefanski, 1990), simulation extrapolation (Cook and Stefanski, 1994), deconvolution methods (Fan and Truong, 1993; Wang and Wang, 2011) and Bayesian analyses (Richardson and Gilks, 1993; Dellaportas and Stephens, 1995; Gustafson, 2004). A thorough overview of the current state of these methods can be found in Carroll et al. (2006) and Buonaccorsi (2010).

In this chapter, we mainly focus on the Bayesian regression models with errors-in-variables using INLA. Our discussion is based on two types of measurement error: *classical measurement error* and *Berkson measurement error*.

A fundamental issue in specifying a measurement error model is whether an assumption is made on the distribution of the observed values given the true values or vice versa. In a classical measurement error model, the predictor of interest X cannot be observed directly but is measured with error. What can be observed is the independent sample $w_1, ..., w_n$. In general, any type of regression model can be used for how W and X are related. Historically, the most commonly used model is the classical measurement error model,

$$w_i = x_i + u_i, \quad ,i = 1, ..., n,$$

where u_i's are the measurement error with mean zero, x_i and u_i are independent of each other. So, $\mathrm{E}(W|X = x) = x$, W is unbiased for the unobserved x.

Berkson measurement error was introduced by Berkson (1950) in cases where an experimenter is trying to achieve a target value w but the true value achieved is x. The model assumed that

$$x_i = w_i + u_i, \quad , i = 1, \ldots, n,$$

where w_i and u_i are independent of each other.

Ignoring measurement error can result in biased estimates and lead to erroneous conclusions in data analysis. Let us look at a simple linear regression with the variable measured with error. Given the observed sample (w_i, y_i), $i = 1, \ldots, n$, the model is given by

$$\begin{cases} y_i = \beta_0 + \beta_1 x_i + \varepsilon_i, \\ w_i = x_i + u_i, \end{cases} \tag{10.1}$$

where u_i's are classical additive measurement errors with $u_i \sim N(0, \sigma_U^2)$, and the ε_i's are the random noise from the regression model $\varepsilon_i \sim N(0, \sigma_\varepsilon^2)$. The regression of Y on W can be obtained by rewriting the model (10.1),

$$\begin{cases} y_i = \beta_0 + \beta_1 w_i + \eta_i, \\ \eta_i = \varepsilon_i - \beta_1 u_i, \end{cases}$$

The least squares estimator for β_1 is

$$\hat{\beta}_1 = \frac{\sum_{i=1}^n (w_i - \bar{w})(y_i - \bar{y})}{\sum_{i=1}^n (w_i - \bar{w})^2},$$

where $\bar{w} = \sum_{i=1}^n w_i / n$, $\bar{y} = \sum_{i=1}^n y_i / n$. Note that W and η are correlated with each other,

$$\text{Cov}(W, \eta) = \text{Cov}(X + U, \varepsilon - \beta_1 U) = -\beta_1 \sigma_u^2 \neq 0.$$

So, the least squares estimator $\hat{\beta}_1$ is inconsistent for β_1. Its probability limit is

$$\text{plim} \hat{\beta}_1 = \beta_1 + \frac{\text{Cov}(X, U)}{\text{Var}(X)} = \beta_1 - \frac{\sigma_U^2}{\sigma_X^2 + \sigma_U^2} \beta_1 = \frac{\sigma_X^2}{\sigma_X^2 + \sigma_U^2} \beta_1,$$

where σ_X^2 is the variance of X. This bias is known as *attenuation bias*. The result could also be extended to a multivariate linear regression model. One should note that, in a multivariate regression case, even if only a single predictor is error-prone, the coefficients on all predictors are generally biased.

In this chapter, we discuss the Bayesian analysis of a generalized linear mixed model (GLMM) with errors-in-variables using INLA. Posterior marginal distributions in such errors-in-variables models have been estimated by employing an MCMC sampler; see for example Richardson and Gilks (1993). However, model-specific implementation is typically challenging, and MCMC computation is very time-consuming. Muff et al. (2015) extended the INLA approach to formulate Gaussian measurement error models within GLMM. They discussed multiple real applications, and showed how parameter estimates were obtained for different measurement error models, including the classical and Berkson error models with heteroscedastic measurement errors. Here we follow Muff et al. (2015)'s modeling framework and their coding skills and features.

10.2 Classical Errors-in-Variables Models

Let us consider a general GLMM with errors-in-variables. Suppose the response $\mathbf{y} = (y_1, ..., y_n)^T$ is of exponential family form with mean $\mu_i = \mathrm{E}(y_i)$ linked to the linear predictor η_i via

$$\begin{cases} \mu_i = h(\eta_i), \\ \eta_i = \beta_0 + \beta_x x_i + \widetilde{\mathbf{x}}_i \alpha + \mathbf{z}_i \gamma, \\ w_i = x_i + u_i. \end{cases} \tag{10.2}$$

Here $h(\cdot)$ is a known monotonic inverse link function, x_i is the error-prone (unobserved) variable, w_i is an observed proxy of x_i, and $\widetilde{\mathbf{x}}_i$ is a vector of error-free predictors. The $\mathbf{z}_i$ is also a vector of error-free predictors, some of which could be in common with $\widetilde{\mathbf{x}}_i$. The parameters $(\beta_0, \beta_x, \alpha)$ are called fixed effects, and the γ are random effects. Without the term $\mathbf{z}_i\gamma$, the GLMM (10.2) becomes a GLM with errors-in-variables. In the following, we discuss two commonly used models for data with classical measurement error.

10.2.1 A Simple Linear Model with Heteroscedastic Errors-in-Variables

Let us go back to the simple linear regression with errors-in-variables. Typically, we assume that the error-prone variable $x_i \sim N(\lambda, \sigma_X^2), i = 1, ..., n$, which can be considered as a special case of the exposure model that Gustafson (2004) proposed. Consider the following model

$$\begin{cases} y_i = \beta_0 + \beta_1 x_i + \varepsilon_i, & \varepsilon_i \sim N(0, \sigma_\varepsilon^2), \\ w_i = x_i + u_i, & u_i \sim N(0, d_i \sigma_u^2), \\ x_i = \lambda + \xi_i, & \xi_i \sim N(0, \sigma_X^2), \end{cases} \tag{10.3}$$

Note here we consider a heteroscedastic error structure $w_i | x_i \sim N(x_i, d_i \sigma_u^2), i = 1, ..., n$, since in practice the distribution of measurement error may vary with each subject or even with each observation so the errors can be heteroscedastic (Wang et al., 2010). The weight d_i, which is known, is proportional to the individual error variance $\sigma_u^2(x_i) = d_i \sigma_u^2$ depending on x_i, which allows for a heteroscedastic error model.

This classical linear model with errors-in-variables (10.1) can be fit using INLA by specifying `model = "mec"` inside the `f()` function, when we define an `inla` model formula. There are 4 special hyperparameters to be defined in `f()`: $\boldsymbol{\theta} = (\theta_1, \theta_2, \theta_3, \theta_4)$, where $\theta_1 = \beta_1$, $\theta_2 = \log(1/\sigma_u^2)$, $\theta_3 = \lambda$, and $\theta_4 = \log(1/\sigma_X^2)$. It is important to select these parameters appropriately in order to achieve the reasonable results.

Let us look at a simulated example and illustrate the use of the INLA approach to fit the errors-in-variables linear model. We assume that the true parameters $\beta_0 = 1$ and $\beta_1 = 5$, the regression noise $\varepsilon_i \sim N(0,1)$, the true unobserved predictor $x_i \sim N(0,1)$, and the heteroscedastic error $u_i \sim N(0, d_i)$ where $d_i \sim \mathrm{Unif}(0.5, 1.5)$. We first set up the parameters to simulate such a dataset:

```
set.seed(5)
n = 100
beta0 = 1
beta1 = 5
prec.y = 1
prec.u = 1
prec.x = 1
```

Then we generate the true unobserved predictor x:

```
x <- rnorm(n, sd = 1/sqrt(prec.x))
```

And we generate the observed predictor w with heteroscedastic error:

```
d <- runif(n, min = 0.5, max = 1.5)
w <- x + rnorm(n, sd = 1/sqrt(prec.u * d))
```

Finally, we generate the response variables from the true model with the unobserved x, and create a data frame for data analysis:

```
y <- beta0 + beta1*x + rnorm(n, sd = 1/sqrt(prec.y))
sim.data <- data.frame(y, w, d)
```

If we ignore the measurement error in the predictor, the simple linear regression model can be fit using INLA:

```
sim.inla <- inla(y ~ w, data = sim.data, family = "gaussian")
summary(sim.inla)
round(sim.inla$summary.fixed, 4)
```

```
              mean     sd 0.025quant 0.5quant 0.975quant   mode kld
(Intercept) 0.8500 0.3639     0.1339   0.8500     1.5653 0.8500   0
w           2.2779 0.2801     1.7267   2.2779     2.8284 2.2779   0
```

The estimate of the slope β_1 is 2.2779. The result is clearly biased, and its 95% credible interval is (1.7267, 2.8284), which does not cover the true slope 5. Now, let us employ the new `mec` model in INLA. We need to define a relatively complex INLA model formula. Let us set the initial values of hyperparameters:

```
init.prec.u <- prec.u
init.prec.x <- var(w) - 1/prec.u
init.prec.y <- sim.inla$summary.hyperpar$mean
```

Then we set prior parameters:

```
prior.beta = c(0, 0.0001)
prior.prec.u = c(10, 10)
prior.prec.x = c(10, 10)
prior.prec.y = c(10, 10)
```

Now we define the `mec` model formula:

```
formula = y ~ f(w, model = "mec", scale = d, values = w,
        hyper = list(
                beta = list(prior = "gaussian", param = prior.beta,
                  ↪ fixed = FALSE),
                prec.u = list(prior = "loggamma", param = prior.prec.u
                  ↪ , initial = log(init.prec.u), fixed = FALSE),
                mean.x = list(prior = "gaussian", initial = 0, fixed =
                  ↪  TRUE),
```

```
                    prec.x = list(prior = "loggamma", param = prior.prec.x
                        ↪ , initial = log(init.prec.x), fixed = FALSE)))
```

In the above formula, the `mec` model contains four hyperparameters:

- `beta` corresponds to the slope coefficient β_1 of the error-prone predictor x, with a Gaussian prior.
- `prec.u` corresponds to $\log(1/\sigma_u^2)$ with log-gamma prior.
- `mean.x` corresponds to the mean parameter of x, with a Gaussian prior, but here it is fixed at 0 due to predictor centering.
- `prec.x` corresponds to $\log(1/\sigma_X^2)$ with log-gamma prior.

The prior settings are defined in the different entries of the list `hyper`. The option `fixed` specifies if the corresponding hyperparameter should be estimated or fixed at the `initial` value. The argument `param` defines the prior parameters of the corresponding prior distribution. When we set up the initial values of hyperparameters, a reasonable guess is important. One may use some information from the naive fit that is from the model ignoring the measurement error.

We also need to define the hyperparameter of the Gaussian regression model σ_ε^2 and the prior distribution for the intercept β_0. These can be specified in the call of the `inla` function via the `control.family` option and the `control.fixed` option. We fit the errors-in-variables model using the following code:

```
sim.mec.inla <- inla(formula, data = sim.data, family = "gaussian",
        control.family = list(hyper = list(prec = list(param = prior.
            ↪ prec.y, initial = log(init.prec.y), fixed=FALSE))),
        control.fixed = list(mean.intercept = prior.beta[1], prec.
            ↪ intercept = prior.beta[2]))
```

Let us output the results:

```
round(sim.mec.inla$summary.fixed, 4)
```

```
                          mean     sd 0.025quant 0.5quant 0.975quant   mode kld
(Intercept)             0.8314 0.3601     0.1202   0.8323     1.5366 0.8342   0
```

```
round(sim.mec.inla$summary.hyperpar, 4)
```

```
                             mean     sd 0.025quant 0.5quant 0.975quant   mode
Precision for the
Gaussian observations      0.9820 0.3177     0.4853   0.9415     1.7203 0.8652
MEC beta for w             5.1508 0.4739     4.2648   5.1321     6.1244 5.0759
MEC prec_u for w           1.0530 0.1416     0.7982   1.0453     1.3532 1.0315
MEC prec_x for w           1.2312 0.2399     0.8376   1.2042     1.7758 1.1496
```

The fitting results show that the corrected estimate for β_1 is 5.151 using the `mec` model, and its 95% confidence interval is (4.2648, 6.125), which does cover the true slope 5. The intercept and other precision parameters are estimated quite well, which are close to the true values. We could plot the estimated regression line and compare it with the true function and the estimate using the naive model, ignoring the measurement error:

```
plot(w, y, xlim=c(-4, 4), col= "grey")
curve(1 + 5*x, -4,4, add=TRUE, lwd=2)
curve(sim.mec.inla$summary.fixed$mean + sim.mec.inla$summary.hyperpar$
    ↪ mean[2]*x, -4,4, add=TRUE, lwd=2, lty = 5)
curve(sim.inla$summary.fixed$mean[1] + sim.inla$summary.fixed$mean[2]*
    ↪ x, -4,4, add=TRUE, lwd=2, lty = 4)
```

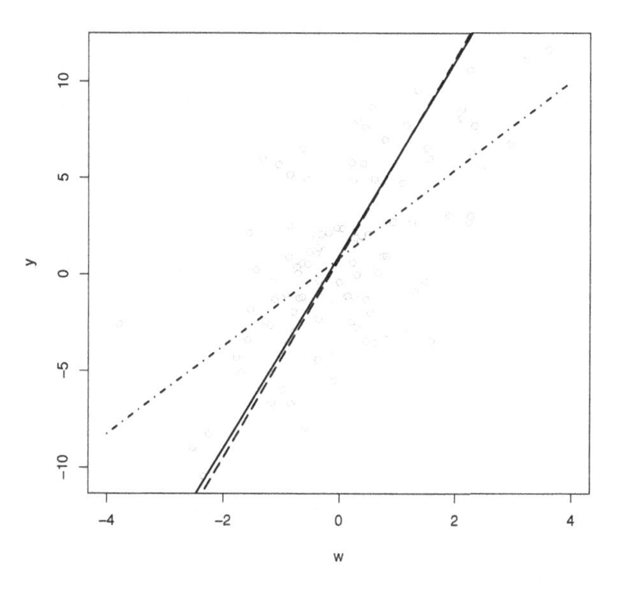

FIGURE 10.1
The simulated example of a simple linear regression with heteroscedastic errors-in-variables: the solid line is the true regression function, the dashed line is the estimate from errors-in-variables regression, the dash-dotted line is the estimate from standard linear regression ignoring the measurement error.

Figure 10.1 shows the comparison of the results: the dashed line is the estimate from errors-in-variables regression, the dash-dotted line is the estimate from standard linear regression that ignores the measurement error, and the solid line is the true regression function. The errors-in-variables regression successfully recovered the underlying true function.

10.2.2 A General Exposure Model with Replicated Measurements

In this subsection, we address the model to deal with the case that the same unobserved variable x_i is independently measured several times, but each measurement is affected by error.

In many applications, repeated measurements $w_{ik}, k = 1,\ldots,m$ of the true value

x_i are available. Data consist of "intrinsic" values that are measured with error a few times. Typically, we assume that

$$w_{ki}|x_i \sim N(x_i, \sigma_u^2).$$

Following Gustafson (2004), the distribution of the x_i's, possibly depending on the error-free predictors $\widetilde{\mathbf{x}}_i$, is specified in the *exposure model*. In the most general situation, the error-prone variable x_i is Gaussian with mean depending on $\widetilde{\mathbf{x}}_i$, i.e.,

$$x_i|\widetilde{\mathbf{x}}_i \sim N(\lambda_0 + \widetilde{\mathbf{x}}_i\boldsymbol{\lambda}, \sigma_X^2),$$

where λ_0 is the intercept and $\boldsymbol{\lambda}$ is a vector of fixed effects in the linear regression of x_i on $\widetilde{\mathbf{x}}_i$. If x_i depends only on some components of $\widetilde{\mathbf{x}}_i$, the corresponding fixed effects in $\boldsymbol{\lambda}$ are set to zero. In the extreme case of $\boldsymbol{\lambda} = 0$, x_i is independent of $\widetilde{\mathbf{x}}_i$, which has been discussed in the previous example.

In the following we show how the above measurement error model fits into the hierarchical structure required by INLA. Based on the model specification, we can have a joint modeling representation, which is hierarchical:

$$\begin{cases} \mathrm{E}(y_i) = h(\beta_0 + \beta_x x_i + \widetilde{\mathbf{x}}_i\boldsymbol{\alpha} + \mathbf{z}_i\boldsymbol{\gamma}), & \\ w_{ki} = x_i + u_{ki}, & u_{ki} \sim N(0, \sigma_u^2), \\ 0 = -x_i + \lambda_0 + \widetilde{\mathbf{x}}_i\boldsymbol{\lambda} + \xi_i, & \xi_i \sim N(0, \sigma_X^2). \end{cases} \tag{10.4}$$

Implementation of this model using the `INLA` library requires a joint model formulation, where the response variable y_i is augmented with pseudo-observation 0, and the observed values w_{ki}.

Let us illustrate the analysis using data from the Framingham Heart Study (Carroll et al., 2006). The Framingham study consists of a series of exams taken two years apart. There are 1615 males aged from 31 to 65 in this dataset, with the outcome indicating the occurrence of coronary heart disease (CHD) within an eight-year period following Exam 3. There were 128 total cases of CHD. Table 10.1 shows the description of variables in the Framingham data.

Predictors employed in this example are the patient's age, smoking status, and transformed systolic blood pressure (SBP), $\log(SBP - 50)$. Following the suggestion of Carroll et al. (2006), we are interested in the variable X, the long-term average transformed SBP, which cannot be observed. In this Framingham heart study data, two SBP measurements from each of two exams were obtained. It was assumed that the two transformed variates-in-errors

$$W_1 = \log((SBP_{21} + SBP_{22})/2 + 50)$$

and

$$W_2 = \log((SBP_{31} + SBP_{32})/2 + 50),$$

where SBP_{ij} is the jth measurement of SBP from the ith exam, $j = 1,2$, $i = 2,3$, were replicate measurements of the the long-term average transformed SBP, X. The following R code is used to set up our initial dataset:

```
data(framingham, package = "brinla")
fram <- subset(framingham, select = c("AGE", "SMOKE", "FIRSTCHD"))
fram$W1 <- log((framingham$SBP21 + framingham$SBP22)/2 - 50)
fram$W2 <- log((framingham$SBP31 + framingham$SBP32)/2 - 50)
fram$W <- (fram$W1 + fram$W2)/2
```

The other error-free variables, SMOKE ($\tilde{X}_1$) and AGE ($\tilde{X}_2$), are considered in both the logistic regression and the exposure model. Thus, the hierarchical model (10.4) is formulated as a joint model by applying the copy feature in INLA. The full model can be written as

$$\begin{bmatrix} y_1 & \text{NA} & \text{NA} \\ \vdots & \vdots & \vdots \\ y_n & \text{NA} & \text{NA} \\ \text{NA} & w_{11} & \text{NA} \\ \vdots & \vdots & \vdots \\ \text{NA} & w_{1n} & \text{NA} \\ \text{NA} & w_{21} & \text{NA} \\ \vdots & \vdots & \vdots \\ \text{NA} & w_{2n} & \text{NA} \\ \text{NA} & \text{NA} & 0 \\ \vdots & \vdots & \vdots \\ \text{NA} & \text{NA} & 0 \end{bmatrix} = \beta_0 \begin{bmatrix} 1 \\ \vdots \\ 1 \\ \text{NA} \\ \vdots \\ \text{NA} \\ \text{NA} \\ \vdots \\ \text{NA} \\ \text{NA} \\ \vdots \\ \text{NA} \end{bmatrix} + \beta_X \begin{bmatrix} 1 \\ \vdots \\ n \\ \text{NA} \\ \vdots \\ \text{NA} \\ \text{NA} \\ \vdots \\ \text{NA} \\ \text{NA} \\ \vdots \\ \text{NA} \end{bmatrix} + \begin{bmatrix} \text{NA} \\ \vdots \\ \text{NA} \\ 1 \\ \vdots \\ n \\ 1 \\ \vdots \\ n \\ -1 \\ \vdots \\ -n \end{bmatrix} + \beta_{smoke} \begin{bmatrix} \tilde{x}_{11} \\ \vdots \\ \tilde{x}_{1n} \\ \text{NA} \\ \vdots \\ \text{NA} \\ \text{NA} \\ \vdots \\ \text{NA} \\ \text{NA} \\ \vdots \\ \text{NA} \end{bmatrix}$$

$$+\beta_{age} \begin{bmatrix} \tilde{x}_{21} \\ \vdots \\ \tilde{x}_{2n} \\ \text{NA} \\ \vdots \\ \text{NA} \\ \text{NA} \\ \vdots \\ \text{NA} \\ \text{NA} \\ \vdots \\ \text{NA} \end{bmatrix} + \lambda_0 \begin{bmatrix} \text{NA} \\ \vdots \\ \text{NA} \\ \text{NA} \\ \vdots \\ \text{NA} \\ \text{NA} \\ \vdots \\ \text{NA} \\ 1 \\ \vdots \\ 1 \end{bmatrix} + \lambda_{smoke} \begin{bmatrix} \text{NA} \\ \vdots \\ \text{NA} \\ \text{NA} \\ \vdots \\ \text{NA} \\ \text{NA} \\ \vdots \\ \text{NA} \\ \tilde{x}_{11} \\ \vdots \\ \tilde{x}_{1n} \end{bmatrix} + \lambda_{age} \begin{bmatrix} \text{NA} \\ \vdots \\ \text{NA} \\ \text{NA} \\ \vdots \\ \text{NA} \\ \text{NA} \\ \vdots \\ \text{NA} \\ \tilde{x}_{21} \\ \vdots \\ \tilde{x}_{2n} \end{bmatrix} + \begin{bmatrix} \varepsilon_1 & \text{NA} & \text{NA} \\ \vdots & \vdots & \vdots \\ \varepsilon_n & \text{NA} & \text{NA} \\ \text{NA} & u_{11} & \text{NA} \\ \vdots & \vdots & \vdots \\ \text{NA} & u_{1n} & \text{NA} \\ \text{NA} & u_{21} & \text{NA} \\ \vdots & \vdots & \vdots \\ \text{NA} & u_{2n} & \text{NA} \\ \text{NA} & \text{NA} & \xi_1 \\ \vdots & \vdots & \vdots \\ \text{NA} & \text{NA} & \xi_n \end{bmatrix}.$$

From this "artificial-looking" model, outcomes are decomposed in terms of the mean and an appropriate error term. The components of the error structure have the appropriate distribution with the variance depending on the mean-variance relationship of the responses. In the left-side matrix of the equation, each column requires specification of a likelihood function. The first follows a Bernoulli distribution, the second is assumed to be Gaussian, and the third component is also Gaussian.

For the parameters $(\beta_0, \beta_X, \beta_{smoke}, \beta_{age}, \lambda_0, \lambda_{smoke}, \lambda_{age})$, we can simply assign

TABLE 10.1
Description of variables in the Framingham data.

Variable Name	Description	Codes/Values
AGE	age at exam 2	Numbers
SBP21	first systolic blood pressure at exam 2	mmHg
SBP22	second systolic blood pressure at exam 2	mmHg
SBP31	first systolic blood pressure at exam 3	mmHg
SBP32	second systolic blood pressure at exam 3	mmHg
SMOKE	present smoking at exam 1	0 = No 1 = Yes
FIRSTCHD	indicator of first evidence of CHD occurring at exam 3 through 6	0 = No 1 = Yes

independent $N(0,10^2)$ priors. To specify the initial values and priors of the hyperparameters $(\tau_u, \tau_X) = (1/\sigma_u^2, 1/\sigma_X^2)$, we could use the information from SBP measurements with error. Note that $w_{ki} = x_i + u_{ki}$, $u_{ki} \sim N(0, \sigma_u^2)$, so,

$$v_i = (w_{1i} - w_{2i}) \sim N(0, 2\sigma_u^2).$$

We suggest the following initial estimator of the precision parameter $\tau_u = 1/\sigma_u^2$,

$$\hat{\tau}_u = 1/\hat{\sigma}_u^2 = 2/S_v^2,$$

where S_v^2 is the sample variance of v_i's. Similarly,

$$r_i = \frac{w_{1i} + w_{2i}}{2} = x_i + \frac{u_{1i} + u_{2i}}{2} \sim N(\lambda_0 + \tilde{\mathbf{x}}_i \boldsymbol{\lambda}, \sigma_X^2 + \frac{\sigma_u^2}{2}).$$

We suggest to estimate $\tau_X = 1/\sigma_X^2$ by

$$\hat{\tau}_X = 1/(S_r^2 - S_v^2/4),$$

where S_r^2 is the sample variance of r_i's. Assuming equal mean and variance for τ_u and τ_X, we specify the priors $\tau_u \sim \text{Gamma}(\hat{\tau}_u, 1)$, and $\tau_X \sim \text{Gamma}(\hat{\tau}_X, 1)$. We first set up the prior parameters in R:

```
prior.beta <- c(0, 0.01)
prior.lambda <- c(0, 0.01)
```

We then estimate initial values of precision parameters:

```
prec.u <- 1/(var(fram$W1 - fram$W2)/2)
prec.x <- 1/(var((fram$W1 + fram$W2)/2) - (1/4)*(var((fram$W1 - fram$
    ↪ W2)/2)))
prior.prec.x <- c(prec.x, 1)
prior.prec.u <- c(prec.u, 1)
```

We need to create a new data frame containing the response matrix Y and the data vectors according to the above joint model equation:

```
Y <- matrix(NA, 4*n, 3)
```

```
Y[1:n, 1] <- fram$FIRSTCHD
Y[n+(1:n), 2] <- rep(0, n)
Y[2*n+(1:n), 3] <- fram$W1
Y[3*n+(1:n), 3] <- fram$W2

beta.0 <- c(rep(1, n), rep(NA, n), rep(NA, n), rep(NA, n))
beta.x <- c(1:n, rep(NA, n), rep(NA, n), rep(NA, n))
idx.x <- c(rep(NA, n), 1:n, 1:n, 1:n)
weight.x <- c(rep(1, n), rep(-1, n), rep(1, n), rep(1,n))
beta.smoke <- c(fram$SMOKE, rep(NA, n), rep(NA, n), rep(NA,n))
beta.age <- c(fram$AGE, rep(NA, n), rep(NA, n), rep(NA,n))
lambda.0 <- c(rep(NA, n), rep(1, n), rep(NA, n), rep(NA, n))
lambda.smoke <- c(rep(NA, n), fram$SMOKE, rep(NA, n), rep(NA, n))
lambda.age <- c(rep(NA, n), fram$AGE, rep(NA, n), rep(NA, n))
Ntrials <- c(rep(1, n), rep(NA, n), rep(NA, n), rep(NA, n))

fram.jointdata <- data.frame(Y=Y,
                       beta.0=beta.0, beta.x=beta.x, beta.smoke=beta
                           ↪ .smoke, beta.age=beta.age,
                       idx.x=idx.x, weight.x=weight.x,
                       lambda.0=lambda.0, lambda.smoke=lambda.smoke,
                           ↪ lambda.age=lambda.age,
                       Ntrials=Ntrials)
```

Next we need to define the INLA formula. In this example, we have six fixed effects $(\beta_0, \beta_{smoke}, \beta_{age}, \lambda_0, \lambda_{smoke}, \lambda_{age})$. Also, the joint model has no common intercept, thus it has to be explicitly removed using `-1` in the formula. We have two special random effects, which are needed to encode that the values of X in the exposure model and the error model are assigned the same values. The first random effect term is defined by `f(beta.x, ...)`, where the `copy="idx.x"` call guarantees the assignment of identical values to X in all components of the joint model. The second random effect term is defined by `f(idx.x, ...)`, where `idx.x` contains the values of the unobserved variable X, encoded as an i.i.d. Gaussian random effect, and weighted with `weight.x` to ensure correct signs in the joint model. The `values` option contains the vector of all values assumed by the covariate for which the effect is estimated. The precision `prec` of the random effect is fixed at $\exp(-15)$ (or other very small number). This is important and necessary since the uncertainty in X has been modeled in the exposure component of the joint model. More details about the INLA formula specification can be found in Muff et al. (2015).

The three options also need to be specified in the call of the `inla` function: the first option is `family`. There are three different likelihoods in the joint model, which correspond to the different columns in the response matrix. Here, we need to specify `family = c("binomial", "gaussian", "gaussian")` to correspond to the binomial likelihood of the regression model, one Gaussian likelihood for the exposure model, and the other Gaussian likelihood for the error model. The second option is `control.family`, which is to specify the hyperparameters for the three likelihoods, in the same order as given in `family`. The binomial likelihood does not contain a hyperparameter, hence the corresponding list is empty. In the second and third likelihoods the hyperparameters σ_X and σ_u need to be specified, respectively. The third option is `control.fixed`, which is to specify priors for the fixed effects. The implementation of fitting the complex measurement error models using INLA is given by the following R code:

```
fram.formula <- Y ~  f(beta.x, copy = "idx.x", hyper = list(beta =
    ↪ list(param = prior.beta, fixed = FALSE)))
      + f(idx.x, weight.x, model = "iid", values = 1:n, hyper = list
          ↪ (prec = list(initial = -15, fixed = TRUE)))
      + beta.0 - 1 + beta.smoke + beta.age + lambda.0 + lambda.smoke
          ↪ + lambda.age

fram.mec.inla <- inla(fram.formula, Ntrials = Ntrials, data = fram.
    ↪ jointdata, family = c("binomial", "gaussian", "gaussian"),
      control.family = list(
        list(hyper = list()),
        list(hyper = list(
          prec = list(initial = log(prec.x),
                      param = prior.prec.x,
                      fixed = FALSE))),
        list(hyper = list(
          prec = list(initial=log(prec.u),
                      param = prior.prec.u,
                      fixed = FALSE)))),
      control.fixed = list(
          mean = list(beta.0=prior.beta[1], beta.smoke=prior.beta[1],
              ↪ beta.age=prior.beta[1],
                  lambda.0=prior.lambda[1], lambda.smoke=prior.
                      ↪ lambda[1], lambda.age=prior.lambda[1]),
          prec = list(beta.0=prior.beta[2], beta.smoke=prior.beta[2],
              ↪ beta.age=prior.beta[2],
                  lambda.0=prior.lambda[2], lambda.smoke=prior.
                      ↪ lambda[2], lambda.age=prior.lambda[2]))
)
```

We often want to improve the estimates of the posterior marginals for the hyperparameters of the measurement error model using the grid integration strategy by the function `inla.hyperpar`, since the coefficient of the covariate measured in error is a hyperparameter in INLA analysis.

```
fram.mec.inla <-inla.hyperpar(fram.mec.inla)
round(fram.mec.inla$summary.fixed, 4)
```

```
                 mean     sd 0.025quant 0.5quant 0.975quant     mode kld
beta.0       -13.9799 2.1885   -18.2224 -14.0133    -9.7163 -14.0132   0
beta.smoke     0.5697 0.2479     0.0999   0.5635     1.0744   0.5510   0
beta.age       0.0515 0.0118     0.0285   0.0514     0.0747   0.0513   0
lambda.0       4.1066 0.0310     4.0458   4.1066     4.1674   4.1066   0
lambda.smoke  -0.0266 0.0125    -0.0512  -0.0266    -0.0019  -0.0266   0
lambda.age     0.0061 0.0006     0.0049   0.0061     0.0073   0.0061   0
```

```
round(fram.mec.inla$summary.hyperpar, 4)
```

```
                                   mean     sd 0.025quant 0.5quant 0.975quant    mode
Precision for the
Gaussian observations[2]        27.4738 1.1244    25.3274  27.4533    29.7357 27.4121
Precision for the
Gaussian observations[3]        78.4611 2.6326    73.3821  78.4328    83.7025 78.3787
Beta for beta.x                  1.9574 0.4800     1.0182   1.9580     2.8986  1.9638
```

From the results, we find that the error free variables, age and smoking status, are positively associated with the occurrence of CHD. The transformed SBP after correcting the measurement error has the coefficient 1.9574 with 95% credible intervals (0.4800, 2.8986). For each increase in 1 unit of transformed SBP, the estimated odds of occurring CHD increases by roughly a factor of 7.0809 (= exp(1.9574)), assuming that all other covariates are fixed. We could compare the measurement error model with the naive model by ignoring measurement error:

```
fram.naive.inla <- inla(FIRSTCHD~ SMOKE + AGE + W, family = "binomial"
    ↪ , data = fram)
round(fram.naive.inla$summary.fixed, 4)
```

```
                 mean     sd 0.025quant 0.5quant 0.975quant     mode kld
(Intercept) -13.2961 1.7984   -16.8379 -13.2925    -9.7776 -13.2851   0
SMOKE         0.5742 0.2484     0.1035   0.5679     1.0802   0.5553   0
AGE           0.0532 0.0116     0.0306   0.0531     0.0761   0.0530   0
W             1.7728 0.4110     0.9662   1.7727     2.5795   1.7724   0
```

The estimates for the error-free covariates are very close, but the estimate for the SBP from the naive model is 1.7728. It clearly underestimates the effect due to measurement error.

10.3 Berkson Errors-in-Variables Models

We now consider a GLMM with Berkson errors-in-variables. Similar to the classical error model, we have the response $\mathbf{y} = (y_1, ..., y_n)^T$ is of exponential family form with mean $\mu_i = \mathrm{E}(y_i)$ linked to the linear predictor η_i via

$$\begin{cases} \mu_i = h(\eta_i), \\ \eta_i = \beta_0 + \beta_x x_i + \widetilde{\mathbf{X}}_i \alpha + \mathbf{Z}_i \gamma, \\ x_i = w_i + u_i. \end{cases} \tag{10.5}$$

In the Berkson error model, $x_i | w_i \sim N(w_i, \sigma_u^2), i = 1, ..., n$. One may generalize to the case of heteroscedastic error, where $x_i | w_i \sim N(w_i, d_i \sigma_u^2), i = 1, ..., n$. Since x_i is defined conditionally on the observation w_i, the exposure model discussed in the classical errors-in-variables problem is not applicable anymore. Again, we can rewrite the model into the hierarchical structure required by INLA:

$$\begin{cases} \mathrm{E}(y_i) = h(\beta_0 + \beta_x x_i + \widetilde{\mathbf{x}}_i \alpha + \mathbf{Z}_i \gamma), \\ -w_i = -x_i + u_i, \qquad\qquad u_i \sim N(0, \sigma_u^2). \end{cases} \tag{10.6}$$

This Berkson errors-in-variables model is termed "meb" in the INLA package. Let us use the bronchitis data example to illustrate the Berkson error model. This example has been discussed in Chapter 8 of Carroll et al. (2006). The bronchitis study was to assess the health hazard of specific harmful substances in a dust-laden mechanical engineering plant in Munich. The outcome of the study is the indicator that the worker has bronchitis. The main covariate of interest is the average dust concentration in the working area over the period of time. In addition, two other variables, the duration of exposure, and the smoking status, are also measured. The description of variables in the bronchitis data is displayed in Table 10.2. Carroll et al. (2006) discussed the estimation of threshold limiting value (TLV) of average dust concentration. The estimated TLV is 1.28, under which there is no risk due to the substance. Here we only focus on the subgroup samples that the dust concentration is greater than the TLV, 1.28.

TABLE 10.2
Description of variables in the bronchitis data.

Variable Name	Description	Codes/Values
cbr	Chronic bronchitis reaction	0 = No
		1 = Yes
dust	Dust concentration at work place	mg/m^3
smoking	Does worker smoke?	0 = No
		1 = Yes
expo	Duration of exposure	years

```
data(bronch, package = "brinla")
bronch1 <- subset(bronch, dust >=1.28)
```

It is impossible to measure the dust concentration exactly, and instead sample dust concentrations were obtained several times between 1960 and 1977. The resulting measurements are `dust` in the dataset. Let us start to explore the data by summarizing the variables:

```
round(prop.table(table(bronch1$cbr)),4)
```

```
     0      1
0.6967 0.3033
```

```
round(prop.table(table(bronch1$smoking)),4)
```

```
     0      1
0.2582 0.7418
```

```
round(c(mean(bronch1$dust), sd(bronch1$dust)), 4)
```

```
[1] 1.9285 0.2118
```

```
round(cor(bronch1$dust, bronch1$expo), 4)
```

```
[1] 0.0119
```

```
round(by(bronch1$dust, bronch1$smoking, mean), 4)
```

```
bronch1$smoking: 0
[1] 1.9258
------------------------------------------------------------------------
bronch1$smoking: 1
[1] 1.9294
```

Among 488 subjects, 30% of the workers are reported as having chronic bronchitis, and 74% are smokers. Measured dust concentration had a mean of 1.93 and a standard deviation of 0.21. The durations are effectively independent of concentrations, with correlation 0.012. Smoking status is also effectively independent of dust concentration, with the smokers having mean concentration 1.926, and the nonsmokers having mean 1.929. In Carroll et al. (2006)'s analysis, they concluded that a reasonably informative prior on σ_u is necessary through numerical experiments. Here we take the initial value of $\sigma_u = 1.3$ based on their analysis. Let us first set up the prior parameters:

```
prior.beta <- c(0, 0.01)
prec.u <- 1/1.3
prior.prec.u <- c(1/1.3, 0.01)
```

Then we fit the Berkson errors-in-variables model with the following code:

```
bronch.formula <- cbr ~  smoking + expo + f(dust, model="meb", hyper =
    ↪  list(beta = list(param = prior.beta, fixed = FALSE), prec.u =
    ↪ list(param = prior.prec.u, initial = log(prec.u), fixed = FALSE
    ↪ )))
bronch.meb.inla <- inla(bronch.formula, data = bronch1, family = "
    ↪ binomial", control.fixed = list(mean.intercept = prior.beta[1],
    ↪  prec.intercept = prior.beta[2], mean = prior.beta[1], prec =
    ↪ prior.beta[2]))
bronch.meb.inla <- inla.hyperpar(bronch.meb.inla)
round(bronch.meb.inla$summary.fixed, 4)
```

```
               mean     sd 0.025quant 0.5quant 0.975quant    mode kld
(Intercept) -5.6646 1.0079    -7.7397  -5.5994    -3.8632 -5.4020   0
smoking1     1.2257 0.2953     0.6635   1.2195     1.8234  1.2069   0
expo         0.0411 0.0117     0.0186   0.0410     0.0645  0.0407   0
```

```
round(bronch.meb.inla$summary.hyperpar, 4)
```

```
                      mean     sd 0.025quant 0.5quant 0.975quant   mode
MEB beta for dust   1.0401 0.4127     0.3970   0.9936     1.9304 0.8429
MEB prec_u for dust 1.0511 0.9889     0.1277   0.7571     3.6641 0.1189
```

Let us also compare the naive model, ignoring the measurement error.

```
bronch.naive.inla <- inla(cbr ~ dust + smoking + expo, data = bronch1,
    ↪  family = "binomial",
control.fixed = list(mean.intercept = prior.beta[1], prec.intercept =
    ↪ prior.beta[2], mean = prior.beta[1], prec = prior.beta[2]))
round(bronch.naive.inla$summary.fixed, 4)
```

```
               mean     sd 0.025quant 0.5quant 0.975quant    mode kld
(Intercept) -4.2577 0.9973    -6.2401  -4.2496    -2.3222 -4.2331   0
dust         0.8420 0.4746    -0.0864   0.8408     1.7762  0.8385   0
smoking1     1.0442 0.2679     0.5344   1.0384     1.5873  1.0267   0
expo         0.0356 0.0096     0.0169   0.0355     0.0546  0.0354   0
```

We notice that the variable of interest, `dust`, is not significant (in the Bayesian sense) in the naive model. With the Berkson errors-in-variables model, the coefficient estimate for `dust` increases to 1.0511 and its 95% credible interval is (0.3970, 1.9304). After correcting the effect of measurement error, the effect for `dust` is positive with very high probability.

As Muff et al. (2015) pointed out, Bayesian methods have provided a very flexible framework for measurement error problems in the past decades. However, Bayesian analysis using MCMC samplers has not become part of standard regression analyses in solving measurement error problems, due to a wide range of problems in terms of convergence and computational time. The implementation of model fitting with MCMC requires careful algorithm construction. It might often be problematic, especially, for end users who might not be experts in programming. INLA provides more user friendly approaches to solve the complex problems.

11

Miscellaneous Topics in INLA

This chapter covers a mix of topics including splines, functional data, extreme values and density estimation.

11.1 Splines as a Mixed Model

Let's consider the simplest nonparametric regression setting

$$y_i = f(x_i) + \varepsilon_i, \quad \varepsilon_i \sim N(0, \sigma_\varepsilon^2),$$

where f is the unknown but smooth function. Although a few methods for estimating f have been given in Chapters 7 and 8, we here introduce two more spline smoothing methods, which are extensively used in practice and can be connected to the linear mixed models introduced in Chapter 5.

11.1.1 Truncated Power Basis Splines

Recall that P-splines (Eilers and Marx, 1996) (see Section 7.5) combine a rich B-spline basis with equally-spaced knots and a simple difference penalty within the regression framework. Ruppert and Carroll (2000) proposed a smoothing method using a similar idea: their basis consists of truncated power functions, the knots are quantiles of x_i, and the penalty is on the size of the coefficients. More specifically, using notation $a_+^p = a^p$ if $a > 0$ or 0 if $a \leq 0$ we expand the function as follows

$$f(x_i) = \beta_0 + \beta_1 x_i + \cdots + \beta_p x_i^p + \sum_{j=1}^{r} u_j (x_i - t_j)_+^p,$$

for $i = 1, \ldots, n$, where f is expanded as a combination of the pth order polynomial and the sum of truncated power basis (TPB) functions given a sequence of r knots t_j. Taking $\boldsymbol{\beta} = (\beta_0, \ldots, \beta_p)^T$ to be the unknown vector of fixed effects and $\boldsymbol{u} = (u_1, \ldots, u_r)^T$ of random effects, this smoothing method can be formulated as a mixed model given by

$$\boldsymbol{y} = \boldsymbol{X}\boldsymbol{\beta} + \boldsymbol{Z}\boldsymbol{u} + \boldsymbol{\varepsilon}, \quad \boldsymbol{u} \sim N(\boldsymbol{0}, \sigma_u^2 \boldsymbol{I}), \quad \boldsymbol{\varepsilon} \sim N\left(\boldsymbol{0}, \sigma_\varepsilon^2 \boldsymbol{I}\right), \tag{11.1}$$

where $\boldsymbol{X}$ is the $n \times (p+1)$ matrix with i^{th} row $(1, x_i, \ldots, x_i^p)$ and $\boldsymbol{Z}$ is the $n \times r$ matrix with i^{th} row $\left((x_i - t_1)_+^p, \ldots, (x_i - t_r)_+^p\right)$. Note that $\lambda = \sigma_\varepsilon^2 / \sigma_u^2$ is the smoothing parameter that controls how smooth the fitted curve is. The best linear unbiased predictors (BLUPs) of $\boldsymbol{\beta}$ and $\boldsymbol{u}$ can be explicitly derived under this framework, and the variance components σ_u^2 and σ_ε^2 are estimated by restricted maximum likelihood (REML). This work has been extended by Ruppert et al. (2003), and also called "P-splines" by some people. To avoid unnecessary confusion we term it here as TPB splines.

From the Bayesian point of view, we need to take priors on the unknown parameters in model (11.1), which are $\boldsymbol{\beta}$, $\boldsymbol{u}$, σ_u^2 and σ_ε^2. To align with mixed model formulation, we use the following Gaussian priors on $\boldsymbol{\beta}$ and $\boldsymbol{u}$:

$$\boldsymbol{\beta} \sim N\left(\mathbf{0},\, \sigma_\beta^2 \boldsymbol{I}\right), \quad \boldsymbol{u} \sim N\left(\mathbf{0},\, \sigma_u^2 \boldsymbol{I}\right),$$

where σ_β^2 is a fixed large number (e.g., 10^6), but σ_u^2 must be random and needs a prior for it. Following the discussion in Section 7.2.2, we may take non-informative or weakly informative priors on σ_u^2 and σ_ε^2.

Finally, we discuss how to choose the knots. The knot t_j is the sample quantile of x_is corresponding to probability $j/(r+1)$. A common default number of knots in the penalized spline literature is $r = \min(n_u/4, 35)$, where n_u is the number of unique x_is (see, e.g., Ruppert et al., 2003). Ruppert (2002) discusses "hi-tech" choices of r, and suggests choosing a large enough number (typically 5 to 20) for the desired flexibility. Also, the distribution of the knots, for a given r, may have some effect on the results, but in most situations this effect will be minor.

11.1.2 O'Sullivan Splines

O'Sullivan splines (O'Sullivan, 1986), abbreviated as O-splines, combine B-splines with the penalty function used in smoothing splines. Given a sequence of knots $x_{min} < t_1 < \cdots < t_r < x_{max}$ we represent $f(x)$ as a sum of cubic B-splines using the expression in (7.11), with $B_1, \ldots, B_p$ $(p = r+4)$ basis functions defined by these knots and the corresponding coefficients $\boldsymbol{\beta} = (\beta_1, \ldots, \beta_p)^T$. Then, the integrated squared second derivative penalty is used, and it can be expressed as a quadratic function of the coefficients

$$\int \left(f''(x)\right)^2 dx = \boldsymbol{\beta}^T \boldsymbol{Q} \boldsymbol{\beta},$$

where $\boldsymbol{Q}$ is the $p \times p$ (banded) matrix with entry $\boldsymbol{Q}[i,j] = \int B_i''(x) B_j''(x) dx$. Finally, the O-splines estimator is obtained by minimizing

$$(\boldsymbol{y} - \boldsymbol{B}\boldsymbol{\beta})^T (\boldsymbol{y} - \boldsymbol{B}\boldsymbol{\beta}) + \lambda \boldsymbol{\beta}^T \boldsymbol{Q} \boldsymbol{\beta}, \tag{11.2}$$

where $\boldsymbol{B}$ is the $n \times p$ design matrix with entry $\boldsymbol{B}[i,j] = B_j(x_i)$, and λ is the smoothing parameter. Although it is not trivial to compute $\boldsymbol{Q}$, Wand and Ormerod (2008) extended the O-splines to higher orders of the derivative, and derived an exact matrix algebraic expression for the corresponding penalty function. They also showed how to derive the design matrices $\boldsymbol{X}$ and $\boldsymbol{Z}$ using the spectral decomposition of $\boldsymbol{Q}$, and represent the O-splines as a mixed model of form in (11.1). It's noteworthy that the simpler specification $\boldsymbol{X} = [1, x_i]$ can be used without affecting the fit, because the original $\boldsymbol{X}$ is a basis for the space of straight lines.

The O-splines are closely related to the P-splines of Eilers and Marx (1996). If the knots are taken to be equally spaced, the family of cubic P-splines is given by (11.2) with $\boldsymbol{Q}$ replaced by $\boldsymbol{D}_2^T \boldsymbol{D}_2$, where $\boldsymbol{D}_2$ is the differencing matrix of RW2 model. Therefore, the P-splines can also be analyzed under mixed model framework (e.g., Eilers et al., 2015). Wand and Ormerod (2008) compared those two spline methods from both theoretical and empirical perspectives. Eilers et al. (2015) conducted a thorough review on the P-splines, and compared it to the O-splines and the TPB splines in detail.

11.1.3 Example: Canadian Income Data

We in this section show how to use INLA to fit the TPB splines and the O-splines to the Canadian income data. The data have 205 pairs of observations on Canadian workers from a 1971 Canadian Census Public Use Tape. We first load the dataset:

```
library(INLA); library(brinla)
data(age.income, package = 'SemiPar')
str(age.income)
```

```
'data.frame':        205 obs. of  2 variables:
 $ age       : int  21 22 22 22 22 22 22 22 22 23 ...
 $ log.income: num  11.2 12.8 13.1 11.7 11.5 ...
```

where the variable `age` is worker's age in years, and `log.income` is the logarithm of worker's income. We would like to see whether there exists a possible nonlinear relationship between `log.income` and `age`.

TPB Splines

To fit TPB splines as a mixed model, we need to first produce two design matrices $\boldsymbol{X}$ and $\boldsymbol{Z}$ as in (11.1):

```
X <- spline.mixed(age.income$age, degree = 2, type = 'TPB')$X
Z <- spline.mixed(age.income$age, degree = 2, type = 'TPB')$Z
```

The function `spline.mixed()` is used to make design matrices for different types of splines. Here `'TPB'` type matrices are based on p (order of the polynomial) and r (number of knots). We here let $p = 2$ specified by `degree=2`, and use the default value for r although other values can be used via option `Nknots` in `spline.mixed()`. Then, we use the so-called "z" model in INLA to specify this mixed model:

```
age.income$ID <- 1:nrow(age.income)
formula <- log.income ~ -1 + X + f(ID, model = 'z', Z = Z, hyper =
    ↪ list(prec = list(param = c(1e-6, 1e-6))))
```

Here `ID` is the sequence of $1, 2, \ldots, n$, and `hyper` is used to specify the hyperprior on the precision $1/\sigma_u^2$. We here use a highly diffuse gamma prior with mean 1 and

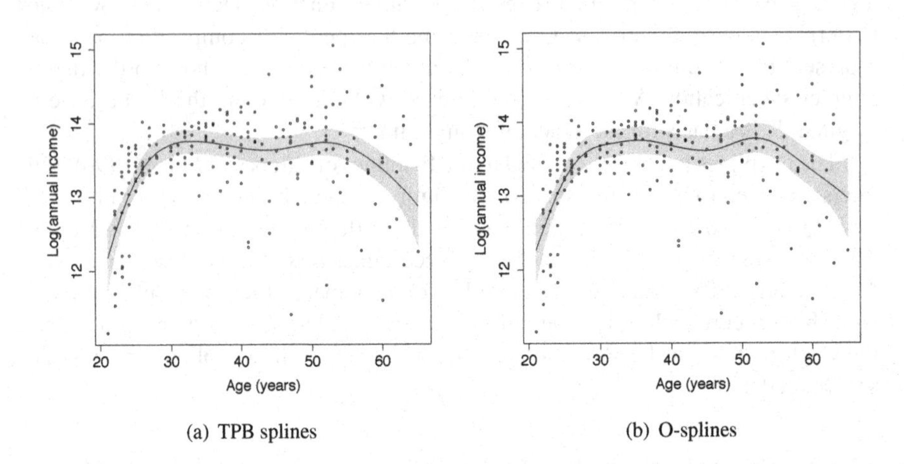

(a) TPB splines (b) O-splines

FIGURE 11.1
Canadian income data: fitted curve (solid) and 95% credible interval (gray) using TPB splines (left) and O-splines (right); black dots are data points.

variance 10^6. Note that we suppress the intercept in `formula` because it is already included in `X` matrix. Finally, let's fit the model with INLA:

```
result.tpb <- inla(formula, data = age.income, control.predictor =
    ↪ list(compute = TRUE))
```

The posterior summary of the function estimates can be retrieved from:

```
result.tpb$summary.fitted.values
```

In Figure 11.1(a) we show the fitted curve (posterior mean) and its 95% credible interval, together with the data points

```
bri.band.plot(result.tpb, x = age.income$age, alpha = 0.05, type = '
    ↪ fitted', xlab = 'Age (years)', ylab = 'Log(annual income)',
    ↪ ylim = range(age.income$log.income))
points(age.income$age, age.income$log.income, cex = 0.5)
```

Although they are not of interest in this case, the posterior quantities of $\boldsymbol{\beta}$ and $\boldsymbol{u}$ are given by `result.tpb$summary.fixed` and
`result.tpb$summary.random$ID[-(1:nrow(age.income)),]`
respectively.

O-splines

It is very similar to fit O-splines in INLA as we do TPB splines. We only need different design matrices:

```
X <- spline.mixed(age.income$age, type = 'OSS')$X
Z <- spline.mixed(age.income$age, type = 'OSS')$Z
```

Here we use `'OSS'` to specify the O-splines type of matrices. Then, we may use the same `formula` and `inla()` for fitting O-splines as for TPB splines:

```
age.income$ID <- 1:nrow(age.income)
formula <- log.income ~ -1 + X + f(ID, model = 'z', Z = Z, hyper =
    ↪ list(prec = list(param = c(1e-6, 1e-6))))
result.oss <- inla(formula, data = age.income, control.predictor =
    ↪ list(compute = TRUE))
```

We present the result in Figure 11.1(b). Compared to the TPB spline fit, the O-spline fit seems to be a little less smooth although the same knots and priors on the variance components are used.

We now present an alternative way to fit O-splines with INLA. Based on equation (11.2), it is intuitive to represent the O-splines as a Bayesian hierarchical model rather than a mixed model:

$$y \mid \beta, \sigma_\varepsilon^2 \sim N\left(B\beta, \sigma_\varepsilon^2 I\right), \quad \beta \mid \sigma_\beta^2 \sim N\left(0, \sigma_\beta^2 Q^{-1}\right), \tag{11.3}$$

where $\lambda = \sigma_\varepsilon^2/\sigma_\beta^2$ is the smoothing parameter. This model is very similar to the P-splines model, where the only difference is in the precision matrix in the prior of β. We therefore follow the INLA procedure used for P-splines to fit O-splines. Let's produce the B-spline basis matrix B in (11.3) and change it to the sparse matrix format as required by INLA:

```
B.tmp <- spline.mixed(age.income$age, type = 'OSS')$B
n.row <- nrow(B.tmp)
n.col <- ncol(B.tmp)
attributes(B.tmp) <- NULL
Bmat <- as(matrix(B.tmp, n.row, n.col), 'sparseMatrix')
```

Although it is not internally coded in INLA, the prior on β in (11.3) can be specified by first making its Q matrix:

```
Q <- as(spline.mixed(age.income$age, type = 'OSS')$Q, 'sparseMatrix')
```

and then using the so-called "`generic0`" model as follows:

```
formula <- y ~ -1 + f(x, model = 'generic0', Cmatrix = Q, hyper = list
    ↪ (prec = list(param = c(1e-6, 1e-6))))
```

Here `hyper` is used to specify the hyperprior for σ_β^2. We are now ready to fit the O-splines using INLA:

```
data.inla <- list(y = age.income$log.income, x = 1:ncol(Bmat))
result.oss2 <- inla(formula, data = data.inla, control.predictor =
    ↪ list(A = Bmat, compute = TRUE))
```

The posterior quantities regarding function estimation are saved in

```
result.oss2$summary.fitted.values[1:n.row, ]
```

We may visualize the fitted curve and the 95% credible interval using the following R commands:

```
bri.band.plot(result.oss2, ind = 1:n.row, x = age.income$age, alpha =
    ↪ 0.05, type = 'fitted', xlab = 'Age (years)', ylab = 'Log(annual
    ↪  income)', ylim = range(age.income$log.income))
points(age.income$age, age.income$log.income, cex = 0.5)
```

We notice that the resulting O-splines fit (not shown) is a little different from that using the mixed model approach. It is because in the two methods parameters σ_u^2 and σ_β^2 shrink different coefficients, and therefore apply different amounts of smoothing to the function fit, although they are put on the same prior (diffuse gamma prior).

It is trivial to apply TPB splines and O-splines to non-Gaussian data using INLA. The only change we need to make to the above R code is to specify the (non-Gaussian) likelihood using the `family` option in `inla()` function.

11.2 Analysis of Variance for Functional Data

Being associated with continuous time monitoring processes, functional data are usually smooth curves or surfaces and are often treated as realizations of underlying random functions. The basic philosophy of functional data is to consider the observed curves (or surfaces) as single entities, rather than only as a sequence of individual observations. Comprehensive surveys of statistical techniques for analyzing functional data can be found in Ramsay and Silverman (2005) and Ferraty and Vieu (2006).

We here focus on the functional analysis of variance (ANOVA) models, which allow us to study how the data differ with certain categorical factors. Let $y_{ik}(\boldsymbol{x})$ denote the kth functional observation in the ith group, and we define a one-way functional ANOVA model given factor A as

$$y_{ik}(\boldsymbol{x}) = \mu(\boldsymbol{x}) + \alpha_i(\boldsymbol{x}) + \varepsilon_{ik}(\boldsymbol{x}), \quad \boldsymbol{x} \in \mathcal{X} \subset \mathbf{R}^d, \tag{11.4}$$

where μ is the grand mean function, α_i is the ith-level main effect function, and ε_{ik} is the Gaussian process with mean zeros, for $i = 1,\ldots,m_A$ and $k = 1,\ldots,n$. To identify α_i from μ, we need certain linear constraints, such as $\sum_i \alpha_i(\boldsymbol{x}) = 0$ or $\alpha_{m_A}(\boldsymbol{x}) = 0$ for all $\boldsymbol{x} \in \mathcal{X}$. Obviously, this model is an extension of the ordinary one-way ANOVA model to the function space.

It is straightforward to generalize model (11.4) for two factors and even for non-Gaussian functional response. For example, let's consider two factors denoted by A and B. Let $y_{ijk}(\boldsymbol{x})$ denote a functional response at the ith level of A and j^{th} level of B from k^{th} subject, and assume it follows a distribution from the exponential family with mean $E(y_{ijk}) = \mu_{ij}$ and a canonical link function g. A (generalized) two-way functional ANOVA model is given by

$$g(\mu_{ijk}) = \mu(\boldsymbol{x}) + \alpha_i(\boldsymbol{x}) + \beta_j(\boldsymbol{x}) + \gamma_{ij}(\boldsymbol{x}), \tag{11.5}$$

for $i = 1,\ldots,m_A$, $j = 1,\ldots,m_B$ and $k = 1,\ldots,n_{ij}$. Here μ is the grand mean function, α_i is the ith-level main effect function of A, β_j is the jth-level main effect function of B, and γ_{ij} is the interaction function. For identifiability, we need a set of linear constraints on α_i, β_j and γ_{ij}, and one possible set is $\alpha_{m_A}(\boldsymbol{x}) = 0$, $\beta_{m_B}(\boldsymbol{x}) = 0$, $\gamma_{m_A,j}(\boldsymbol{x}) = 0$ and $\gamma_{i,m_B}(\boldsymbol{x}) = 0$ for all $\boldsymbol{x} \in \mathcal{X}$. It is straightforward to extend model (11.5) for more than two factors.

To do ANOVA for functional data, we need to first estimate the effect functions, and then test whether or not those functions are significantly different from 0. By taking on each effect function a Gaussian prior mentioned in Chapter 7 or 8, the functional ANOVA models fall in the class of latent Gaussian models, and hence can be fitted by INLA. We are also able to use the excursion method described in Section 7.8 to find which part of the function is non-zero with a significantly high probability. Below we present an example where we use INLA to fit a one-way ANOVA model to the continuous functional data with one-dimensional domain. However, INLA is able to deal with more complicated ANOVA models with more factors and/or discrete functional data (see Yue et al., 2018).

Example: Diffusion Tensor Imaging (DTI)

Goldsmith et al. (2012) study DTI metrics of multiple sclerosis (MS) patients over multiple clinical visits. The data consist of 100 subjects, aged between 21 and 70 years at first visit. The number of visits per subject ranges from 2 to 8, and a total of 340 visits were recorded. For each visit the DTI scans were obtained and used to create tract profiles. The derived outcome is known as fractional anisotropy (FA), which describes the degree of anisotropy of a diffusion process in a brain and has been routinely used as an index of white matter integrity. The FA tract profiles of two particular brain regions are contained in the data: the corpus callosum area (CCA) and the right cortico-spinal tract (RCST). The damage to either region may be linked with a decline in cognitive performance in MS patients. The same data have been analyzed in Yue et al. (2018).

Let's load the dataset and take a look at its structure:

```
data(DTI, package = 'refund')
str(DTI)
```

```
'data.frame':        382 obs. of  9 variables:
 $ ID        : num  1001 1002 1003 1004 1005 ...
 $ visit     : int  1 1 1 1 1 1 1 1 1 1 ...
 $ visit.time: int  0 0 0 0 0 0 0 0 0 0 ...
 $ Nscans    : int  1 1 1 1 1 1 1 1 1 1 ...
 $ case      : num  0 0 0 0 0 0 0 0 0 0 ...
 $ sex       : Factor w/ 2 levels "male","female": 2 2 1 1 1 1 1 1 1 1 ...
 $ pasat     : int  NA NA NA NA NA NA NA NA NA NA ...
 $ cca       : num [1:382, 1:93] 0.491 0.472 0.502 0.402 0.402 ...
  ..- attr(*, "dimnames")=List of 2
  .. ..$ : chr  "1001_1" "1002_1" "1003_1" "1004_1" ...
  .. ..$ : chr  "cca_1" "cca_2" "cca_3" "cca_4" ...
 $ rcst      : num [1:382, 1:55] 0.257 NaN NaN 0.508 NaN ...
  ..- attr(*, "dimnames")=List of 2
  .. ..$ : chr  "1001_1" "1002_1" "1003_1" "1004_1" ...
  .. ..$ : chr  "rcst_1" "rcst_2" "rcst_3" "rcst_4" ...
```

The data frame is made up of `ID` = subject ID numbers, `visit` = subject-specific visit numbers, `Nscans` = total number of visits for each subject, `case` = multiple sclerosis case status (0-control, 1-case), `cca` = 382×93 matrix of FA tract profiles from the CCA, and `rcst` = 382 × 55 matrix of FA tract profiles from the RCST. The descriptions of other variables can be found in the R document.

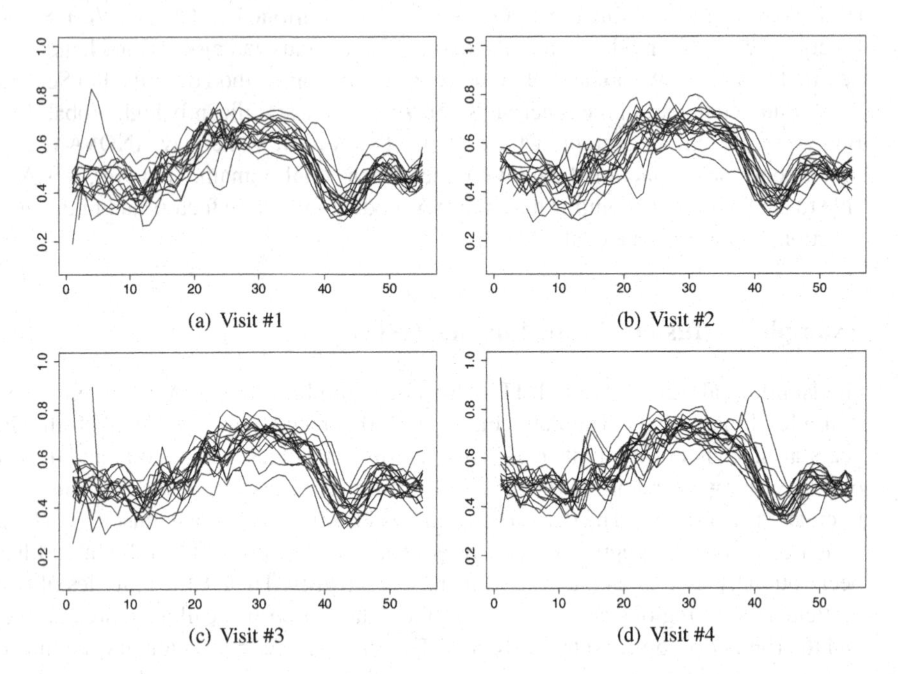

(a) Visit #1 (b) Visit #2

(c) Visit #3 (d) Visit #4

FIGURE 11.2
DTI of MS patients: FA tract profile of RCST for each patient at each visit.

There are two research questions we try to answer in this case study. The first question is: "Is there a significant functional effect of 'visit' on FA measures of MS patients?", and the second one is: "How does the significance of 'visit' change dynamically over the trace locations if it exists?" We therefore restrict our attention to the 18 patients, who had completed 4 visits within approximately one year:

```
DTI.sub <- DTI[DTI$Nscans==4 & DTI$case == 1, ]
```

We here only present the analyses of RCST for demonstration purposes. The CCA can be analyzed similarly. Let's plot the FA tract profiles of RCST for the first visit, which is presented in Figure 11.2(a):

```
dat.v1 <- DTI.sub[DTI.sub$visit==1,]$rcst
plot(dat.v1[1,], type = "n", ylim = c(0.1, 1))
for(i in 1:dim(dat.v1)[1]) lines(dat.v1[i,])
```

The plots for the other 3 visits are made similarly and shown in Figures 11.2(b), 11.2(b) and 11.2(d). From the discontinuities in the curves we know there are some missing values in the profiles. We take each FA tract profile as a functional observation and would like to study whether those functions have different mean patterns across the visits.

Let $y_{ij}(x)$ denote the FA measure at location x for j^{th} subject on i^{th} visit. Since

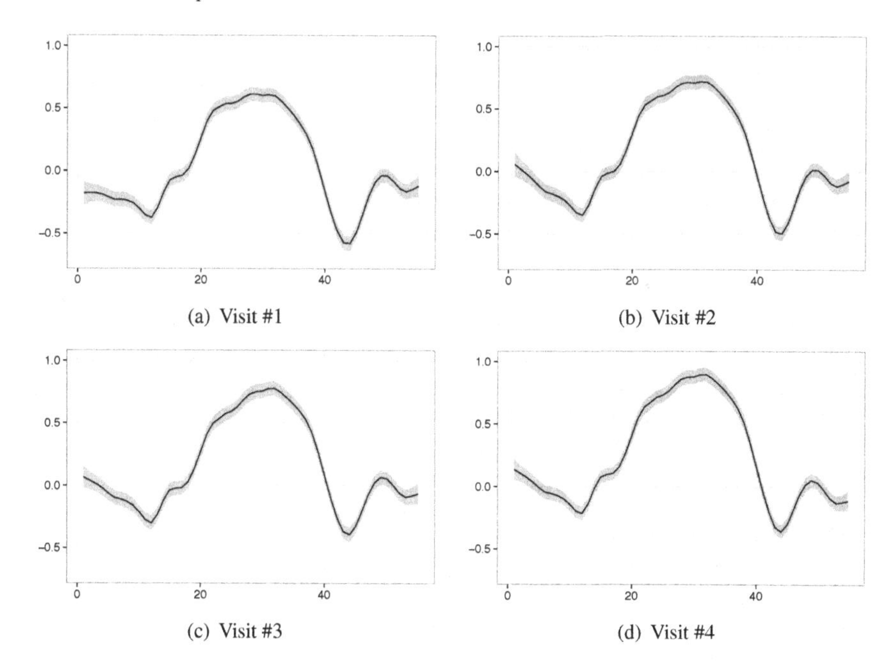

(a) Visit #1 (b) Visit #2

(c) Visit #3 (d) Visit #4

FIGURE 11.3
DTI of MS patients: mean function and 95% credible band of the tract profile of RCST for each visit.

it is bounded between 0 and 1, we assume $y_{ij}(x)$ follows a beta distribution, denoted by $Beta(p_{ij}, \tau)$, with mean p_{ij} and variance $p_{ij}(1-p_{ij})/(1+\tau)$. To investigate the "visit" effect, we use the following one-way ANOVA model:

$$\text{logit}(p_{ij}) = \mu(x) + \alpha_i(x), \quad i = 1,2,3,4, \tag{11.6}$$

where μ is the grand mean function and α_i is the main effect function of ith visit. For identifiability we here let $\alpha_4(x) = 0$ for all x, and therefore μ becomes the mean function for Visit #4. We can write down model (11.6) in matrix form:

$$\text{logit}(\boldsymbol{p}) = \boldsymbol{A}_\mu \boldsymbol{\mu} + \boldsymbol{A}_\alpha \boldsymbol{\alpha} = \boldsymbol{A}\boldsymbol{f},$$

where $\boldsymbol{A} = [\boldsymbol{A}_\mu, \boldsymbol{A}_\alpha]$ and $\boldsymbol{f} = [\boldsymbol{\mu}', \boldsymbol{\alpha}']'$. Here $\boldsymbol{A}_\mu$ and $\boldsymbol{A}_\alpha$ are incidence matrices used to map vectors $\boldsymbol{\mu}$ and $\boldsymbol{\alpha}$ to the vector $\boldsymbol{p}$. Since the locations of each tract profile are discrete, it is intuitive to assign RW2 priors for μ and α_i. Regarding precision τ, we use the default gamma prior as specified in INLA.

To fit the model let's first extract necessary information from the data:

```
ns <- dim(DTI.sub$rcst)[2]
ng <- length(unique(DTI.sub$visit))
n <- length(unique(DTI.sub$ID))
```

Here ns is the number of locations, ng is the number of visits, and n is the number of subjects. We produce matrices $\boldsymbol{A}_\mu$ and $\boldsymbol{A}_\alpha$:

```
D1 <- Matrix(rep(1,ng*n),ng*n,1)
A.mu <- kronecker(D1, Diagonal(n=ns, x=1))
D1 <- Diagonal(n = ns, x = 1)
D2 <- Diagonal(n = (ng-1), x = 1)
D3 <- Matrix(rep(0, ng-1), 1, ng-1)
D4 <- kronecker(rBind(D2, D3), D1)
A.a <- kronecker(Matrix(rep(1, n), n, 1), D4)
```

We combine them into matrix $\boldsymbol{A}$

```
A <- cBind(A.mu, A.a)
```

Then, we make the index vectors for $\boldsymbol{\mu}$, $\boldsymbol{\alpha}$ and $\boldsymbol{\alpha}$'s replicates required by INLA:

```
mu <- 1:ns
alpha <- rep(1:ns, ng-1)
alpha.rep <- rep(1:(ng-1), each = ns)
```

In addition, we need to add a few NA's to the vectors made above in order to match their lengths to the dimension of $\boldsymbol{A}$:

```
mu2 <- c(mu, rep(NA, length(alpha)))
alpha2 <- c(rep(NA, length(mu)), alpha)
alpha2.rep <- c(rep(NA, length(mu)), alpha.rep)
```

Finally, we may collect the vectors we need in a list, and express this one-way functional ANOVA model in INLA as follows:

```
y <- as.vector(t(DTI.sub$rcst))
data.inla <- list(y=y, mu=mu2, alpha=alpha2, alpha.rep=alpha2.rep)
formula <- y ~ -1 + f(mu, model = 'rw2', constr = FALSE, scale.model =
    ↪ T) + f(alpha, model = 'rw2', constr = FALSE, scale.model =
    ↪ TRUE, replicate = alpha.rep)
```

The estimated functions from this model are the mean function of Visit #4 (μ), and main effect functions of Visit #1, #2, and #3 (α_1, α_2 and α_3). If one is also interested in the mean functions of the first three visits, one may write down such functions as linear combinations of $\boldsymbol{\mu}$ and $\boldsymbol{\alpha}$, i.e., $\boldsymbol{A}_1\boldsymbol{\mu} + \boldsymbol{A}_2\boldsymbol{\alpha}$, and build them in INLA:

```
A1.lc <- kronecker(Matrix(rep(1,ng-1),ng-1,1), Diagonal(n=ns, x=1))
A2.lc <- Diagonal(n = (ng - 1)*ns, x = 1)
lc <- inla.make.lincombs(mu = A1.lc, alpha = A2.lc)
```

We then use `lincomb` option in `inla()` to estimate the linear combinations `lc` defined above, as well as functions $\boldsymbol{\mu}$ and $\boldsymbol{\alpha}$:

```
result <- inla(formula, data = data.inla, family = 'beta', control.
    ↪ predictor = list(A = A, compute = TRUE), control.compute = list
    ↪ (config = TRUE), lincomb = lc)
```

Now let's display the result. We first plot the fitted mean function and its 95% credible band for each visit:

```
bri.band.ggplot(result, ind = 1:ns, type = 'lincomb')
bri.band.ggplot(result, ind = 1:ns + ns, type = 'lincomb')
bri.band.ggplot(result, ind = 1:ns + 2*ns, type='lincomb')
bri.band.ggplot(result, name = 'mu', type = 'random')
```

Figure 11.3 shows the plots. As we can see, the tract profiles show similar patterns

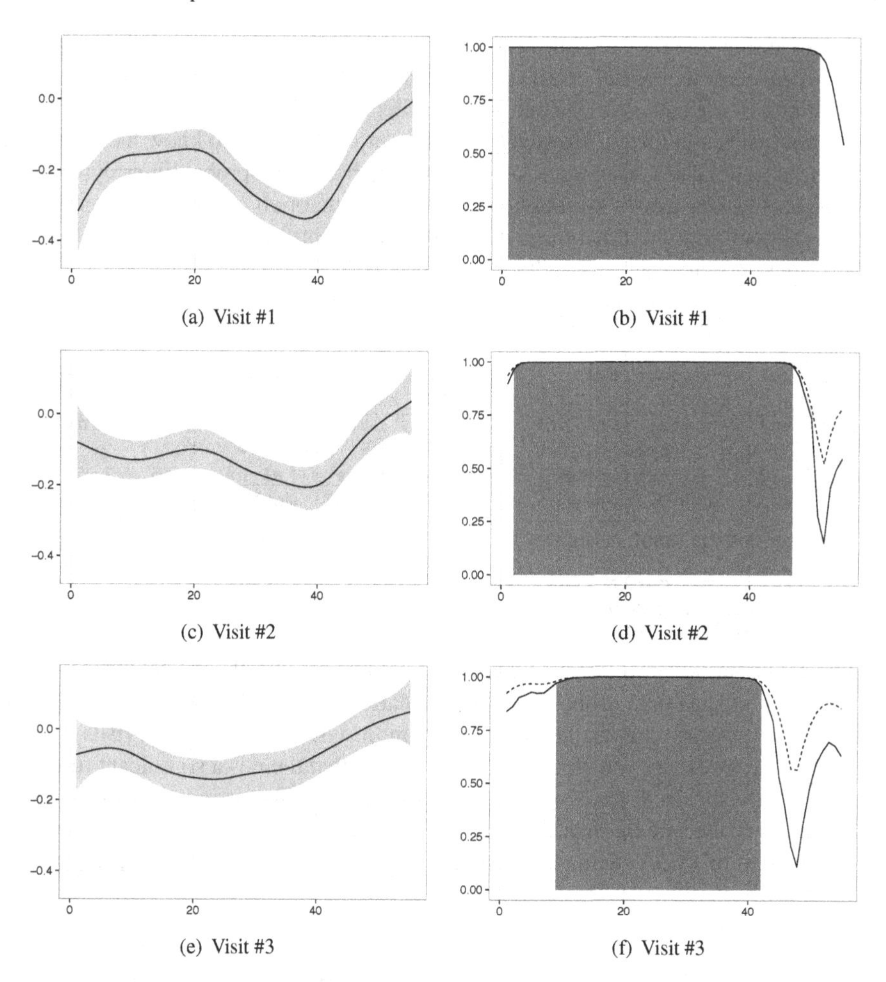

FIGURE 11.4
DTI of MS patients: main effect function and its 95% credible interval for each visit (left panels) and corresponding joint probabilities (solid) and marginal probabilities (dashed) that the main effect functions are non-zeroes, and the set of locations where the joint probabilities are at least 0.95 (gray) for each visit.

across the visits, but seem to shift upward as the number of visits increases. We also plot the main effect function for each of the first 3 visits to see how different they are from the last visit:

```
bri.band.ggplot(result, name='alpha', ind=1:ns, type='random')
bri.band.ggplot(result, name='alpha', ind=1:ns+ns, type='random')
bri.band.ggplot(result, name='alpha', ind=1:ns+2*ns, type='random')
```

From Figures 11.4(a), 11.4(c) and 11.4(e) we see the functions get more and more flattened, indicating the difference between each of the first three visits and the last visit becomes smaller as the patients pay more visits.

To answer the research questions mentioned earlier, we must understand how the significance of the "visit" effect on FA dynamically measures changes over the trace locations if it exists. We can see some evidence in Figures 11.3 and 11.4 that the main effect is significant. However, it will be more desirable to know the probability of the significance at each trace location and see how those probabilities dynamically change. In other words, we need to find for each main effect function a set of locations D such that the function is not zero with a significant joint probability for all locations in that region, i.e., $P(\alpha_i(x) \neq 0) \geq 0.95$ for all $x \in D$ and $i = 1,2,3$. We therefore implement the excursion method introduced in Section 7.8 to each main effect function:

```
res.exc1 <- excursions.brinla(result, name = 'alpha', ind = 1:ns, u =
    ↪ 0, type = '!=', alpha = 0.05, method = 'NIQC')
res.exc2 <- excursions.brinla(result, name = 'alpha', ind = 1:ns + ns,
    ↪  u = 0, type = '!=', alpha = 0.05, method = 'NIQC')
res.exc3 <- excursions.brinla(result, name = 'alpha', ind = 1:ns + 2*
    ↪ ns, u = 0, type = '!=', alpha = 0.05, method = 'NIQC')
```

And we plot the results (Figures 11.4(b), 11.4(d) and 11.4(f)):

```
bri.excursions.ggplot(res.exc1)
bri.excursions.ggplot(res.exc2)
bri.excursions.ggplot(res.exc3)
```

For each visit the gray region shows the set of locations such that the function is not zero with at least 0.95 probability for all locations in that region. As we can see, the region keeps shrinking as the patients pay more visits, and the three visits have the common region between 10 and 40. It concludes that there is a significant functional effect of "visit." In terms of its dynamic change, the effect is more significant at the locations in the middle than those at two ends. It indicates that for MS patients the middle part of RCST seems to be damaged more quickly than the two end parts.

11.3 Extreme Values

Most statistical modeling is concerned with the mean response but in some applications, the extreme values of the response are the main interest. For example, when considering insurance claims, the maximum claim is of particular interest since the insurance company must have ready access to funds to pay such a claim. Extreme values, as the name implies, are usually generated from the maximum (or minimum) of a large collection of random variables. The *generalized extreme value* distribution is a flexible way to model extreme values with distribution function:

$$F(y|\mu,\tau,\xi) = \exp[-(1+\sqrt{\tau s}(y-\mu))^{-1/\xi}].$$

This is defined for values of y such that $1+\sqrt{\tau s}(y-\mu) > 0$. The μ is a location parameter which we might link to a linear predictor. τ is the precision and ξ is a shape parameter. The scale s is not a parameter but is a fixed value which we will need for scaling purposes. The shape parameter ξ determines the particular type of extreme value distribution:

1. In the limit as $\xi \to 0$, we have a *Gumbel* distribution. This distribution can arise from the maximum of a large number of exponential random variables.
2. For $\xi > 0$, we have a *Fréchet* distribution.
3. For $\xi < 0$, we have a *Weibull* distribution. But the shape will be reversed from the usual Weibull in that there will be no lower bound but there will be an upper bound.

The Gumbel distribution is unrestricted while the Fréchet distribution has a lower bound.

Extreme flows in rivers are of special interest since flood defenses must be designed with these in mind. The National River Flow Archive provides data about river flows in the United Kingdom. For this example, we consider data on annual maximum flow rates from the River Calder in Cumbria, England, from 1973 to 2014.

```
data(calder, package="brinla")
plot(Flow ~ WaterYear, calder)
```

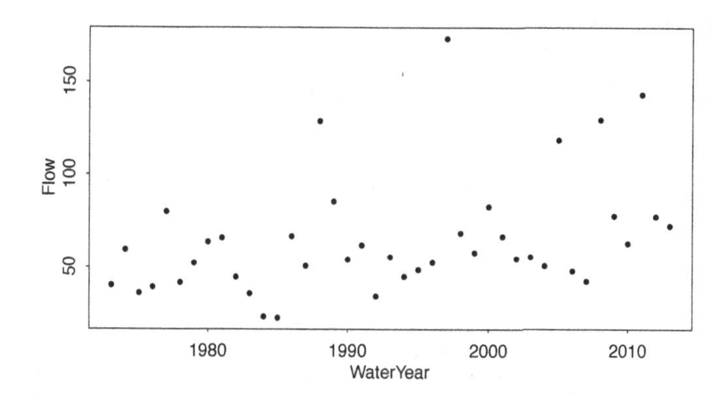

FIGURE 11.5
Annual maximum flow rates on the River Calder in Cumbria.

The data, as plotted in Figure 11.5, show the maximum flows during the period of observation. There is some indication that maximum flows may be increasing, perhaps due to changing land use in the catchment area of the river. It would not be sensible to use a Gaussian linear model here as extreme values do not follow a Gaussian distribution. In this case, we can see that the response has a skewed distribution. Furthermore, we want to make predictions about future extreme flows and the Gaussian distribution is unlikely to be suitable for the tail probabilities. The generalized extreme value distribution can be used for the observed responses y_i, $i = 1, \ldots, n$ with

$$\mu_i = \eta_i = \beta_0 + \beta_1 \text{year}_i.$$

The default priors for the fixed effects, β, are the standard flat distributions while the precision τ has a diffuse gamma distribution. The default prior on ξ is $N(0,16)$. An SD of 4 is appropriate for ξ as we expect this parameter to be moderate in size. We accept these default priors here. For convenience, we shift the water year predictor to start from 1973 as year zero. We fit the INLA model, taking care to first load the packages we will need for this section:

```
library(INLA); library(brinla)
calder$year <- calder$WaterYear-1973
imod <- inla(Flow ~ 1+year, data=calder, family="gev",scale=0.1)
```

In this case, the scale, s, needs to be used — we have set `scale=0.1`. Without this, the fitting algorithm fails to converge. The generalized extreme value distribution is problematic to fit so the need for some data conditioning is not surprising. We recommend you choose the scale to reduce the response to single digits. Some experimentation may be necessary.

First we consider the fixed effects:

```
imod$summary.fixed
```

```
                mean      sd 0.025quant 0.5quant 0.975quant     mode        kld
(Intercept) 36.71617 5.82063   25.18051 36.72028    48.1901 36.71110 6.1575e-13
year         0.75902 0.24656    0.26552  0.76149     1.2404  0.76786 4.6615e-12
```

The slope parameter has a posterior which is quite clearly positive. We can compute the probability of it being negative as:

```
inla.pmarginal(0, imod$marginals.fixed$year)
```

```
[1] 0.0019135
```

So there is strong evidence that maximum flows are increasing for this river over time.

We can display the posterior distributions for the hyperparameters, as seen in Figure 11.6:

```
plot(bri.hyper.sd(imod$marginals.hyperpar$`precision for GEV
    ↪ observations`), type="l", xlab="SD", ylab="density")
plot(imod$marginals.hyperpar$`shape-parameter for gev observations`,
    ↪ type="l", xlim=c(-0.2,0.5), xlab="xi", ylab="density")
```

The shape parameter is most interesting as this has a large impact on the shape of the distribution. We see that the posterior is mostly concentrated on positive values of ξ but there is a small chance that ξ is negative.

The maximum flow over the period of observation occurred in the 1997 water year measuring 173.17 m^3/s. Under our fitted model, what was the probability of observing such a flow (or greater)? This will give us a measure of how unusual this event was. First we need an R function to compute $P(Y < y)$ for the generalized extreme value distribution:

```
pgev <- function(y,xi,tau,eta,sigma=1){
  exp(-(1+xi*sqrt(tau*sigma)*(y-eta))^(-1/xi))
}
```

Now we can make the calculation using the posterior means of the parameters:

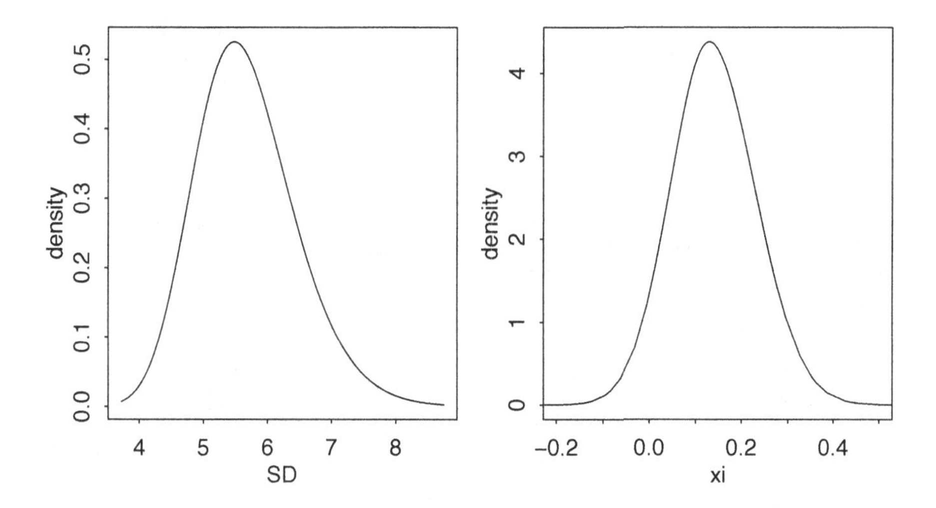

FIGURE 11.6
Posterior distribution of the hyperparameters. The SD (derived from the precision τ) is shown on the left while the shape ξ is shown on the right.

```
yr <- 1997-1973
maxflow <- 173.17
eta <- sum(c(1,yr)*imod$summary.fixed$mean)
tau <- imod$summary.hyperpar$mean[1]
xi <- imod$summary.hyperpar$mean[2]
sigma <- 0.1
pless <- pgev(maxflow, xi, tau, eta,sigma)
1-pless
```

```
[1] 0.0085947
```

We can see the observed event should be considered quite rare. Hydrologists often work with the expected time for the event to occur called the *recurrence interval.* In this case, the value is:

```
1/(1-pless)
```

```
[1] 116.35
```

Hence, we would expect such a flood to recur about every 116 years. There is concern that river flooding is becoming more common in the UK with global warming and changing land use. For this river, there is an increasing trend in flow rates. If we recompute the recurrency interval for 2017, we find this drops to 74 years which is no longer quite so exceptional.

These are point estimates but we should view the recurrency interval as a random variable whose distribution we can construct from the posterior distribution of the parameters. The computation of the recurrency interval requires all four parameters so we must consider the full joint posterior distribution. We cannot get by with just the marginal distributions because the parameters are likely to be correlated. In principle, the computation could be done exactly using the joint posterior but this cannot be produced from INLA. An easier solution is to generate samples from the joint posterior and use these to estimate the distribution of the recurrency interval. The function `inla.posterior.sample()` can be used to achieve this. We generate 999 samples:

```
nsamp <- 999
imod <- inla(Flow ~ 1+year, data=calder, family="gev",scale=0.1,
    ↪ control.compute = list(config=TRUE))
postsamp <- inla.posterior.sample(nsamp, imod)
pps <- t(sapply(postsamp, function(x) c(x$hyperpar, x$latent[42:43])))
colnames(pps) <- c("precision","shape","beta0","beta1")
```

We get a sample of all the latent variables — 41 for each of the cases and two more which are β_0 and β_1. We only want the β here. The matrix `pps` has four columns for the four parameters. We plot the hyperparameters, ξ and τ only in Figure 11.7. The multiplication by $0.01 = 0.1^2$ is required due to the internal scaling applied to ξ.

```
plot(shape*0.01 ~ precision, pps, ylab="shape")
```

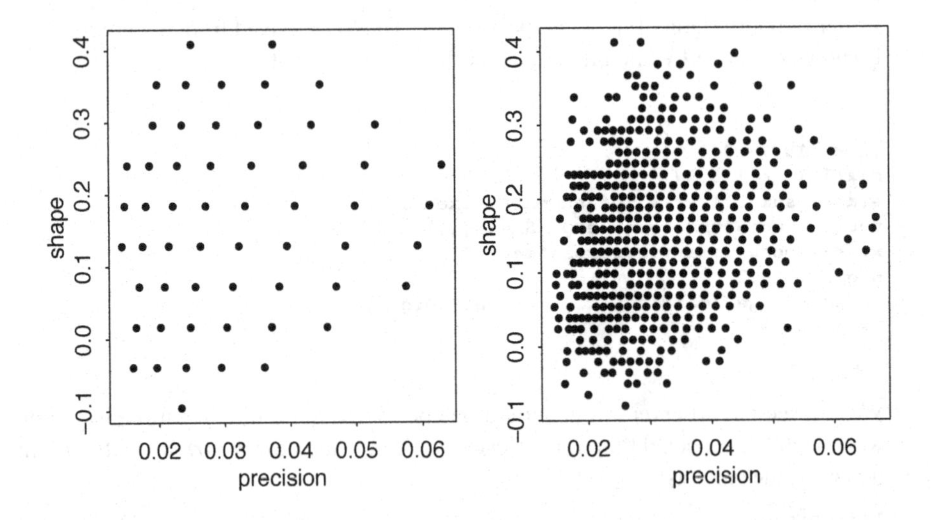

FIGURE 11.7
Samples from the posterior of the hyperparameters. The default is shown on the left while the shorter step length is shown on the right.

We are surprised to find that the samples are confined to a grid of points. This is explained by the approximation methods used by INLA. The posteriors for the latent variables are computed on a grid of points in the space of the hyperparameters. We can obtain a standard sample from the posterior of the hyperparameters using `inla.hyperpar.sample()` but this would not allow for the dependence with the latent variables. Since the gridded sample of the hyperparameters properly reflects the posterior on the hyperparameters, computations using this sample will not be as inaccurate as one might fear. Nevertheless, one might hope to do better. We can use a finer grid in the INLA computation. The value dz is the step-length in the standarized scale for the integration of the hyperparameters. We reduce the default choice of 0.75 to 0.2:

```
imod <- inla(Flow ~ 1+year, data=calder, family="gev",scale=0.1,
   ↪ control.compute = list(config=TRUE),control.inla=list(int.
   ↪ strategy='grid', dz=0.2))
postsamp <- inla.posterior.sample(nsamp, imod)
pps <- t(sapply(postsamp, function(x) c(x$hyperpar, x$latent[42:43])))
colnames(pps) <- c("precision","shape","beta0","beta1")
plot(shape*0.01 ~ precision, pps, ylab="shape")
```

The plot shown in the second panel of Figure 11.7 is more dense. We now compute the recurrency interval for each sample and find some quantiles of interest:

```
sigma <- 0.1
maxflow <- 173.17
retp <- numeric(nsamp)
for(i in 1:nsamp){
  eta <- sum(c(1,yr)*pps[i,3:4])
  tau <- pps[i,1]
  xi <- 0.01*pps[i,2]
  pless <-  pgev(maxflow, xi, tau, eta,sigma)
  retp[i] <- 1/(1-pless)
}
quantile(retp, c(0.025, 0.5, 0.975))
```

```
   2.5%     50%   97.5%
 20.908  70.351 512.848
```

We see that 95% credible interval runs from 21 to 513 years which is rather wide. Nevertheless, the lower end of this interval is quite low, causing some concern if we are worried about flooding. If we still have some anxiety about the accuracy of this interval, we have two ways to improve it. We can take more samples — this is relatively inexpensive. We can further reduce the step length dz, which will increase the computation time significantly. For a single small dataset like this, we can easily afford this but in larger problems, this may be an obstacle.

11.4 Density Estimation Using INLA

Nonparametric density estimation can also be implemented using INLA. Brown et al. (2010) proposed a "root-unroot" density estimation procedure, which turned density estimation into a nonparametric regression problem. The regression problem was created by binning the original observations into suitable size of bins and applying a mean-matching variance stabilizing root transform to the binned data counts. Then, a wavelet block thresholding regression was used to obtain the density estimate. Here we adopt Brown et al. (2010)'s root-unroot procedure but use a second-order random walk model with INLA for the regression step. The second-order random walk model is particularly suitable for an equi-spaced nonparametric time series regression problem (Fahrmeir and Knorr-Held, 2000). See more discussions about random walk models in Chapter 7. There are two advantages to use the Bayesian nonparametric approach. First, we avoid the smoothing parameter selection, where the smoothness of curve is automatically determined by the Bayesian model fitting. Second, it is straightforward to construct the credible bound of the regression curve. As a result, constructing the credible band for the probability density function becomes a natural by-product in the density estimation. Let $\{x_1,...,x_n\}$ be a random sample from a distribution with the density function f_X. The estimation algorithm is summarized as follows.

1. *Poissonization.* Divide $\{x_1,...,x_n\}$ in T equal length intervals. Let $C_1,...,C_T$ be the count of observations in each of the intervals.
2. *Root Transformation.* Apply the mean-matching variance stabilizing root transform, $y_j = \sqrt{C_j + 1/4}, j = 1,...,T$.
3. *Bayesian Smoothing with INLA.* Consider the time series $y = (y_1,...,y_T)$ to be the sum $y_j = m_j + \varepsilon_j, j = 1,...,T$ of a smooth trend function $m(\cdot)$ and a noise component ε. Fit a second-order random walk model with INLA for the equi-spaced time series to obtain a posterior mean estimate $\hat{m}$ of m, and $\alpha/2$ and $1-\alpha/2$ quantiles, $\hat{m}_{\alpha/2}$ and $\hat{m}_{1-\alpha/2}$.
4. *Unroot Transformation and Normalization.* The density function f_X is estimated by
$$\hat{f}_X(x) = \gamma[\hat{m}(x)]^2,$$
and the $100(1-\alpha)\%$ credible bands of $f(x)$ is
$$(\gamma[\hat{m}_{\alpha/2}(x)]^2, \gamma[\hat{m}_{1-\alpha/2}(x)]^2)$$
where $\gamma = (\int \hat{f}_X dx)^{-1}$ is a normalization constant.

In Step 3, we need to fit a nonparametric smooth function $m(\cdot)$ from the "pseudo" time series $y = (y_1,...,y_T)$. Wahba (1978) showed that the smoothing spline is equivalent to Bayesian estimation with a partially improper prior. $m(z)$ has the prior distribution which is the same as the distribution of the stochastic process

$$S(z) = \theta_0 + \theta_1 z + b^{1/2} V(z),$$

where $\theta_0, \theta_1 \sim N(0,\zeta)$, $b = \sigma^2/\lambda$ is fixed, and $V(z)$ is the one-fold integrated Wiener process,

$$V(z) = \int_0^z (z-t)dW(t).$$

Thus, estimating $m(z)$ becomes to seek the solution to the stochastic differential equation

$$\frac{d^2 m(z)}{dz^2} = \lambda^{-1/2}\sigma\frac{dW(z)}{dz}, \tag{11.7}$$

as a prior over m. Note that such a differential equation of order two is the continuous-time version of a second-order random walk. However, the solution of (11.7) does not have any Markov properties. The precision matrix is dense, hence it is computationally intensive. Lindgren and Rue (2008b) suggested a Galerkin approximation to $m(z)$, as the solution of (11.7). To be specific, let $z_1 < z_2 < ... < z_n$ be the set of fixed points, a finite element representation of $m(z)$ is constructed as

$$\tilde{m}(z) = \sum_{i=1}^{n} \psi_i(z) w_i,$$

for the piecewise linear basis functions ψ_i's and random weights w_i's.

In order to estimate the smooth function $m(z)$, one needs to determine the joint distribution of the weights $\boldsymbol{w} = (w_1, ..., w_n)^T$. Using the Galerkin method, $\boldsymbol{w}$ is derived as a GMRF with mean zero and precision matrix $\boldsymbol{G}$. Let $d_i = z_{i+1} - z_i$ for $i = 1, ..., n-1$, and $d_{-1} = d_0 = d_n = d_{n+1} = \infty$. The $n \times n$ symmetric matrix $\boldsymbol{G}$ is defined as

$$\boldsymbol{G} = \begin{pmatrix} g_{11} & g_{12} & g_{13} & & & 0 & \\ g_{21} & g_{22} & g_{23} & g_{24} & & & \\ g_{31} & g_{32} & g_{33} & g_{34} & g_{35} & & \\ & \ddots & \ddots & \ddots & \ddots & \ddots & \\ & & g_{n-2,n-4} & g_{n-2,n-3} & g_{n-2,n-2} & g_{n-2,n-1} & g_{n-2,n} \\ & 0 & & g_{n-1,n-3} & g_{n-1,n-2} & g_{n-1,n-1} & g_{n-1,n} \\ & & & & g_{n,n-2} & g_{n,n-1} & g_{n,n} \end{pmatrix},$$

where the non-zero elements of row i are given by

$$g_{i,i-2} = \frac{2}{d_{i-2}d_{i-1}(d_{i-2}+d_{i-1})},$$

$$g_{i,i-1} = \frac{-2}{d_{i-1}^2}\left(\frac{1}{d_{i-2}} + \frac{1}{d_i}\right),$$

$$g_{i,i} = \frac{2}{d_{i-1}^2(d_{i-2}+d_{i-1})} + \frac{2}{d_{i-1}d_i}\left(\frac{1}{d_{i-1}} + \frac{1}{d_i}\right) + \frac{2}{d_i^2(d_i+d_{i+1})},$$

with $g_{i,i+1} \equiv g_{i+1,i}$ and $g_{i,i+2} \equiv g_{i+2,i}$ due to symmetry. $\boldsymbol{G}$ is a sparse matrix with rank $n-2$, making the model computationally effective.

If we assign $\tilde{m}$ as a smoothness prior over m, the cubic smoothing spline $\hat{m}(z)$ at z coincides with the posterior expectation of $m(z)$ given the data, i.e., $\hat{m}(z) \approx E(m(z)|y)$. Therefore, the nonparametric regression problem becomes to fit a latent Gaussian model. It can be accomplished using INLA since $\boldsymbol{w}$ is a GMRF. The implementation of the method needs some extra programming in `R` to define a user-specified GMRF.

The R function, `bri.density` in our library `brinla` implements the above root-unroot algorithm. The argument `x` is a numeric vector of data values, `m` is the number of equally spaced points at which the density is to be estimated, `from` and `to` are the left- and right-most points of the grid at which the density is to be estimated. If `from` and to are missing, from equals the minima of the data values minus cut times the range of the data values and to equals the maxima of the data values plus cut times the range of the data values. In the following, we show two simulated examples to compare the INLA method and conventional kernel density estimation. In the first example data are generated from the standard normal distribution, $X \sim N(0,1)$, with sample size $n = 500$:

```
library(brinla)
set.seed(123)
n <- 500
x <- rnorm(n)
x.den1 <- bri.density(x, cut = 0.3)
x.den2 <- density(x, bw = "SJ")

curve(dnorm(x, mean = 0, sd = 1), from = -4, to = 4, lwd = 3, lty = 3,
    ↪ xlab = "x", ylab = "f(x)", cex.lab = 1.5, cex.axis = 1.5, ylim
    ↪ =c(0, 0.45))
lines(x.den1, lty = 1, lwd = 3, ylim = c(0, 0.25))
lines(x.den1$x, x.den1$y.upper, lty = 2, lwd = 3)
lines(x.den1$x, x.den1$y.lower, lty = 2, lwd = 3)
lines(x.den2, lty = 4, lwd = 3)
```

In the second example data are generated from a normal mixture model, $X \sim 0.5N(-1.5,1) + 0.5N(2.5, 0.75^2)$, with sample size $n = 1000$:

```
set.seed(123)
n <- 1000
x <- c(rnorm(n/2, mean = -1.5, sd = 1), rnorm(n/2, mean = 2.5, sd =
    ↪ 0.75))
x.den1 <- bri.density(z, cut = 0.3)
x.den2 <- density(z, bw = "SJ")

curve(dnorm(x, mean = -1.5, sd = 1)/2 + dnorm(x, mean = 2.5, sd =
    ↪ 0.75)/2, from = -6, to = 6, lwd = 3, lty = 3, xlab = "x", ylab
    ↪ = "f(x)", ylim=c(0, 0.3), cex.lab = 1.5, cex.axis = 1.5)
lines(x.den1, lty = 1, lwd = 3)
lines(x.den1$x, x.den1$y.upper, lty = 2, lwd = 3)
lines(x.den1$x, x.den1$y.lower, lty = 2, lwd = 3)
lines(x.den2, lty = 4, lwd = 3)
```

Figure 11.8 displays the estimation results. The estimates using INLA are denoted by the solid lines, and the 95% credible bands are denoted by the dashed lines. The kernel density estimates with Sheather and Jones (1991)'s plug-in bandwidth are denoted by the dash-dotted lines. The true functions are denoted by the dotted lines. We note that the INLA estimates are very close to the kernel density estimates. The INLA approach allows us to compute the credible bands of the density function without additional computational effort.

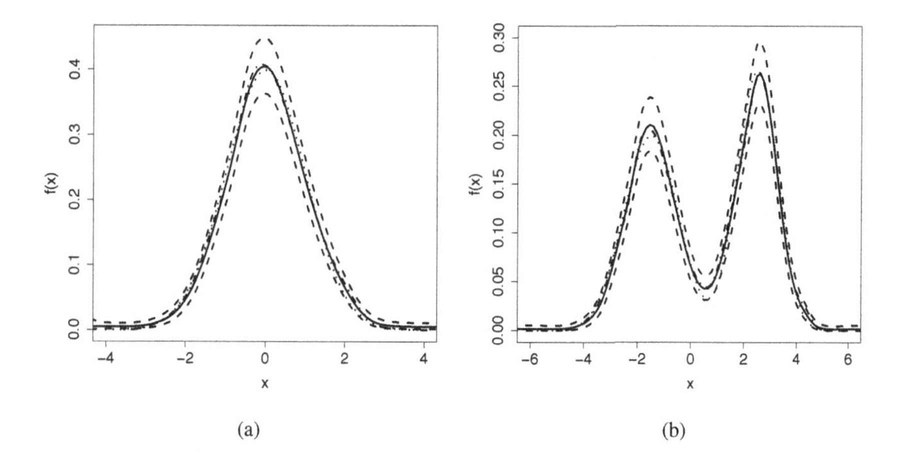

(a) (b)

FIGURE 11.8
Simulated examples for density estimation: (a) $X \sim N(0,1)$, $n = 500$; (b) $X \sim 0.5N(-1.5,1) + 0.5N(2.5,0.75^2)$, $n = 1000$. The estimates (solid lines) using INLA are compared to the estimates (dash-dotted lines) using kernel density estimation with Sheather and Jones (1991)'s plug-in bandwidth. The true functions are denoted by the dotted lines.

A

Installation

R

In this book, we prepare the data, call INLA and process the results using R. Hence you need R installed to get started. You can obtain it from:

```
www.r-project.org
```

You can also obtain a wide selection of introductory guides to R at the same site. Although R comes with its own GUI, you may find that the Rstudio interface offers many more features obtainable from:

```
www.rstudio.com
```

R-INLA

The heart of INLA is written in C but these routines are accessed using the `INLA` package. The `INLA` package is not available from CRAN where most R packages may be found. Instead it must be downloaded from the R-INLA website at:

```
www.r-inla.org
```

where installation instructions may be found. A large amount of information regarding the use of R-INLA may be found at this website. In particular, there is a Google group entitled "R-inla discussion group" in which users may find the answer to many questions.

brinla

We have packaged the functions we have written and some of the data used in this book into an R package called `brinla`. You can install it from within R using:

```
library(devtools)
install_github("julianfaraway/brinla")
```

You will need to do this only once per computer but in each R session you run on the examples in this book, you will need:

```
library(brinla)
```

We have prefaced our functions with `bri` so you know where they come from. If you receive a message saying that one of these functions has "not been found," it means you have not installed `brinla` or not run the `library(brinla)` command in the current session.

Book website

You can find the website for the book at

`julianfaraway.github.io/brinla/`

There you can find the R scripts and errata (or updates) for the book. If you spot an error or have a question about the content, please contact us depending on the chapter. Xiaofeng Wang for Chapters 3, 4, 6, 10 and 11.4; Ryan Yue for Chapters 2, 7, 9, 11.1 and 11.2; Julian Faraway for Chapters 1, 5, 8 and 11.3.

Other R packages

We use a number of other R packages in this book. You may choose to install these as needed but here is a list: `GGally MASS R2jags betareg dplyr faraway fields ggplot2 gridBase gridExtra lme4 mgcv reshape2 splines tidyr survival`

INLA special features

INLA has a number of features which are not linked to any particular model, likelihood or prior but which can be useful in many different situations.

Define a prior The INLA website lists a good number of priors which are usually adequate for most situations but it is possible to define your own prior. An example may be found in Section 5.2.1.

Copy Sometimes you have two random components in a model that need to be linked together. You can achieve this with the so-called "copy" feature. In truth, the components are linked not copied. You can find examples in Sections 3.7 and 5.3.2.

Replicate If you have two random components that depend on the same hyperparameter, you need the "replicate" feature as seen in Section 5.7.

Linear combination of predictor When the linear predictor for the response is formed from another linear combination of predictors, INLA provides a way for achieving this. An example is seen in Section 5.4.1.

B

Uninformative Priors in Linear Regression

Uninformative priors are commonly used in Bayesian linear regression. In this appendix, we discuss the uninformative priors and compare different modeling approaches. We begin our discussion by assuming that the errors are normal and homoscedastic, i.e., the error term ε in (3.2) is assumed to be distributed as $N(0,\sigma^2\mathbf{I})$ with an unknown parameter σ. The likelihood function for the model (3.2) is

$$L(\beta,\sigma^2|\mathbf{X},\mathbf{y}) = \left(\frac{1}{\sqrt{2\pi}\sigma}\right)^n \exp\left[-\frac{1}{2\sigma^2}(\mathbf{y}-\mathbf{X}\beta)^T(\mathbf{y}-\mathbf{X}\beta)\right]. \tag{B.1}$$

In the frequentist approach to the model (3.2), the well-known ordinary least squares estimator of β is

$$\hat{\beta} = (\mathbf{X}^T\mathbf{X})^{-1}\mathbf{X}^T\mathbf{y}, \tag{B.2}$$

and the estimator of σ^2 is given by

$$\hat{\sigma}^2 = \frac{(\mathbf{y}-\mathbf{X}\hat{\beta})^T(\mathbf{y}-\mathbf{X}\hat{\beta})}{n-p-1}. \tag{B.3}$$

We can plug the two estimates (B.2) and (B.3) into (B.1) and process according to:

$$\begin{aligned}
L(\beta,\sigma^2|\mathbf{X},\mathbf{y}) &\propto \sigma^{-n}\exp\left[-\frac{1}{2\sigma^2}\left(\mathbf{y}^T\mathbf{y}-2\beta^T\mathbf{X}^T\mathbf{y}+\beta^T\mathbf{X}^T\mathbf{X}\beta\right)\right]\\
&= \sigma^{-n}\exp\left[-\frac{1}{2\sigma^2}\left(\mathbf{y}^T\mathbf{y}-2\beta^T\mathbf{X}^T\mathbf{y}+\beta^T\mathbf{X}^T\mathbf{X}\beta\right.\right.\\
&\quad\left.\left.-2((\mathbf{X}^T\mathbf{X})^{-1}\mathbf{X}^T\mathbf{y})^T\mathbf{X}^T\mathbf{y}+2((\mathbf{X}^T\mathbf{X})^{-1}\mathbf{X}^T\mathbf{y})^T(\mathbf{X}^T\mathbf{X})(\mathbf{X}^T\mathbf{X})^{-1}\mathbf{X}^T\mathbf{y}\right)\right]\\
&= \sigma^{-n}\exp\left[-\frac{1}{2\sigma^2}\left((\mathbf{y}-\mathbf{X}\hat{\beta})^T(\mathbf{y}-\mathbf{X}\hat{\beta})\right.\right.\\
&\qquad\left.\left.+\hat{\beta}^T\mathbf{X}^T\mathbf{X}\hat{\beta}+\beta^T\mathbf{X}^T\mathbf{X}\beta-2\beta^T\mathbf{X}^T\mathbf{X}\hat{\beta}\right)\right]\\
&= \sigma^{-n}\exp\left[-\frac{1}{2\sigma^2}\left(\hat{\sigma}^2(n-p-1)+(\beta-\hat{\beta})^T\mathbf{X}^T\mathbf{X}(\beta-\hat{\beta})\right)\right].
\end{aligned} \tag{B.4}$$

To conduct Bayesian inference for the linear model, we are required to specify the prior distributions for the unknown parameters β and σ^2. The simple priors are the improper uninformed priors

$$p(\beta)\propto c, \quad p(\sigma^2)\propto 1/\sigma^2.$$

Here we assume β and σ^2 are independent. Therefore, the joint posterior distribution $\pi(\beta,\sigma^2|\mathbf{X},\mathbf{y})$ is provided by

$$\begin{aligned}\pi(\beta,\sigma^2|\mathbf{X},\mathbf{y}) &\propto L(\beta,\sigma^2|\mathbf{X},\mathbf{y})p(\beta)p(\sigma^2)\\ &\propto \sigma^{-n-2}\exp\left[-\frac{1}{2\sigma^2}\left(\hat{\sigma}^2(n-p-1)+(\beta-\hat{\beta})^T\mathbf{X}^T\mathbf{X}(\beta-\hat{\beta})\right)\right].\end{aligned} \tag{B.5}$$

Integrating (B.5) with respect to σ^2 to get the marginal distribution for β, we have

$$\pi(\beta|\mathbf{X},\mathbf{y}) \propto \int \sigma^{-n-2}\exp\left[-\frac{1}{2\sigma^2}\left(\hat{\sigma}^2(n-p-1)+(\beta-\hat{\beta})^T\mathbf{X}^T\mathbf{X}(\beta-\hat{\beta})\right)\right]d\sigma^2.$$

Note that the integrand here is the kernel of an inverse gamma distribution for σ^2. So, after certain simplification, one can obtain

$$\begin{aligned}\pi(\beta|\mathbf{X},\mathbf{y}) &\propto \Gamma(n/2)\left[\frac{1}{2}\left(\hat{\sigma}^2(n-p-1)+(\beta-\hat{\beta})^T\mathbf{X}^T\mathbf{X}(\beta-\hat{\beta})\right)\right]^{-n/2}\\ &\propto \left[1+\frac{(\beta-\hat{\beta})^T\hat{\sigma}^2\mathbf{X}^T\mathbf{X}(\beta-\hat{\beta})}{n-p-1}\right]^{-\left((n-p-1)+(p+1)\right)/2}.\end{aligned} \tag{B.6}$$

It is easy to recognize that (B.6) is the kernel of a $(p+1)$-dimensional t distribution with location $\hat{\beta}$, scale matrix $\hat{\sigma}^2(\mathbf{X}^T\mathbf{X})^{-1}$, and $n-p-1$ degrees of freedom, i.e.,

$$\beta|\mathbf{X},\mathbf{y} \sim T_{p+1}\left((\mathbf{X}^T\mathbf{X})^{-1}\mathbf{X}^T\mathbf{y},(\mathbf{X}^T\mathbf{X})^{-1}\hat{\sigma}^2,n-p-1\right).$$

We see that, under the improper uninformed priors, the Bayesian posterior mean $E(\beta|\mathbf{X},\mathbf{y})$ is identical to the ordinary least squares estimate. And $100(1-\alpha)\%$ credible intervals are also the same as the $100(1-\alpha)\%$ confidence intervals in the least squares approach, although the two intervals could have quite different interpretations Tiao and Zellner (1964); Wakefield (2013).

We can also derive the marginal posterior distribution of σ^2. Specifically,

$$\begin{aligned}\pi(\sigma^2|\mathbf{X},\mathbf{y}) &\propto \int \sigma^{-n-2}\exp\left[-\frac{1}{2\sigma^2}\left(\hat{\sigma}^2(n-p-1)+(\beta-\hat{\beta})^T\mathbf{X}^T\mathbf{X}(\beta-\hat{\beta})\right)\right]d\beta\\ &= \sigma^{-n-2}\exp\left[-\frac{1}{2\sigma^2}\hat{\sigma}^2(n-p-1)\right]\\ &\qquad\times\int\exp\left[-\frac{1}{2\sigma^2}(\beta-\hat{\beta})^T\mathbf{X}^T\mathbf{X}(\beta-\hat{\beta})\right]d\beta\\ &\propto \sigma^{-n-2}\exp\left[-\frac{1}{2\sigma^2}\hat{\sigma}^2(n-p-1)\right](2\pi\sigma^2)^{(p+1)/2}\\ &\propto (\sigma^2)^{-\frac{1}{2}(n-p-1)-1}\exp\left[-\frac{1}{2\sigma^2}\hat{\sigma}^2(n-p-1)\right].\end{aligned} \tag{B.7}$$

It is obvious that (B.7) is a kernel of a scaled inverse chi-squared distribution. That is,

$$\sigma^2|\mathbf{X},\mathbf{y} \sim (n-p-1)\hat{\sigma}^2 \times \chi^{-2}_{n-p-1}. \tag{B.8}$$

We obtain the analytic forms of the marginal posterior distributions of $\boldsymbol{\beta}$ and σ^2, however the close form for the parameters of interest is still not available. An algorithm of direct sampling from the posteriors could be applied here in the case of uninformative priors, without involving MCMC iterations Wakefield (2013). Note that

$$\pi(\boldsymbol{\beta},\sigma^2|\mathbf{X},\mathbf{y}) = \pi(\sigma^2|\mathbf{X},\mathbf{y})\pi(\boldsymbol{\beta}|\sigma^2,\mathbf{X},\mathbf{y}),$$

where $\sigma^2|\mathbf{X},\mathbf{y}$ is given by (B.8) and $\boldsymbol{\beta}|\sigma^2,\mathbf{X},\mathbf{y} \sim N_{p+1}\left(\hat{\boldsymbol{\beta}},\sigma^2(\mathbf{X}^T\mathbf{X})^{-1}\right)$. We can generate independent samples from the pair of distributions,

$$\begin{cases} \sigma^{2(b)} \sim \pi(\sigma^2|\mathbf{X},\mathbf{y}), \\ \boldsymbol{\beta}^{(b)} \sim \pi(\boldsymbol{\beta}|\sigma^{2(b)},\mathbf{X},\mathbf{y}), \end{cases}$$

for $b = 1,...,B$. From the samples, one can form various summaries for the model parameters including point estimates such as posterior means, medians, percentiles, and interval estimates such as credible intervals.

To illustrate many aspects of Bayesian techniques, let us start exploring the simple linear regression with one of the standard sets of data available in R. In the 1920s, braking distances were recorded for cars traveling at different speeds. Analyzing the relationship between speed and braking distance can influence the lives of great number of people, via changes in speeding laws, car design, and other factors (McNeil, 1977). The `cars` data contain two variables with 50 observations:

```
str(cars)
plot(cars$speed,cars$dist,ylab="Stopping Distance",xlab="Speed")
```

```
'data.frame':        50 obs. of  2 variables:
 $ speed: num  4 4 7 7 8 9 10 10 10 11 ...
 $ dist : num  2 10 4 22 16 10 18 26 34 17 ...
```

We shall begin by looking at the scatterplot of the two variables.

```
plot(cars$speed,cars$dist, xlab = "Speed (mph)", ylab = "Stopping
    ↪ distance (ft)", main = "Speed and Stopping Distances of Cars")
```

The plot is shown in Figure B.1, which indicates that there is a linear trend between the `speed` variable and the `dist` variable. The linear regression with the frequentist least-squares approach can simply be implemented using the following R code:

```
cars.lm <- lm(dist ~ speed, data=cars)
round(summary(cars.lm)$coeff, 4)
```

```
            Estimate Std. Error t value Pr(>|t|)
(Intercept) -17.5791     6.7584 -2.6011   0.0123
speed         3.9324     0.4155  9.4640   0.0000
```

```
round(summary(cars.lm)$sigma, 4)
```

```
[1] 15.3796
```

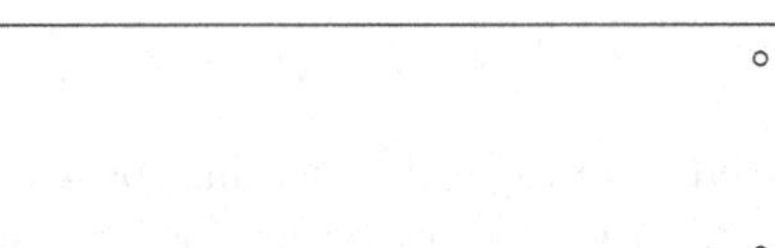
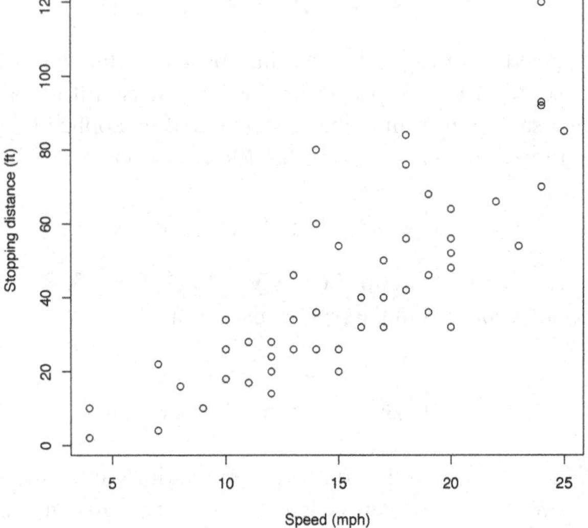

FIGURE B.1
The scatterplot for the `cars` data.

The results show the coefficient of `speed` is 3.93 and is significantly different from zero. It indicates that for every additional mph in speed of cars we can expect the stopping distance to increase by an average 3.93 feet.

Let us now consider a Bayesian linear regression with uninformative priors for the data. The model can be implemented by direct sampling without involving MCMC iterations. We have written an R function, `BayesLM.nprior`, implementing the direct sampling algorithm in our `brinla` package. The arguments in the function `BayesLM.nprior` include "`lmfit`", an object of class `"lm"` which is the output from R`lm` function, and "`B`", the size of direct sampling. We fit the model for the `car` data with the size `B = 10,000`:

```
cars.blm <- BayesLM.nprior(cars.lm,10000)
round(cars.blm$summary.stat, 4)
```

```
                mean     se 0.025quant   median 0.975quant
(Intercept) -17.6523 6.9373   -31.4474 -17.7014    -4.3668
speed         3.9359 0.4265     3.1067   3.9400     4.7869
sigma        15.6207 1.6423    12.8480  15.4920    19.2183
```

The results are very close to the least squared approach. A nice feature of the Bayesian analysis is that we can easily obtain the standard deviation and confidence levels for the parameter σ.

The linear regression can also be fitted through MCMC with normal priors. The following R commands implement the MCMC algorithm using the `R2jags` package.

```
cars.inits <- list(list("beta0" = 0, "beta1" = 0, "tau" = 1))
cars.dat <- list(x = cars$speed, y = cars$dist, n = length(cars$speed)
    ↪ )
parameters <- c("beta0", "beta1", "sigma")

SimpleLinearReg <- function(){
        for(i in 1:n){
                y[i] ~ dnorm(mu[i], tau)
                mu[i] <- beta0 + beta1*x[i]
        }
        beta0 ~ dnorm(0.0,1.0E-6)
        beta1 ~ dnorm(0.0,1.0E-6)
        tau ~ dgamma (0.001, 0.001)
        sigma <- sqrt(1/tau)
}
cars.blm2 <- jags(data = cars.dat, inits = cars.inits, parameters.to.
    ↪ save = parameters, model.file = SimpleLinearReg, n.chains = 1,
    ↪ n.iter = 5000, n.burnin = 2000, n.thin = 1)
cars.blm2
```

```
          mean  sd  2.5%   25%   50%   75% 97.5%
beta0    -17.5 7.0 -30.7 -22.2 -17.6 -12.8  -3.1
beta1      3.9 0.4   3.1   3.6   3.9   4.2   4.7
deviance 416.3 2.6 413.4 414.4 415.6 417.4 423.1
sigma     15.6 1.6  12.9  14.5  15.5  16.7  19.1

DIC info (using the rule, pD = var(deviance)/2)
pD = 3.4 and DIC = 419.6
DIC is an estimate of expected predictive error (lower deviance is better).
```

There is not much difference in the results compared to the two previous methods.

Finally, we fit the linear regression model using INLA:

```
cars.inla <- inla(dist ~ speed, data=cars, control.predictor = list(
    ↪ compute = T))
round(cars.inla$summary.fixed, 4)
```

```
                mean    sd 0.025quant 0.5quant 0.975quant     mode kld
(Intercept) -17.5686 6.767   -30.9117 -17.5690    -4.2415 -17.5693   0
speed         3.9317 0.416     3.1113   3.9317     4.7510   3.9318   0
```

```
round(cars.inla$summary.hyperpar, 4)
```

```
                                  mean    sd 0.025quant 0.5quant 0.975quant   mode
Precision
for the Gaussian observations 0.0044 9e-04     0.0029   0.0043     0.0063 0.0042
```

We want the hyperparameter summarized as the SD rather than precision. So, we apply the `bri.hyperpar.summary` function in our `brinla` library for producing this summary:

```
bri.hyperpar.summary(cars.inla)
```

```
                                  mean       sd   q0.025     q0.5   q0.975     mode
SD
for the Gaussian observations 15.29788 1.553266 12.61431 15.16697 18.70892 14.90626
```

We also obtain very similar results. We want a plot of the fitted line and its credible interval and the marginal posterior densities for intercept, `speed` and the precision parameter:

```
plot(cars$speed,cars$dist,ylab="Stopping Distance",xlab="Speed")
lines(cars$speed, cars.inla$summary.linear.predictor[,1], lwd=2)
lines(cars$speed, cars.inla$summary.linear.predictor[,3],lty=2, lwd=2)
lines(cars$speed, cars.inla$summary.linear.predictor[,5],lty=2, lwd=2)
plot(inla.smarginal(cars.inla$marginals.fixed[[1]]), type="l", xlab=""
    ↪ ,ylab="", main="Marginal posterior: Intercept")
plot(inla.smarginal(cars.inla$marginals.fixed[[2]]), type="l", xlab=""
    ↪ ,ylab="", main="Marginal posterior: Speed")
plot(inla.smarginal(cars.inla$marginals.hyperpar[[1]]), type="l", xlab
    ↪ ="",ylab="", main="Marginal posterior: Hyperparamter")
```

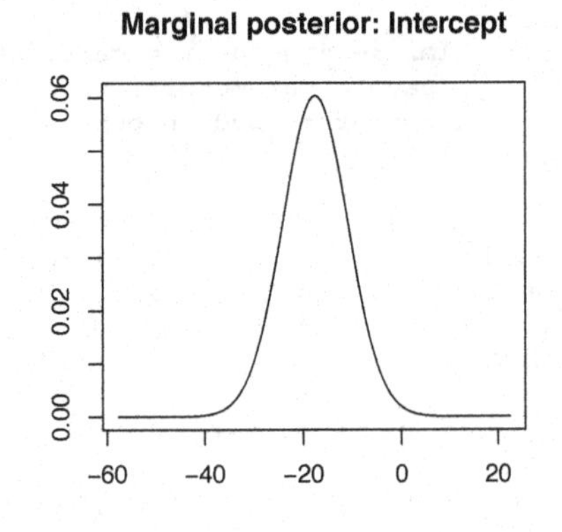

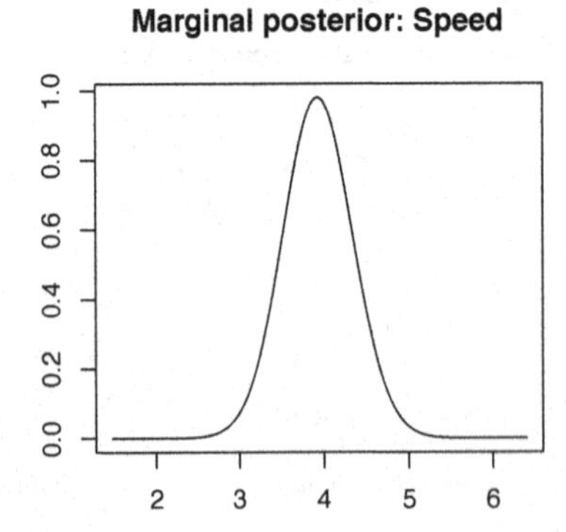

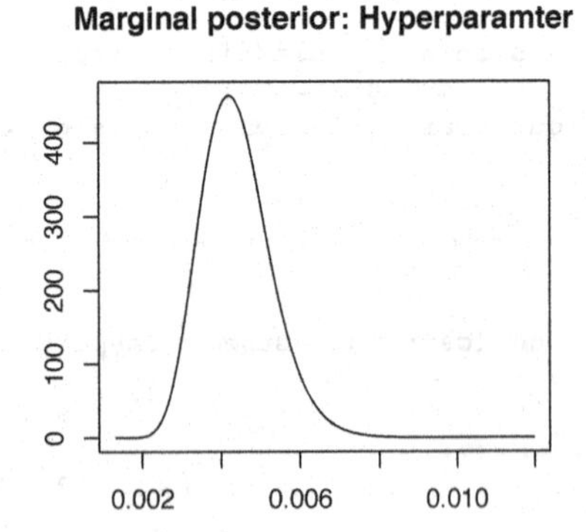

FIGURE B.2
Bayesian regression with INLA for the `cars` data: the upper left panel shows the plot of fitted line and its credible interval. The other three panels show the marginal posterior densities for intercept, `speed` and the precision parameter.

Figure B.2 shows the INLA method performs very well in fitting the linear model for the `cars` data.

Bibliography

Abrams, D., A. Goldman, C. Launer, J. Korvick, J. Neaton, L. Crane, M. Grodesky, S. Wakefield, K. Muth, S. Kornegay, D. L. Cohn, A. Harris, R. Luskin-Hawk, N. Markowitz, J. H. Sampson, M. Thompson, and L. Deyton (1994). Comparative trial of didanosine and zalcitabine in patients with human immunodeficiency virus infection who are intolerant of or have failed zidovudine therapy. *New England Journal of Medicine 330*, 657–662.

Adler, R. J. (1981). *The Geometry of Random Fields*. New York: Wiley.

Agresti, A. (2012). *Categorical Data Analysis* (3rd ed.). New York: Wiley.

Aitkin, M. A., B. Francis, and J. Hinde (2005). *Statistical Modelling in GLIM 4*. Oxford University Press, New York.

Barlow, W. E. and R. L. Prentice (1988). Residuals for relative risk regression. *Biometrika 75*(1), 65–74.

Bauwens, L. and A. Rasquero (1993). Approximate hpd regions for testing residual autocorrelation using augmented regressions. In *Computer Intensive Methods in Statistics*, pp. 47–61. Berlin: Springer.

Bell, D. F., J. L. Walker, G. O'Connor, and R. Tibshirani (1994). Spinal deformity after multiple-level cervical laminectomy in children. *Spine 19*, 406–411.

Berkson, J. (1950). Are there two regressions? *Journal of the American Statistical Association 45*(250), 164–180.

Besag, J., P. Green, D. Higdon, and K. Mengersen (1995). Bayesian computation and stochastic systems. *Statistical Science 10*, 3–41.

Besag, J. and C. Kooperberg (1995). On conditional and intrinsic autoregressions. *Biometrika 82*, 733–746.

Bickel, P. J. and K. A. Doksum (2015). *Mathematical Statistics: Basic Ideas and Selected Topics* (2nd ed.). Boca Raton: CRC Press.

Blangiardo, M. and M. Cameletti (2015). *Spatial and Spatio-Temporal Bayesian Models with R-INLA*. Chichester: John Wiley & Sons.

Bogaerts, K. and E. Lesaffre (2004). A new, fast algorithm to find the regions of possible support for bivariate interval-censored data. *Journal of Computational and Graphical Statistics 13*(2), 330 – 340.

Bolin, D. and F. Lindgren (2015). Excursion and contour uncertainty regions for latent Gaussian models. *Journal of the Royal Statistical Society: Series B (Statistical Methodology) 77*(1), 85–106.

Bralower, T., P. Fullagar, C. Paull, G. Dwyer, and R. Leckie (1997). Mid-cretaceous strontium-isotope stratigraphy of deep-sea sections. *Geological Society of America Bulletin 109*, 1421–1442.

Breslow, N. (1972). Discussion on regression models and life-tables. *Journal of the Royal Statistical Society: Series B (Methodology) 34*(2), 216 – 217.

Brockmann, H. J. (1996). Satellite male groups in horseshoe crabs, limulus polyphemus. *Ethology 102*(1), 1–21.

Brown, L., T. Cai, R. Zhang, L. Zhao, and H. Zhou (2010). The root–unroot algorithm for density estimation as implemented via wavelet block thresholding. *Probability Theory and Related Fields 146*(3-4), 401–433.

Buja, A., T. Hastie, and R. Tibshirani (1989). Linear smoothers and additive models. *The Annals of Statistics 17*(2), 453–510.

Buonaccorsi, J. P. (2010). *Measurement Error: Models, Methods, and Applications*. Boca Raton: Chapman & Hall.

Carlin, B. P. and T. A. Louis (2008). *Bayesian Methods for Data Analysis*. Boca Raton: CRC Press.

Carroll, R. J. and D. Ruppert (1988). *Transformation and Weighting in Regression*. New York: CRC Press.

Carroll, R. J., D. Ruppert, L. A. Stefanski, and C. Crainiceanu (2006). *Measurement Error in Nonlinear Models: A Modern Perspective* (2nd ed.). New York: Chapman & Hall/CRC Press.

Carroll, R. J. and L. A. Stefanski (1990, September). Approximate quasi-likelihood estimation in models with surrogate predictors. *Journal of the American Statistical Association 85*(411), 652–663.

Chaloner, K. (1991). Bayesian residual analysis in the presence of censoring. *Biometrika 78*(3), 637–644.

Chaloner, K. and R. Brant (1988). A bayesian approach to outlier detection and residual analysis. *Biometrika 75*, 651–659.

Chatterjee, S. and A. S. Hadi (2015). *Regression Analysis by Example* (5th ed.). New York: John Wiley & Sons.

Chaudhuri, P. and J. S. Marron (1999). SiZer for exploration of structures in curves. *Journal of the American Statistical Association 94*(447), 807–823.

Comte, F. (2004, July). Kernel deconvolution of stochastic volatility models. *Journal of Time Series Analysis 25*(4), 563–582.

Cook, J. R. and L. A. Stefanski (1994). Simulation extrapolation estimation in parametric measurement error models. *Journal of American Statistical Association 89*, 1314–1328.

Cox, D. R. (1972). Regression models and life-tables. *Journal of the Royal Statistical Society: Series B (Methodology) 34*(2), 187 – 220.

Cox, D. R. and E. J. Snell (1968). A general definition of residuals (with discussion). *Journal of the Royal Statistical Society - Series B (Methodology) 30*, 248–275.

Davison, A. C. and D. V. Hinkley (1997). *Bootstrap Methods and Their Application*. New York: Cambridge University Press.

Dawid, A. (1984). Present Position and Potential Developments: Some Personal Views Statistical Theory The Prequential Approach. *Journal of the Royal Statistical Society: Series A (Statistics in Society) 147*, 278–292.

De Boor, C. (1978). *A Practical Guide to Splines*. New York: Springer.

Dellaportas, P. and D. Stephens (1995). Bayesian analysis of errors-in-variables regression models. *Biometrics 51*(3), 1085–1095.

Diebold, F. X., T. A. Gunther, and A. S. Tay (1998). Evaluating density forecasts with applications to financial risk management. *International Economic Review 39*(4), 863–883.

Dobson, A. J. and A. Barnett (2008). *An Introduction to Generalized Linear Models*. Boca Raton: CRC press.

Draper, D. (1995). Assessment and propogation of model uncertainty. *Journal of the Royal Statistical Society: Series B (Methodology) 57*, 45–97.

Dreze, J. H. and M. Mouchart (1990). Tales of testing Bayesians. In *Contributions to Econometric Theory and Application*, pp. 345–366. New York: Springer.

Eilers, P. and B. Marx (1996). Flexible smoothing with B-splines and penalties (with discussion). *Statistical Science 11*, 89–121.

Eilers, P. H., B. D. Marx, and M. Durbán (2015). Twenty years of P-splines. *SORT-Statistics and Operations Research Transactions 39*(2), 149–186.

Evans, M., H. Moshonov, et al. (2006). Checking for prior-data conflict. *Bayesian Analysis 1*(4), 893–914.

Everitt, B. S. (2006). *An R and S-PLUS Companion to Multivariate Analysis*. London: Springer.

Fahrmeir, L. and T. Kneib (2009). Propriety of posteriors in structured additive regression models: Theory and empirical evidence. *Journal of Statistical Planning and Inference 139*(3), 843–859.

Fahrmeir, L. and L. Knorr-Held (2000). Dynamic and semiparametric models. In *Smoothing and Regression: Approaches, Computation, and Application*, pp. 513 – 544. New York: John Wiley & Sons.

Fahrmeir, L. and S. Lang (2001). Bayesian inference for generalized additive mixed models based on Markov random field priors. *Journal of the Royal Statistical Society: Series C (Applied Statistics) 50*(2), 201–220.

Fahrmeir, L. and G. Tutz (2001). *Multivariate Statistical Modeling Based on Generalized Linear Models*. Berlin: Springer.

Fan, J. and Y. K. Truong (1993). Nonparametric regression with errors in variables. *The Annals of Statistics 21*(4), 1900–1925.

Faraway, J. (2014). *Linear Models with R* (2nd ed.). Boca Raton: Chapman & Hall/CRC.

Faraway, J. (2016a). Confidence bands for smoothness in nonparametric regression. *Stat 5*(1), 4–10.

Faraway, J. (2016b). *Extending the Linear Model with R: Generalized Linear, Mixed Effects and Nonparametric Regression Models* (2nd ed.). London: Chapman & Hall.

Ferkingstad, E. and H. Rue (2015). Improving the INLA approach for approximate bayesian inference for latent gaussian models. *Electronic Journal of Statistics 9*(2), 2706–2731.

Ferrari, S. L. P. and F. Cribari-Neto (2004). Beta regression for modelling rates and proportions. *Journal of Applied Statistics 31*(7), 799–815.

Ferraty, F. and P. Vieu (2006). *Nonparametric Functional Data Analysis: Theory and Practice*. New York: Springer.

Fitzmaurice, G. M. and N. M. Laird (1993). A likelihood-based method for analysing longitudinal binary responses. *Biometrika 80*(1), 141–151.

Fong, Y., H. Rue, and J. Wakefield (2010). Bayesian inference for generalized linear mixed models. *Biostatistics 11*(3), 397–412.

Friedman, J. H. and W. Stuetzle (1981). Projection pursuit regression. *Journal of the American Statistical Association 76*(376), 817–823.

Fuglstad, G.-A., D. Simpson, F. Lindgren, and H. Rue (2015). Constructing Priors that Penalize the Complexity of Gaussian Random Fields. *arXiv preprint arXiv:1503.00256*, 1 – 44.

Fuller, W. A. (1987). *Measurement Error Models*. New York: John Wiley & Sons.

Geisser, S. and W. F. Eddy (1979). A predictive approach to model selection. *Journal of the American Statistical Association 74*(365), 153–160.

Gelfand, A. E. and A. F. Smith (1990). Sampling-based approaches to calculating marginal densities. *Journal of the American Statistical Association 85*(410), 398–409.

Gelman, A. (2006). Prior distributions for variance parameters in hierarchical models. *Bayesian Analysis 1*(3), 515–533.

Gelman, A., J. B. Carlin, H. S. Stern, and D. B. Rubin (2014). *Bayesian Data Analysis* (3rd ed.). New York: Chapman & Hall/CRC Press.

Gelman, A., J. Hwang, and A. Vehtari (2014). Understanding predictive information criteria for Bayesian models. *Statistics and Computing 24*(6), 997–1016.

Gelman, A., X.-L. Meng, and H. Stern (1996). Posterior predictive assessment of model fitness via realized discrepancies. *Statistica Sinica 6*, 733–760.

Gneiting, T., F. Balabdaoui, and A. E. Raftery (2007). Probabilistic forecasts, calibration and sharpness. *Journal of the Royal Statistical Society: Series B (Methodology) 69*(2), 243–268.

Godfrey, P. J., A. Ruby, and O. T. Zajicek (1985). The Massachusetts acid rain monitoring project: Phase 1. In *Water Resource Research Center*. University of Massachusetts.

Goldsmith, J., C. M. Crainiceanu, B. Caffo, and D. Reich (2012). Longitudinal penalized functional regression for cognitive outcomes on neuronal tract measurements. *Journal of the Royal Statistical Society: Series C (Applied statistics) 61*(3), 453–469.

Gomez, G., M. L. Calle, R. Oller, and K. Langohr (2009). Tutorial on methods for interval-censored data and their implementation in R. *Statistical Modelling 9*(4), 259–297.

Green, P. J. and B. W. Silverman (1994). *Nonparametric Regression and Generalized Linear Models: a Roughness Penalty Approach*. Boca Raton: Chapman & Hall.

Guo, X. and B. P. Carlin (2004). Separate and joint modeling of longitudinal and event time data using standard computer packages. *The American Statistician 58*(1), 16–24.

Gustafson, P. (2004). *Measurement Error and Misclassification in Statistics and Epidemiology: Impacts and Bayesian Adjustments*. Boca Raton: Chapman & Hall.

Hadfield, J. D. (2010). MCMC methods for multi-response generalized linear mixed models: The MCMCglmm R package. *Journal of Statistical Software 33*(2), 1–22.

Hammer, S. M., K. E. Squires, M. D. Hughes, J. M. Grimes, L. M. Demeter, J. S. Currier, J. J. Eron Jr., J. E. Feinberg, H. H. Balfour Jr., L. R. Deyton, et al. (1997). A controlled trial of two nucleoside analogues plus indinavir in persons with human immunodeficiency virus infection and CD4 cell counts of 200 per cubic millimeter or less. *New England Journal of Medicine 337*(11), 725–733.

Hastie, T. and R. Tibshirani (1990). *Generalized Additive Models.* New York: Chapman & Hall.

Hastie, T. and R. Tibshirani (2000). Bayesian backfitting. *Statistical Science 15*(3), 196–223.

Hastings, W. K. (1970). Monte Carlo sampling methods using Markov chains and their applications. *Biometrika 57*(1), 97–109.

Hawkins, D. (2005). *Biomeasurement.* New York: Oxford University Press.

Held, L., B. Schrödle, and H. Rue (2010). Posterior and cross-validatory predictive checks: a comparison of MCMC and INLA. In *Statistical Modelling and Regression Structures*, pp. 91–110. New York: Springer.

Henderson, C. R. (1982). Analysis of covariance in the mixed model: Higher-level, nonhomogeneous, and random regressions. *Biometrics 38*, 623–640.

Henderson, R., P. Diggle, and A. Dobson (2000). Joint modelling of longitudinal measurements and event time data. *Biostatistics 1*(4), 465–480.

Hjort, N. L., F. A. Dahl, and G. H. Steinbakk (2006). Post-processing posterior predictive *p* values. *Journal of the American Statistical Association 101*(475), 1157–1174.

Hoerl, A. E., R. W. Kannard, and K. F. Baldwin (1975). Ridge regression: Some simulations. *Communications in Statistics: Theory and Methods 4*(2), 105–123.

Hoerl, A. E. and R. W. Kennard (1970). Ridge regression: Biased estimation for nonorthogonal problems. *Technometrics 12*(1), 55–67.

Hosmer, D. W. and S. Lemeshow (2004). *Applied Logistic Regression.* New York: John Wiley & Sons.

Hosmer, D. W., S. Lemeshow, and S. May (2008). *Applied Survival Analysis: Regression Modelling of Time to Event Data* (2nd ed.). New York: Wiley-Interscience.

Hsiang, T. (1975). A Bayesian view on ridge regression. *The Statistician 24*(4), 267–268.

Jolliffe, I. T. (1982). A note on the use of principal components in regression. *Journal of the Royal Statistical Society: Series C (Applied Statistics) 31*(3), 300–303.

Kardaun, O. (1983). Statistical survival analysis of male larynx-cancer patients: A case study. *Statistica Neerlandica 37*(3), 103–125.

Klein, J. P. and M. L. Moeschberger (2005). *Survival Analysis: Techniques for Censored and Truncated Data.* New York: Springer.

Kutner, M. H., C. J. Nachtsheim, J. Neter, and W. Li (2004). *Applied Linear Statistical Models* (5th ed.). New York: McGraw-Hill Irwin.

Lambert, D. (1992). Zero-inflated Poisson regression models with an application to defects in manufacturing. *Technometrics 34*(1), 1 – 14.

Lang, S. and A. Brezger (2004). Bayesian P-splines. *Journal of Computational and Graphical Statistics 13*(1), 183–212.

Lange, K. L., R. J. Little, and J. M. Taylor (1989). Robust statistical modeling using the t distribution. *Journal of the American Statistical Association 84*(408), 881–896.

Le Cam, L. (2012). *Asymptotic Methods in Statistical Decision Theory.* New York: Springer.

Lindgren, F. and H. Rue (2008a). On the second-order random walk model for irregular locations. *Scandinavian Journal of Statistics 35*(4), 691–700.

Lindgren, F. and H. Rue (2008b). On the second-order random walk model for irregular locations. *Scandinavian Journal of Statistics 35*(4), 691–700.

Lindgren, F. and H. Rue (2015). Bayesian spatial modelling with R-INLA. *Journal of Statistical Software 63*(19), 1–25.

Lindgren, F., H. Rue, and J. Lindström (2011). An explicit link between gaussian fields and gaussian markov random fields: the stochastic partial differential equation approach (with discussion). *Journal of the Royal Statistical Society: Series B (Methodology) 73*(4), 423–498.

Lindsey, J. K. (1997). *Applying Generalized Linear Models.* New York: Springer.

Liu, C. and D. B. Rubin (1995). ML estimation of the t distribution using EM and its extensions, ECM and ECME. *Statistica Sinica 5*(1), 19–39.

Long, J. S. (1997). *Regression Models for Categorical and Limited Dependent Variables.* Thousand Oaks: Sage Publications.

Lunn, D., C. Jackson, N. Best, A. Thomas, and D. Spiegelhalter (2012). *The BUGS Book: A Practical Introduction to Bayesian Analysis.* Boca Raton: CRC Press.

Marshall, E. C. and D. J. Spiegelhalter (2007). Identifying outliers in Bayesian hierarchical models: A simulation-based approach. *Bayesian Analysis 2*(2), 409–444.

Martino, S., R. Akerkar, and H. Rue (2011). Approximate Bayesian inference for survival models. *Scandinavian Journal of Statistics 38*(3), 514–528.

Martins, T. G., D. Simpson, F. Lindgren, and H. Rue (2013). Bayesian computing with INLA: New features. *Computational Statistics and Data Analysis 67*, 68–83.

McCullagh, P. and J. A. Nelder (1989). *Generalized Linear Models* (2nd ed.). London: CRC Press.

McGilchrist, C. and C. Aisbett (1991). Regression with frailty in survival analysis. *Biometrics 47*(2), 461–466.

McNeil, D. R. (1977). *Interactive Data Analysis.* New York: Wiley.

Meng, X.-L. (1994). Posterior predictive *p*-values. *The Annals of Statistics 22*(3), 1142–1160.

Metropolis, N., A. W. Rosenbluth, M. N. Rosenbluth, A. H. Teller, and E. Teller (1953). Equation of state calculations by fast computing machines. *The Journal of Chemical Physics 21*(6), 1087–1092.

Montgomery, D. C. (2013). *Design and Analysis of Experiments* (8th ed.). New York: John Wiley & Sons.

Morrison, H. L., M. Mateo, E. W. Olszewski, P. Harding, et al. (2000). Mapping the galactic halo I: The "spaghetti" survey. *The Astronomical Journal 119*, 2254–2273.

Muff, S., A. Riebler, L. Held, H. Rue, and P. Saner (2015). Bayesian analysis of measurement error models using integrated nested laplace approximations. *Journal of the Royal Statistical Society: Series C (Applied Statistics) 64*(2), 231–252.

Myers, R. and D. Montgomery (1997). A tutorial on generalized linear models. *Journal of Quality Technology 29*(3), 274 – 291.

Nelder, J. A. and R. J. Baker (2004). Generalized linear models. In *Encyclopedia of Statistical Sciences*. New York: John Wiley & Sons.

O'Sullivan, F. (1986). A statistical perspective on ill-posed inverse problems. *Statistical Science 1*(4), 502–527.

Pettit, L. (1990). The conditional predictive ordinate for the normal distribution. *Journal of the Royal Statistical Society: Series B (Methodological) 52*, 175–184.

Plummer, M. et al. (2003). JAGS: A program for analysis of Bayesian graphical models using Gibbs sampling. In *Proceedings of the 3rd International Workshop on Distributed Statistical Computing*, Volume 124, pp. 125. Vienna.

Prater, N. (1956). Estimate gasoline yields from crudes. *Petroleum Refiner 35*(5), 236 – 238.

Ramsay, J. O. and B. W. Silverman (2005). *Functional Data Analysis* (2nd ed.). New York: Springer.

Rasmussen, C. and C. Williams (2006). *Gaussian Processes for Machine Learning*. Cambridge, MA: The MIT Press.

Richardson, S. and W. Gilks (1993). Conditional independence models for epidemiological studies with covariate measurement error. *Statistics in Medicine 12*(18), 1703–1722.

Rue, H. and L. Held (2005). *Gaussian Markov Random Fields: Theory and Applications*. London: Chapman & Hall.

Rue, H. and S. Martino (2007). Approximate Bayesian inference for hierarchical Gaussian Markov random field models. *Journal of Statistical Planning and Inference 137*(10), 3177–3192.

Rue, H., S. Martino, and N. Chopin (2009). Approximate Bayesian inference for latent Gaussian models using integrated nested Laplace approximations (with discussion). *Journal of the Royal Statistical Society: Series B (Methodological) 71*(2), 319–392.

Rue, H., A. Riebler, S. H. Sørbye, J. B. Illian, D. P. Simpson, and F. Lindgren (2017). Bayesian computing with INLA: a review. *Annual Review of Statistics and Its Application 4*, 395–421.

Ruppert, D. (2002). Selecting the number of knots for penalized splines. *Journal of Computational and Graphical Statistics 11*, 735–757.

Ruppert, D. and R. J. Carroll (2000). Spatially-adaptive penalties for spline fitting. *Australian and New Zealand Journal of Statistics 42*(2), 205–223.

Ruppert, D., M. P. Wand, and R. J. Carroll (2003). *Semiparametric Regression*. New York: Cambridge University Press.

Sheather, S. J. and M. C. Jones (1991). A reliable data-based bandwidth selection method for kernel density estimation. *Journal of the Royal Statistical Society: Series B (Methodological) 53*, 683–690.

Shumway, R. H. and D. S. Stoffer (2011). *Time Series Analysis and Its Applications: with R Examples* (3rd ed.). New York: Springer.

Simpson, D., F. Lindgren, and H. Rue (2012). Think continuous: Markovian Gaussian models in spatial statistics. *Spatial Statistics 1*, 16–29.

Simpson, D. P., T. G. Martins, A. Riebler, G.-A. Fuglstad, H. Rue, and S. H. Sørbye (2017). Penalising model component complexity: A principled, practical approach to constructing priors. *Statistical Science 32*(1), 1–28.

Singer, J. D. and J. B. Willett (2003). *Applied Longitudinal Data Analysis: Modeling Change and Event Occurrence*. London: Oxford University Press.

Sørbye, S. H. and H. Rue (2014). Scaling intrinsic Gaussian Markov random field priors in spatial modelling. *Spatial Statistics 8*, 39–51.

Speckman, P. L. and D. Sun (2003). Fully Bayesian spline smoothing and intrinsic autoregressive priors. *Biometrika 90*(2), 289–302.

Spiegelhalter, D. J., N. G. Best, B. P. Carlin, and A. Linde (2014). The deviance information criterion: 12 years on. *Journal of the Royal Statistical Society: Series B (Methodological) 76*(3), 485–493.

Spiegelhalter, D. J., N. G. Best, B. P. Carlin, and A. Van Der Linde (2002). Bayesian measures of model complexity and fit. *Journal of the Royal Statistical Society: Series B (Methodological) 64*(4), 583–639.

Stan Development Team (2016). *Stan Modeling Language: User's Guide and Reference Manual*. Stan Development Team.

Sun, D. and P. L. Speckman (2008). Bayesian hierarchical linear mixed models for additive smoothing splines. *Annals of the Institute of Statistical Mathematics 60*(3), 499–517.

Sun, D., R. K. Tsutakawa, and P. L. Speckman (1999). Posterior distribution of hierarchical models using CAR(1) distributions. *Biometrika 86*, 341–350.

Therneau, T. M., P. M. Grambsch, and T. R. Fleming (1990). Martingale-based residuals for survival models. *Biometrika 77*(1), 147–160.

Thodberg, H. H. (1993). Ace of Bayes: Application of neural networks with pruning. Technical report, The Danish Meat Research Institute, Maglegaardsvej 2, DK-4000.

Tiao, G. C. and A. Zellner (1964). On the bayesian estimation of multivariate regression. *Journal of the Royal Statistical Society: Series B (Methodological) 26*(2), 277–285.

Tierney, L. and J. B. Kadane (1986). Accurate approximations for posterior moments and marginal densities. *Journal of the American Statistical Association 81*(393), 82–86.

Tsiatis, A. A. and M. Davidian (2004). Joint modeling of longitudinal and time-to-event data: An overview. *Statistica Sinica 14*(3), 809–834.

Umlauf, N., D. Adler, T. Kneib, S. Lang, and A. Zeileis (2015). Structured additive regression models: An R interface to BayesX. *Journal of Statistical Software 63*(1), 1–46.

Vaupel, J. W., K. G. Manton, and E. Stallard (1979). The impact of heterogeneity in individual frailty on the dynamics of mortality. *Demography 16*(3), 439–454.

Wahba, G. (1978). Improper priors, spline smoothing and the problem of guarding against model errors in regression. *Journal of the Royal Statistical Society: Series B (Methodology) 40*(3), 364–372.

Wahba, G. (1990). *Spline Models for Observational Data.* Philadelphia: SIAM [Society for Industrial and Applied Mathematics].

Wakefield, J. (2013). *Bayesian and Frequentist Regression Methods.* New York: Springer.

Wand, M. P. and J. T. Ormerod (2008). On semiparametric regression with O'Sullivan penalized splines. *Australian and New Zealand Journal of Statistics 50*(2), 179–198.

Wang, X.-F. (2012). Joint generalized models for multidimensional outcomes: A case study of neuroscience data from multimodalities. *Biometrical Journal 54*(2), 264–280.

Wang, X.-F., Z. Fan, and B. Wang (2010). Estimating smooth distribution function in the presence of heteroscedastic measurement errors. *Computational Statistics and Data Analysis 54*(1), 25–36.

Wang, X.-F. and B. Wang (2011). Deconvolution estimation in measurement error models: The R package decon. *Journal of Statistical Software 39*(10), 1–24.

Watanabe, S. (2010). Asymptotic equivalence of bayes cross validation and widely applicable information criterion in singular learning theory. *Journal of Machine Learning Research 11*, 3571–3594.

Whittle, P. (1954). On stationary processes in the plane. *Biometrika 41*, 434–449.

Whyte, B., J. Gold, A. Dobson, and D. Cooper (1987). Epidemiology of acquired immunodeficiency syndrome in Australia. *The Medical Journal of Australia 146*(2), 65–69.

Wood, S. N. (2003). Thin plate regression splines. *Journal of the Royal Statistical Society: Series B (Methodological) 65*(1), 95–114.

Wood, S. N. (2006). *Generalized Additive Models: An Introduction with R.* New York: Chapman & Hall/CRC Press.

Wood, S. N. (2008). Fast stable direct fitting and smoothness selection for generalized additive models. *Journal of the Royal Statistical Society: Series B (Methodological) 70*(3), 495–518.

Yue, Y. and P. L. Speckman (2010). Nonstationary spatial Gaussian Markov random fields. *Journal of Computational and Graphical Statistics 19*(1), 96–116.

Yue, Y. R., D. Bolin, H. Rue, and X.-F. Wang (2018). Bayesian generalized two-way ANOVA modeling for functional data using INLA. *Statistica Sinica*, in press.

Yue, Y. R., D. Simpson, F. Lindgren, and H. Rue (2014). Bayesian adaptive smoothing spline using stochastic differential equations. *Bayesian Analysis 9*(2), 397–424.

Yue, Y. R., P. Speckman, and D. Sun (2012). Priors for Bayesian adaptive spline smoothing. *Annals of the Institute of Statistical Mathematics 64*(3), 577–613.

Index

Printed in the United States
by Baker & Taylor Publisher Services